ENGINEERING INTELLIGENT
HYBRID MULTI-AGENT SYSTEMS

ENGINEERING INTELLIGENT HYBRID MULTI-AGENT SYSTEMS

by

Rajiv Khosla
La Trobe University, Melbourne
Australia

Tharam Dillon
La Trobe University, Melbourne
Australia

KLUWER ACADEMIC PUBLISHERS
Boston / Dordrecht / London

Distributors for North America:
Kluwer Academic Publishers
101 Philip Drive
Assinippi Park
Norwell, Massachusetts 02061 USA

Distributors for all other countries:
Kluwer Academic Publishers Group
Distribution Centre
Post Office Box 322
3300 AH Dordrecht, THE NETHERLANDS

Library of Congress Cataloging-in-Publication Data

A C.I.P. Catalogue record for this book is available
from the Library of Congress.

Printed on acid-free paper.

Printed in the United States of America

This book is dedicated by Rajiv
Khosla to his wife Anu, son
Anurag and daughter Sukriti.

Contents

Preface

This book is about building intelligent hybrid systems, problem solving, and software modeling. It is relevant to practitioners and researchers in the areas of intelligent hybrid systems, control systems, multi-agent systems, knowledge discovery and data mining, software engineering, and enterprise-wide systems modeling. The book in many ways is a synergy of all these areas. The book can also be used as a text or reference book for postgraduate students in intelligent hybrid systems, software engineering, and system modeling.

On the intelligent hybrid systems front, the book covers applications and design concepts related to fusion systems, transformation systems and combination systems. It describes industrial applications in these areas involving hybrid configurations of knowledge based systems, case-based reasoning, fuzzy systems, artificial neural networks, and genetic algorithms.

On the problem solving front, the book describes an architectural theory for engineering intelligent associative hybrid multi-agent systems. The architectural theory is described at the task structure level and the computational level. From an organizational context the problem solving architecture is not only relevant at the knowledge engineering layer for developing knowledge agents but also at the information engineering layer for developing information agents.

On the software modeling front the book describes the role of objects, agents and problem solving in knowledge engineering, information engineering, data engineering, and software modeling of intelligent hybrid systems. Based on concepts developed in the book, an enterprise-wide systems modeling framework is described to facilitate forward and backward integration of systems developed in the knowledge, information, and data engineering layers of an organization. In the modeling process, agent oriented analysis, design, and reuse aspects of software engineering are also discussed.

The book consists of four parts:

Part I: introduces various methodologies and their hybrid applications in the industry.

Part II: describes a multi-agent architectural theory of associative intelligent hybrid systems at the task structure level and the computational level. It covers various aspects related to knowledge modeling of hybrid systems.

Part III: describes the software engineering aspects of the architecture. It does that by describing a real-time alarm processing application of the architecture.

Part IV: takes a broader view of the various concepts and theories developed in Part II and III of the book respectively in terms of enterprise-wide systems modeling, multi-agent systems, control systems, and software engineering and reuse.

Part I is described through chapters 1, 2, 3, 4 and 5 respectively.

Chapter 1: primarily discusses the evolution of hybrid systems from a real world applications viewpoint and a human information processing system viewpoint. It also outlines the four classes of intelligent hybrid systems, namely, fusion systems, transformation systems, combination systems, and associative systems.

Chapter 2: describes various methodologies which have been used in the book. These include expert systems, artificial neural networks, fuzzy systems, genetic algorithms, object-oriented methodology, and agent and agent architectures.

Chapter 3: describes various types of fusion and transformation hybrid systems based on their knowledge engineering strategy and types of problems addressed by them. A number of fusion and transformation system models of symbolic knowledge representation, symbolic reasoning, fuzzy reasoning, fuzzy clustering, rule extraction, design automation of fuzzy systems, optimization of fuzzy and neural networks, and others are covered.

Chapter 4: describes various types of combination hybrid systems based on synergistic combination of symbolic knowledge based systems, case-based reasoning, fuzzy logic, artificial neural networks, and genetic algorithms. A number of industrial applications are described in the areas of process control, scheduling, diagnosis, design, forecasting, and condition based maintenance.

Chapter 5: Knowledge discovery and data mining has become a significant area of intelligent hybrid system applications. This chapter outlines the process of Knowledge Discovery in Databases (KDD). It illustrates the application of the KDD process through applications in forecasting, financial trading, learning concept hierarchies, and computer network diagnosis.

Part II is described through chapters 6, 7, and 8 respectively.

Chapter 6: Associative hybrid systems associate fusion, transformation, and combination systems in a manner so as to maximize the quality as well as range of tasks that can be covered by fusion, transformation, and combination architectures on their own. Chapter 6 develops a Task Structure Level (TSL) associative hybrid architecture based on a number of perspectives including philosophical, neurobiological, cognitive science, levels of intelligence, forms of knowledge, learning, user, and others. The TSL architecture is defined in terms of the information processing phases, tasks and task constraints in each phase, knowledge engineering strategy, intelligent methodologies for task accomplishment, and hybrid configuration of these methodologies.

Chapter 7: develops the computational level of the TSL architecture The computational architecture, Intelligent Multi-Agent Hybrid Distributed Architecture (IMAHDA), is realized through integration of TSL architecture with object-oriented, agent, and operating system process models. The role and knowledge content of IMAHDA in terms of its four layers, namely, object, software agent, intelligent agent, and the problem solving agent respectively is described in this chapter. dynamic analysis and verification constructs of IMAHDA are also outlined.

Chapter 8: continues with the computational level description of IMAHDA. It describes the communication constructs, learning knowledge and strategy, and dynamic analysis and verification knowledge constructs of IMAHDA. A comprehensive view of IMAHDA is provided through an agent based description of its problem solving agents. The emergent behavioral characteristics of IMAHDA are also described.

Part III is described through chapters 9, 10, 11, and 12 respectively.

Chapter 9: introduces the reader to the problems associated with real-time alarm processing and fault diagnosis (an application of IMAHDA) in a power system control center. It surveys some of the existing methods of alarm processing and their problems. It outlines the objectives of the Real-Time Alarm Processing System (RTAPS) in the Thomastown Terminal Power Station (TTS).

Chapter 10: Analysis and design precede machine implementation of any large scale problem. This chapter outlines the steps for carrying out the agent oriented analysis of the RTAPS. It outlines the complementary role of objects and agents and how they enrich each other in the analysis process. The agent oriented analysis steps described in this chapter include understanding the domain, object-oriented problem domain structure analysis, agent analysis of problem solving behavior, association of problem domain structure & agent analysis with the problem solving agents of IMAHDA, and object vocabulary of the RTAPS objects.

Chapter11: continues with the agent oriented analysis started in chapter 10. It defines the vocabulary of the RTAPS problem solving agents in terms of contextual rules, learning knowledge and strategy, and dynamic analysis. The agent oriented design described in this chapter among other aspects maps the software and intelligent agents to the RTAPS problem solving agents and other software objects identified in the analysis stage. Some of the emergent behavioral characteristics of the RTAPS design architecture are also described.

Chapter 12: describes the implementation of the RTAPS. It describes the realization of the RTAPS design architecture in terms of methodology related objectives (e.g. IMAHDA), domain related objectives, and management related objectives. Various real-time issues like response time and temporal reasoning are also discussed.

Part IV is described through chapters 13 and 14 respectively.

Chapter 13: Chapters 6 to 12 describe various aspects related to architecture of intelligent associative hybrid systems, problem solving and software engineering of intelligent hybrid systems. This chapter takes a broader view of IMAHDA (e.g. its problem solving aspects) and the various concepts (e.g. role of objects and agents) described in chapters 6 to 12 respectively. The chapter provides a framework for enterprise-wide systems modeling. It describes a unified approach for developing databases, database agents, information agents, and knowledge agents.

Chapter 14: This final chapter of the book revisits some of the contributions made in this book in the areas of problem solving, intelligent hybrid systems, software modeling of hybrid systems, control systems, multi-agent systems, software engineering, and enterprise-wide systems modeling.

RAJIV KHOSLA

THARAM DILLON

Acknowledgments

The authors would like to acknowledge the support of Demi Dallikavak for drawing the diagrams. The authors would also like to acknowledge the assistance of Marcus J. de Rijk, Sue Lane and Mary Witten for the front cover page design.

1 WHY INTELLIGENT HYBRID SYSTEMS

1.1 INTRODUCTION

The need for intelligent systems has grown in the past decade because of the increasing demand on humans and machines for better performance. One of the reasons for this increasing demand is that we are passing through an era of information explosion, information globalization and consequently ever increasing competition. The information explosion has put huge constraints on time in which decisions have to be made. Today knowledge has become an important strategic resource to help humans deal with the complexity and sheer quantum of information. This strategic resource will help them to enhance their own performance and that of the machines they work with. The areas like knowledge discovery and data mining and soft computing are the latest manifestations of intelligent systems and knowledge in general.

In business, intelligent systems are needed to deal with competition, markets, and customers. In manufacturing, intelligent systems are needed for optimization of processes and systems related to monitoring, control, troubleshooting and diagnosis. The need for intelligent systems in business and manufacturing comes with many aspirations. Some problem solvers simply want machines to do various sorts of things that people call intelligent. These are applications oriented or applied intelligent systems. It usually involves getting a machine to do what previously was only done by humans (Rich, 1986; Schank, 1990). Others hope to understand what enables people to do such things. This generally falls within the purview of cognitive science and psychology as it relates to the study of human mind and brain. Yet other problem solvers want to write programs that would allow machines to grow and learn from experience.

These different aspirations or perceptions have arisen because intelligent systems are not defined through some physical law such as Kirchoff's law in electric circuit theory but are defined through the type of problem addressed by them (Schank 1990; Minsky 1991). These problems range from psychology, linguistics, business to engineering. The one thing common in these aspirations is that all of them address one or more aspects of human cognition, namely, information processing, knowledge representation and learning.

The increasing significance of intelligent systems has put more pressure on the intelligent systems community to come out with more powerful and flexible problem solving strategies which will provide a better understanding of the human information processing system and at the same time help to build complex real world applications.

The intelligent systems community has responded to these needs with the development of intelligent hybrid systems. In this introductory chapter the evolution of hybrid systems is discussed in order to establish their role in a proper perspective. This is followed by outlining the different classes of intelligent hybrid systems and the link between these classes and the other chapters in the book.

1.2   EVOLUTION OF HYBRID SYSTEMS

The evolution of intelligent hybrid systems can be discussed from two viewpoints

- Hybrid Systems and Real World Applications

- Hybrid Systems and Human Information Processing

1.2.1   Hybrid Systems and Real World Applications

The four most commonly used intelligent methodologies in the 90's are symbolic knowledge based systems (e.g. expert systems), artificial neural networks, fuzzy systems and genetic algorithms.

Symbolic knowledge based systems have served varied purposes in industry and commerce during the last three decades. The most widely used versions are expert systems. Many expert systems have found their way into industry and commerce. Today expert systems are applied in all areas of business including manufacturing, planning, scheduling, design, diagnosis, sales, finance, etc.. In these applications the unitary architecture of production rules has been used exhaustively to capture human expertise and experience to solve different problems. The different knowledge representation techniques like semantic networks, frames, scripts and objects (Barr, et al., 1982; Feigenbaum et al., 1989) have been able to capture the ways in which humans store knowledge. As a consequence, a manufacturing planning task that took hours to do is now done in 15 minutes (Kraft, 1984; Feigenbaum, 1989). A scheduling activity that once took eight man weeks is done in half a day. A design task that took a month of work is done in two days. A product configuration task that normally takes a manufacturing engineer three hours is done in half a minute. As a consequence we have come to appreciate a number of advantages associated with the use of this methodology

Symbolic systems with case based reasoning have also been used in areas where there are no given rules or rules cannot be used directly in the decision making process. A number of design problems (e.g. fashion shoe design), and legal domain problems employ case based reasoning.

However, practioners have also identified some of the limitations of symbolic knowledge based systems. These include among others, slow and constricted knowledge acquisition processes, an inability to properly deal with imprecision in data, inability to process incomplete information, combinatorial explosion of rules, retrieval problems in recovering past cases, and inability to reason under time constraints on occasions.

People deal with imprecision in data every day. This imprecision is represented by linguistic statements like "It is a hot day," "Mary has a big house,". In these statements "hot" and "big" are imprecise or fuzzy concepts with fuzzy boundaries.That is, when does a day go from very warm to hot in temperature or when does a medium house become big or large in size. Expert systems have problems with such fuzzy concepts and fuzzy boundaries. A number of fuzzy systems have been built based on fuzzy concepts and human like imprecise reasoning. Fuzzy systems have been used in a number of areas including control of trains in Japan, sales predictions, and stock market risk analysis.

A major disadvantage of fuzzy systems and expert systems is their heavy reliance on the human experts for knowledge acquisition. This knowledge may be in the form of rules used to solve a problem and/or the shape of the membership functions used for modeling a fuzzy concept "hot". Besides the knowledge acquisition problem, these systems are restricted in terms of their adaptive and learning capabilities.

The limitations in knowledge based systems and fuzzy systems have been primarily responsible for the resurgence of artificial neural networks. Applications using artificial neural networks can be found in in various industrial sectors (Davlo, et al. 1991; Beale, et al. 1990). In the financial sector neural networks are used for prediction and modeling of markets, signature analysis, automatic reading of handwritten characters (cheques), assessment of credit worthiness and selection of investments. In the telecommunication sector, applications can be found in signal analysis, noise elimination and data compression. Similarly, in the environment sector, neural networks have been used for risk evaluation, chemical analysis, weather forecasting and resource management. Other applications can be found in quality control, production planning and load forecasting in power systems. In these applications the inherent parallelism in artificial neural networks and their capacity to learn, process incomplete information and generalize have been exploited. However, the stand alone approaches of artificial neural networks have exposed some limitations such as the, problems associated with lack of structured knowledge representation, inability to interact with conventional symbolic databases and inability to explain the reasons for conclusions reached. Their inability to explain their conclusions have limited their applicability in high risk domains (e.g. real time alarm processing), and other training related problems like local minima. Another major limitation associated with neural networks is the problem of scalability. For large and complex problems, difficulties exist in training the networks and also in assessing generalization capabilities of the networks.

Optimization of manufacturing processes is another area where intelligent methodologies like artificial neural networks and genetic algorithms have been used. Genetic algorithms which are largely unconstrained by the limitations of neural networks (e.g. local minima) and other search methods, and do not require any mathematical modeling are being used for solving scheduling and control problems in industry. They have also being successfully used for optimization of symbolic, fuzzy and neural network based intelligent systems because of their modeling convenience. One of the problems associated with genetic algorithms is that they are computationally expensive which can restrict their on line use in real time systems where time and space are at a premium.

In fact, real time systems add another dimension to the problems associated with various intelligent methodologies described in this section. These problems are largely associated with the time and space constraints of real-time systems. Real-time systems are characterized by the fact that severe consequences will result if logical as well as timing correctness properties of the systems are not satisfied. Some examples of real-time systems are command and control systems, process control systems, flight control systems, and alarm processing systems. The additional constraints like response time, reasoning under time constraints,and reliability imposed by real time systems further expose the limitations of expert systems and artificial neural networks. For example, expert systems with unpredictable search time are not good at coping with applications which involve temporal constraints and a fast response time. On the other hand, the state of the art of artificial neural networks does not generate confidence in an operator in the control center to take action in real time based on the generalized results of an artificial neural network.

These computational and practical issues associated with the four intelligent methodologies have made the practiners and researchers look at ways of hybridizing the different intelligent methodologies from an applications viewpoint. However, as mentioned earlier, the evolution of hybrid systems is not only an outcome of the practical problems encountered by these intelligent methodologies but is also is an outcome of different aspects of human information processing system modeled by these methodologies. In the next section, the evolution of hybrid systems based on the underlying human information processing modeling aspects of intelligent methodologies are discussed.

1.2.2 Hybrid Systems and Human Information Processing

Symbolic knowledge based systems, fuzzy systems, artificial neural networks and genetic algorithms have their underpinnings in cognitive psychology, cognitive science, structure and behavior of human brain, and evolution.

These systems represent a rich set of methodologies that have been developed for characterizing human cognition with respect to information processing, learning and knowledge representation. In this chapter human information processing aspects related to evolution of hybrid systems are only discussed. Issues related to learning and knowledge representation are discussed in chapter 5.

Symbolic knowledge based systems (particularly expert systems) are an outcome of studies in applied Artificial Intelligence (AI), cognitive psychology and cognitive science, and also practical considerations for solving real world problems

Cognitive psychology has relied on purely experimental approach of hypothesize and test, whereas cognitive science employs protocol analysis and computer simulations of mental processes as primary techniques for studying human information processing systems. Practical considerations refer to those ingredients which need to be incorporated in the computational models based on these two cognitive branches in order to ground the theories in the real world and solve real world problems in approximately real time.

Various theories/models for simulating human information processing (e.g., Newell and Simon 1972; Simon 1979; Hayes-Roth et al. 1979; Newell 1980; McClelland,1986a, 86b; Smolensky 1988, 90) have been suggested in the cognitive psychology and cognitive science literature.

There are basically two models of information processing which represent fundamentally different views. One developed by Newell and Simon (1972) represents the traditional symbolic AI view and forms the basis of the Von Neumann computer architecture. The other developed by McClelland and his colleagues (1986a) represents the parallel distributed processing view (also known as the connectionist view) which draws its inspiration from the human brain.

Newell and Simon (1972) assert that the human mind is essentially a serial information processing machine where information is processed in a production rules like fashion. McClelland, et al. (1986a) reject the notion that intelligent processes consist of sequential application of explicit rules. Instead, they posit a large number of simple processing units, operating in parallel, where each unit sends excitatory and/or inhibitory signals to other units. Information processing takes place through the interactions of a large number of units.

These two models represent the two different schools of thought among researchers in intelligent systems today. At the same time, these two models present an opportunity to derive computationally efficient and robust models for complex applications and also develop a better understanding of the human information processing system. The underlying features of the two models are briefly described now.

Newell and Simon (1972) and Newell (1977), describe the characteristics of human Information Processing System (IPS) in their work on the theory of human problem solving. The IPS best articulates the traditional Von Neumann architecture where sets of symbols are moved about from one memory store to another, and are processed by explicit rules applied in a sequence.

Newell and Simon postulate that human IPS is a serial system with input (sensory) patterns and output (motor) patterns. It consists of long term memory (LTM), short term memory (STM) and External memory (EM). LTM has unlimited capacity and is organized associatively. Associativity is achieved by storing information in LTM as symbol structures, each consisting of a set of symbols connected by relations. As new symbol structures are stored in LTM, they are designated by symbols drawn from a potentially infinite vocabulary of symbols. These new names can, in turn, be embedded as symbols in other structures. On the other hand, STM has limited capacity and holds about seven symbols based on digit-span experiments. Each of these symbols can designate an entire structure of arbitrary size and complexity in LTM. Only two symbols can be retained for one task while another unrelated task is being performed. EM is the immediately available visual field related to a particular task and has unlimited capacity. The STM and LTM are homogeneous in that sensory and motor patterns are symbolized and handled identically in STM and LTM. The

STM consists of number of elementary processes with processing time largely dependent on the read rates from LTM and STM.

Their work postulates that problem solving takes place by search in problem space (characterized by STM, LTM, and EM). The search is characterized by considering one knowledge state after another until the desired knowledge state is reached. The structure of information in the problem space is determined by the structure of information in the task environment. The function program used to search the problem space is determined by the amount of information available in the problem space.

The two invariant aspects of the the function program are its production-like and goal-like character. The production-like character is based on circumstantial evidence from various experiments in chess, symbolic logic, and cryptarithmetic puzzles. That is, production systems are capable of expressing arbitrary calculations, they encode homogeneously the information that instructs the human information processing system (IPS) how to behave. They have a strong stimulus-response (If-Then) flavor, and the dynamic memory for a production system is the STM on which its productions are contingent, and which they modify. The goal-like character of the function program is reflected in the movement from one state (subgoal) to another in the search space till the desired state (goal) is achieved.

The production-like and goal-like character suggests a deliberate reasoning nature of the human information processing system and a reductionist philosophy at work.

In relation to these invariants is the serial character of the IPS. A serial IPS as per Newell and Simon is one that can execute a single elementary information process at one time. They use time required for the division of one and two numbers as an experimental measure of seriality of the human mind. Seriality is also expressed by the fact that time taken to recognize a character (say 'R') when it is inverted is usually more than when it is upright. Nonetheless, Newell and Simon do concede that parallelism does exist for highly automated activities and can exist in motor activities which are independent in nature.

The work done by Newell and Simon on human IPS has led to the development of symbolic knowledge based systems like expert systems (Hayes-Roth et al. 1983) which compete with human specialists in different problem domains.

There is a basic structural similarity in the system architecture of performance-oriented expert systems and computer simulations of human cognition developed by cognitive scientists. This is shown in Table 1.1. This structural comparison is based on the human IPS proposed by Newell and Simon. However, the concept of working memory has gone through many transformations since then. The more recent view (Anderson, 1983, 88; Card et. al.., 1983) is that short-term memory is the currently activated portion of long-term memory and

Table 1.1.   Comparison of Cognitive and Expert System Architecture

COGNITIVE ARCHITECTURE	EXPERT SYSTEM ARCHITECTURE
SHORT TERM MEMORY (OR WORKING MEMORY)	CURRENT STATUS DATA BASE (OR DYNAMIC DATA BASE)
LONG TERM MEMORY (OR PERMANENT MEMORY)	KNOWELDGE BASE (OR STATIC DATA BASE)
COGNITIVE PROCESSOR (OR MENTAL OPERATIONS)	INFERENCE ENGINE (OR INTERPRETER)

is thus nested in the long-term memory. While this clearly weakens the claim that the modular system architecture of expert systems emulates the human cognitive architecture, a certain structural similarity remains.

However, certain concerns still remain. The main is whether serial processing programs - however fast - can ever approach the power of human thinking. For example, several natural cognitive tasks like perceiving objects in natural scenes, understanding language are accomplished faster by humans than by computers employing traditional Von Neumann architecture and production system software. Another major concern is the impracticality of writing all the detailed instructions needed for a program to respond intelligently to unexpected events.

These concerns have provided the impetus for an alternative parallel distributed processing approach, which is influenced by the computational properties of the human brain. These properties (Feldman 1985; Hinton 1985) include:

- a human can perform a simple task as picture naming in around 500 milliseconds, or about 100 steps (the best available AI programs require millions of time steps to perform comparable tasks)

- the cortex of the human brain contains some 100 billion neurons

- each neuron is connected with up to about 10000 others

- the human brain is a massively parallel natural computer

This has lead to the emergence of parallel distributed processing models primarily developed by McClelland and his colleagues (1986a).

1.2.3   Parallel Distributed Processing Models

According to McClelland et al. (1986a) people are smarter than today's computers because the brain employs a basic computational architecture that is more suited to dealing with a central aspect of natural information processing tasks that people are so good at. Many information processing tasks generally require the simultaneous consideration of many pieces of information or constraints. This is so in motor skill related mechanisms like reaching and grasping where constraints like structure and musculature of the arm as well as the constraints imposed by the environment have to be considered simultaneously in order to successfully grasp a particular object. In natural language processing numerous mutual constraints operate not only between syntactic and semantic processing, but also with each of these domains as well. In these domains numerous constraints have to be kept in mind at once where each constraint affects the other constraints and at the same time is influenced by them. Further, most everyday situations generally involve an interplay between a number of different sources of information rather than being assigned to single episodes or frames. Like what a child's birthday party in a restaurant would be like will involve interaction between two types of knowledge sources i.e. birthday party and restaurant respectively. This, sort of interplay also helps to capture the generative capacity of human beings in novel situations.

The PDP models assume that information processing involving multiple constraints takes place through the interactions of a large number of simple processing elements called units, each sending excitory and inhibitory signals to other units. In some cases, the units stand for possible hypotheses about such things as letters in a particular display or the syntactic roles of words in a particular sentence. In these cases, the activations stand roughly for the strengths associated with different possible hypotheses, and the interconnections among the units stand for the constraints the system knows to exists between the hypotheses. In other cases, the units stand for possible goals and actions, such as a goal of typing a letter, or the action of moving the left index finger, and the connections relate goals to subgoals, subgoals to actions, and actions to muscle movements.

In these models, memory and processing are diffusely distributed throughout the network, with little control. Problem solving then takes place by way of evolutionary rules which are formed by trying out different connections in the network until the network settles into a stable or goal state. The multiple constraints are represented in the form of weight links connecting input and output units. A range of PDP models and others of its kind have been developed for motor control, perception, content addressable memory and natural language and other applications. Besides the cognitive science perspective of these models, they also claim similarity to the physiological structure of the human brain which gives them an extra appeal. In fact these these models are more popularly known as artificial neural networks to reflect the neural hardware of the human brain.

In domains like the ones mentioned above that have multiple constraints, do not go well with the sequential or serial processing nature of human cognition when examined at the microstructure level on a time scale of seconds or minutes. Serious attempts to model even the simplest macrosteps of cognition, say, recognition of single words require vast numbers of microsteps if they are implemented sequentially. The biological hardware as pointed out by Feldman and Ballard (1982), is too sluggish for sequential models of the microstructure to provide a plausible account, at least of the microstructure of human thought. The time limitation only gets worse, not better, when sequential mechanisms try to take large number of constraints into account. Each additional constraint requires more time in a sequential machine, if the constraints are imprecise, the constraints can lead to a computational explosion. This is in strong contrast to humans, who get faster, and not slower, when they are able to exploit additional constraints.

This leads one to believe that though at the macrostructure level there is a sequential interaction, say, between two sentences of a story, at the microstructure level of cognition, Parallel Distributed Models(PDP) offer alternatives to the serial models. According to Chandrasekaran (1990), a duration of 50-100 milliseconds has often been suggested as the size of temporal "grain" of processes at the microstructure level. On the other hand, macro phenomena takes place over seconds if not minutes in case of a human.

Symbols, according to Chandrasekaran (1990) are types about which abstract rules of behavior are known. This leads to symbols being labels which are interpreted during processing, whereas no such interpretations can be directly associated with processing in artificial neural networks. Processing in artificial neural networks can be said to be evolutionary in nature where the weight connections are constantly changing till the desired mapping from input to output is accomplished.

Interpretation in symbolic systems is a consequence of the algorithmic nature of symbolic processing whereas non-interpretation in artificial neural network systems is a consequence of multiple constraints and highly parallel and distributed processing. This view is supported by Smolensky (1990) and also explains the problems associated with knowledge acquisition in expert systems.

One of the bottlenecks in knowledge acquisition (an integral part of expert systems) from domain experts is to extract those rules or heuristics which are based on intuition of the domain expert and not based on logic. Extraction of these heuristics becomes increasingly difficult as the number of constraints and task complexity increases. Smolensky (1990) refers to the microstructure level as being representative of this unconscious or intuitive knowledge, whereas macrostructure level as being representative of conscious or algorithmic knowledge.

The intuitive knowledge which involves multiple constraints and corresponds to the sub-conceptual level does not necessarily come from conscious application of rules (Smolensky 1990). Further, the intuitive processing need not have the same semantics and syntax as conscious rule application (Smolensky 1990). This intuitive processing, according to Smolemsky (1990) is responsible for perception, practiced or automated motor behavior, intuition in problem-solving, etc. or in other words skilled behavior. This intuitive knowledge becomes more distributed with increasing number of constraints and task complexity.

Thus symbolic knowledge based systems and artificial neural networks can be said to represent two ends of the human information processing spectrum as shown in Figure 1.1. One end of the spectrum represents analytical, interpretative and deliberate processing behavior. The other end of the spectrum represents intuitive, non-interpretive and automated processing behavior. On one end of the spectrum information processing takes place in a matter of few seconds or minutes, whereas on the other end of the spectrum the information processing takes place in a matter of few milliseconds.

Fuzzy systems which like artificial neural networks engage in approximate reasoning but which unlike artificial neural networks are more interpretive in nature have been placed in the middle. In Figure 1.1 it is also shown that the symbolic knowledge based systems satisfy hard constraints, whereas fuzzy systems and artificial neural networks satisfy soft constraints. This is because these systems vary with respect to the type of solution they provide for a given task. Logic and provability which dominate symbolic systems modeling behavior make provision for more discrete and precise solutions. In comparison, artificial neural and fuzzy systems which model approximate behavior provide imprecise solutions. As against all (1) or nothing (0) result of a symbolic system, artificial neural networks and fuzzy systems provide approximate solutions to a given task with values varying anywhere between 0 and 1. The term soft constraints (which has now taken up a distinct name, namely, soft computing) has been referred and used by Smolensky (1990). The weight connections in the models are referred to as soft constraints, where a positive connection from unit a to unit b represents a soft constraint to the effect that if unit a is active then unit b should be too and the intensity of activation of unit b, is proportional to the numerical magnitude of the connection. Thus, at the sub-conceptual level solution is achieved by satisfaction of large amount of soft constraints, whereas, at the symbolic or macro level the solution is achieved by satisfaction of hard constraints(e.g., Ohm's Law, Kirchoff's Law) under ideal conditions.

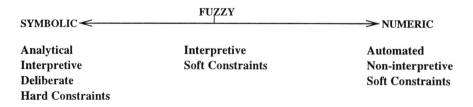

Figure 1.1.   Human Information Processing Spectrum

Smolensky (1990) calls them hard constraints, because they are formalized as hard rules which can be applied singly, independently and sequentially with an all or nothing result, whereas, the soft constraints have no implications singly and can be overridden by others.

In more recent times the information processing distinction on the macro and microstructure level has led two levels of intelligence, namely Artificial intelligence and computational intelligence as described by Bezdek (1994).

In Bezdek (1994) a model outlining three levels of intelligent activity namely, computational, artificial, and biological is described. The computational level (or the lowest level), according to the model deals with numeric data and involves pattern recognition, whereas the artificial intelligence level augments the computational level by adding small pieces of knowledge to the computational processes. The biological intelligence level (the highest level) processes sensory inputs and, through associate memory, links many subdomains of biological neural networks to recall knowledge (Bezdek 1994; Medskar 1995). The computational and the artificial levels according to (Bezdek 1994) lead to the biological level. An integration of the computational and artificial intelligence levels can thus be considered as a way of modeling biological intelligence. The computational intelligence is also known as the soft computing level with intelligent soft computing methodologies like fuzzy systems, artificial neural networks and genetic algorithms. These different levels of intelligence, in fact have provided the motivation for hybridization of various intelligent methodologies from a human information processing viewpoint. However, the hybrid approach adopted to realize the macro and microstructure levels of information processing and the different levels of intelligence have been varied. Broadly speaking, these varied hybrid approaches are realized either through implicit and explicit hybridization or both. The next section outlines these varied hybrid approaches by grouping them into four classes of hybrid systems.

1.3   CLASSES OF HYBRID SYSTEMS

Intelligent hybrid systems can be grouped into four classes, namely, fusion systems, transformation systems, combination systems, and associative systems.

1.3.1   Fusion Systems

In fusion systems representation and/or information processing features of intelligent methodology A are fused into the representation structure of another intelligent methodology B. In this way the intelligent methodology B augments its information processing in a manner which can cope with different levels of intelligence and information processing. From an application or practical viewpoint, this augmentation can be seen as a way by which an intelligent methodology addresses its weaknesses and exploits its existing strengths to a solve a particular real world problem. The hybrid systems based on the fusion approach revolve around artificial neural networks and genetic algorithms. In artificial neural network based fusion systems representation and/or information processing features of other intelligent methodologies like symbolic knowledge based systems and fuzzy systems are fused into artificial neural networks. Genetic algorithm based fusion systems involve fusion of intelligent methodologies like knowledge based systems, fuzzy systems, and artificial neural networks. Chapter 3 describes in detail the motivation behind the fusion systems and the types of problems addressed by these systems. A number of fusion systems are described in chapter 3 to illustrate the type of problems addressed by them.

1.3.2   Transformation Systems

Transformation systems are used to transform one form of representation into another. They are used to alleviate the knowledge acquisition problem as highlighted by Smolensky (1990) by transforming distributed or continuous representations into discrete representations. From a practical perspective, they are used in situations where knowledge required to accomplish the task is not available and one intelligent methodology depends upon another intelligent methodology for its reasoning or processing. For example, artificial neural networks are used for transforming numerical/continuous data into symbolic rules which can then be used by a symbolic knowledge based system for further processing. Transformation systems in conjunction with fusion systems are also used for knowledge refinement or optimization.

Transformation systems are described in chapter 3 and 5 respectively. In chapter 3 they are covered in the conjunction with fusion systems for purpose of optimization. In chapter 5 they are covered in the context of knowledge discovery and data mining.

1.3.3   Combination Systems

Combination systems involve explicit hybridization. Instead of fusion, they model the different levels of information processing and intelligence by using intelligent methodologies which best model a particular level. Intelligent combination systems unlike fusion systems retain the separate identity of an intelligent methodology.These systems involve a modular arrangement of two or more intelligent methodologies to solve real world problems. Chapter 4 outlines the motivation behind these systems and describes a number of real world applications involving different permutation and combinations of the four intelligent methodologies.

1.3.4   Associative Systems

The constraints imposed by the fusion, transformation and combination systems in terms of their knowledge representation, and processing capabilities limit the the range of tasks or problems and the quality of solution of those tasks or problems. Association systems (or Associative systems)[1] attempt to associate fusion, transformation and combination systems in a manner so as to maximize the range of tasks or problems as well as the quality of solution of those tasks.

It needs to be noted here that the class of hybrid systems described in this section have not been used in many real world problems because at the moment associative hybrid architectures have not been designed. Unlike, intelligent methodologies like knowledge based which have well defined architectures for modeling solutions to real world problems, hybrid system architectures have not fully emerged to help a problem solver model real world problems in these systems. Knowledge modeling or Knowledge content issues, division of labor between different methodologies with respect to information processing stages and tasks in these stages have not been addressed. The associative systems box shown in dotted lines in Figure 1.2 is indicative of this fact. Chapters 6, 7 and 8 of this book describe an intelligent associative hybrid architecture at the task structure level and computational level. The task structure level describes the information processing phases, tasks to be accomplished in each information processing phase, task constraints, intelligent methodologies to be used and knowledge engineering strategy employed to accomplish the tasks. The computational level architecture described in chapter 7 and 8 respectively is realized in an agent oriented framework and is called Intelligent Multi-Agent Hybrid Distributed Architecture (IMAHDA). Chapters 6 and 7 address knowledge modeling issues and various other issues of the associative hybrid architecture before its application in chapters 9, 10, 11, and 12.

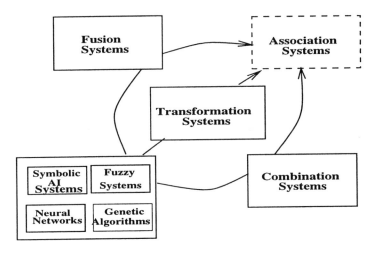

Figure 1.2.   Classes of Intelligent Hybrid Systems

Chapter 13 places IMAHDA in an enterprise-wide system modeling context. It looks at IMAHDA as a problem solving architecture employing objects and agents as the vehicles for developing database systems, information systems and knowledge systems. It describes an application of IMAHDA in the sales management functional area. The chapter proposes a unified approach to enterprise-side system modeling through use of objects, agents, data warehouse, and IMAHDA as shown in Figure 1.3.

Finally, chapter 14 reviews the contributions of this book in areas like problem solving, hybrid systems, control systems, multi-agent systems, software engineering and reuse, and enterprise-wide system modeling.

In the next chapter software and intelligent methodologies including agents which are used in this book are introduced to the reader.

1.4 SUMMARY

Intelligent systems have assumed significance in this era of information explosion and globlization. In this era of information explosion intelligent systems are needed to help humans deal with the complexity and sheer quantum of information and enhance their own performance and that of the machines they work with. In the 70's and 80's information was a strategic resource. In the 90's knowledge has become a strategic resource. As a result there is a greater emphasis on developing more powerful and versatile intelligent problem solving strategies.

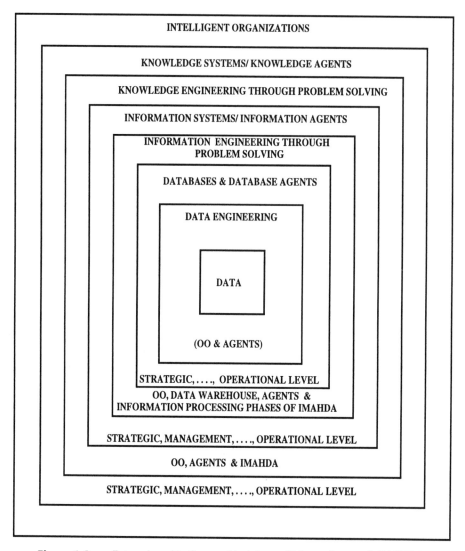

Figure 1.3.   Enterprise-wide System Modeling - Objects, Agents & IMAHDA

Intelligent hybrid systems have evolved in the past seven years in order to develop more powerful problem solving strategies and gain a better understanding of the human information processing system. Under the banner of real world applications, knowledge based systems, fuzzy systems, artificial neural networks and genetic algorithms have been used in various industrial applications. These applications however, have also highlighted limitations of these methodologies which have encouraged their hybridization. On the human information processing front, studies in cognitive psychology, cognitive science and artificial intelligence indicate that information processing takes place at a macrostructure level and a microstructure level. Information processing at macrostructure level takes place in the order of few seconds to minutes, whereas at the microstructure level it takes in the order of few milliseconds to a few hundred milliseconds. In recent times the macrostructure and microstructure levels have also come to be known as the artificial intelligence and computational intelligence levels respectively. The hybrid approaches adopted to model these two levels can be grouped into intelligent fusion systems, transformation systems, combination systems and associative systems.

Notes

1. The word Association and Associative are used interchangeably throughout the book.

References

Anderson, J. R. (1983), *The Architecture of Cognition*, Harvard University Press, Cambridge, Massachusetts.

Anderson, J. R. (1988), "Acquisition of Cognitive Skill," *Readings in Cognitive Science: A Perspective from Psychology and Artificial Intelligence*, pp. 362-80.

Barr, A., and Feigenbaum, E. A. (1982), *The Handbook of Artificial Intelligence*, HeirisTech press, Stanford, California.

Beale, R., and Jackson, T. (1990), *Neural Computing: An Introduction*, Bristol: Hilger, U.K.

Bezdek, J.C. (1994) "What is Computational Intelligence?," Computational Intelligence: Imitating Life, IEEE Press, New York.

Chandrasekaran, B. (1990), "What Kind of Information Processing is Intelligence," *The Foundations of AI: A Sourcebook*, Cambridge, UK: Cambridge University Press, pp. 14-46.

Card, S. K., Moran, T. P., and Newell, A. (1983), *The Psychology of Human-computer Interaction*, Hillsdale, NJ: Erlbaum.

Davlo, E., and Naim, P. (1991), *Neural Networks*, Macmillan Computer Science Series, U.K.

Feigenbaum, E., McCorduck, P., and Penny N. H. (1989), *Rise of the Expert Company : How Visionary Companies are Using Artificial Intelligence to Achieve Higher Productivity and Profits*, Vintage Books, New York.

Feldman, J. A., and Ballard, D. H. (1982), "Connectionist Models and their Properties," *Cognitive Science* vol. 6, pp. 205-264.

Feldman, J. A. (1985), "Connections: Massive Parallelism in Natural and Artificial Intelligence," *Byte*, April, pp. 277-284.

Hayes-Roth, B., and Hayes-Roth, F. (1979), "A Cognitive Model of Planning," *Cognitive Science*, vol. 3, pp. 275-310.

Kraft, A. (1984), "XCON: An Expert Configuration System at Digital Equipment Corporation," *The AI Business: Commercial Uses of Artificial Intelligence*, Cambridge, MA: MIT Press.

McClelland, J. L., Rumelhart, D. E., and Hinton, G.E. (1986a), "The Appeal of Parallel Distributed Processing," *Parallel Distributed Processing*, vol.1, Cambridge , MA: The MIT Press, pp. 3-40.

McClelland, J. L., Rumelhart, D. E., and Hinton, G.E., (1986b), *Parallel Distributed Processing: Explorations in the Microstructure of Cognition*, vol. 2, Cambridge, MA: The MIT Press.

Minsky, M. (1991), "Logical Versus Analogical or Symbolic Versus Connectionist or Neat Versus Scruffy," *AI Magazine*, 12(2), pp34-51.

Newell, A., and Simon, H. A. (1972), "The Theory of Human Problem Solving," *Human Problem Solving*, Englewood Cliffs, NJ:Prentice Hall.

Newell, A. (1977), "On Analysis of Human Problem Solving," *Thinking: Readings in Cognitive Science*, Cambridge UK: Cambridge University Press.

Newell, A. (1979), "Reasoning, Problem Solving, and Decision Processes: The Problem space as a Fundamental Category," *Technical Report - (CMU-CS-79-133)*, Department of Computer Science, Carnegie-Mellon University, Pittsburgh.

Newell, A. (1980), Physical Symbol Systems," *Cognitive Science*, vol. 4, pp. 135-183.

Newell, A., and Rosenbloom, P. S. (1981), "Mechanisms of Skill Acquisition and Learning in Practice," *Cognitive Skills and their Acquisition*, J. R. Anderson (Ed.), Hillsdale, NJ:Erlbaum.

Rich, E. (1983), *Artificial Intelligence*, McGraw Hill, New York.

Schank, R. C. (1990), "What is AI anyway?," *The Foundations of Artificial Intelligence: A Source Book*, Cambridge Unive rsity Press, UK, pp. 3-13.

Simon, H. A. (1979), *Models of Thought*, New Haven and London: Yale University Press.

Smolensky, P. (1988), "On the Proper Treatment of Connectionism," *The Behavioral and Brain Sciences*, vol. 11, no. 1, pp. 1-23.

Smolensky, P. (1990), "Connectionism and Foundations of AI," *The Foundations of AI: A Sourcebook*, Cambridge, UK: Cambridge University

2 METHODOLOGIES

2.1 INTRODUCTION

This book is a synergy of various methodologies used in intelligent systems, software engineering, enterprise modeling, and multi-agent systems. This chapter introduces the reader to these methodologies. The breadth and depth of description of these methodologies has been governed by the extent of their use in this book. The various methodologies used are:

- Expert Systems
- Case-based Reasoning
- Artificial Neural Networks
- Fuzzy Systems
- Genetic Algorithms
- Knowledge Discovery and Data Mining
- Object-Oriented Methodology

- Agents and Agent Architectures

2.2 EXPERT SYSTEMS

Expert Systems handle a whole array of interesting tasks that require a great deal of specialized knowledge and these tasks can only be performed by experts who have accumulated the required knowledge. These specialized tasks are performed in a variety of areas including Diagnosis, Classification, Prediction, Scheduling and Decision Support (Hayes-Roth 1983). The various expert systems developed in these areas can be broadly grouped under four architectures:

- Rule Based Architecture.

- Rule and Frame Based Architecture.

- Model Based Architecture.

- Blackboard Architecture.

These four architectures use a number of symbolic knowledge representation formalisms developed in the last thirty years. Thus before describing these architectures the symbolic knowledge representation formalisms are briefly described.

2.2.1 Symbolic Knowledge Representation

Symbolic Artificial Intelligence (AI) has developed a rich set of representational formalisms which have enabled cognitive scientists to characterize human cognition. The symbolic representational power has in fact been seen for a long time as an advantage over the connectionist representations for human problem solving. The real reason according to Chandrasekaran (1990), for loss of interest in the perceptrons by Minsky and Papert (1969), was not due to limitations of the single layer perceptrons, but for the lack of powerful representational and representation manipulation tools. These AI symbolic knowledge representational formalisms are briefly overviewed.

Knowledge representation schemes have been used to represent the semantic content of natural language concepts, as well as to represent psychologically plausible memory models. These schemes facilitate representation of semantic and episodic memory.

Human semantic memory is the memory of facts we know, arranged in some kind of hierarchical network (Quillian 1968; Kolodner 1984). For example, in a semantic memory "stool" may be defined as a type of "chair", in turn defined as a type of "furniture". Properties and relations are handled within the overall

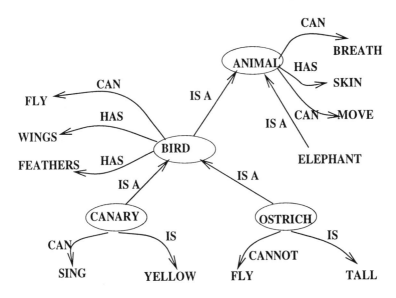

Figure 2.1.   A Semantic Network

hierarchical framework. Semantic networks are a means of representing semantic memory in which any concept (e.g Bird in Figure 2.1) is represented by a set of properties, which in turn consists of pointers to other concepts as shown in Figure 2.1 (Quillian 1968). The properties are made up of attribute-value pairs. Semantic networks also introduce property inheritance as a means to establish hierarchies and a form of default reasoning. Whereas, semantic nets provide a form of default reasoning, first order predicate calculus and production systems (Newell 1977) provide a means of deductive reasoning observed in humans. First order predicate calculus and production systems are both a representational and processing formalism. A predicate calculus expression like:

$$\forall X \ \exists Y \ (student(X) \ \Rightarrow \ AI\_subject(Y) \ \land \ likes(X, Y))$$

provides a clear semantics for the symbols and expressions formed from objects and relations. It also provides a means for representing connectives, variables, and universal and existenial quantifiers (like forall ($\forall$) and forsome or there exists ($\exists$)). The existensial and universal quantifiers provide a powerful mechanism for generalization which is difficult to model in semantic networks.

Knowledge in production systems is represented as condition-action pairs called production rules (e.g., "if it has a long neck and brown blotches, infer that it is a giraffe").

If semantic memory encodes facts, then episodic encodes experience. An episode is a record of an experienced event like visiting a restaurant or a diagnostic consultation. Information in episodic memory is defined and organized in accordance with its intended uses in different situations or operations. Frames and scripts (Schank 1977; Minsky 1981) which are extensions of semantic networks are used to represent complex events (e.g., like going to a restaurant) in terms of structured units with specific slots (e.g., being seated, ordering), with possible default values (e.g., ordering from a menu), and with a range of possible values associated with any slot. These values are either given or computed with help of demons (procedures) installed in slots. Schank's (1972) earlier work in this direction on conceptual dependencies, involves the the notion of representing different actions or verbs in terms of language independent primitives (e.g., object transfer, idea transfer). The idea is to represent all the paraphrases of a single idea with the same representation (e.g., Mary gave the ball to John; John got the ball from Mary).

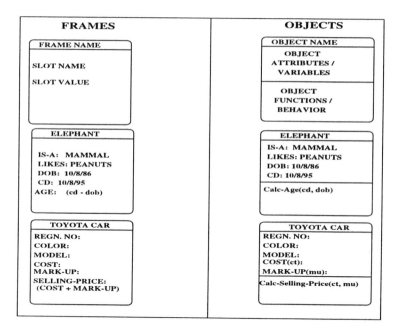

Figure 2.2.   Frames and Objects

Object-Oriented representation, a recent knowledge representational formalism from research in artificial intelligent and software engineering has some similarities with frames as shown in Figure 2.2. It is a highly attractive idea, combining as it does both development from the theory of programming languages, and knowledge representation. The object-oriented representational formalism identifies the real-world objects relevant to a problem as humans do, the attributes of those objects, and the processing operations (methods) in which they participate. Some similarities with frame-based class hierarchies and various procedures (methods in objects) attached to the slots are evident. However, demons or procedures in frames are embedded in the slots, whereas in objects procedures or methods and attributes are represented separately. This delineation of methods from attributes provides them strong encapsulation properties which makes them attractive from a software implementation viewpoint.

The four expert system architectures are now briefly described in rest of the subsections.

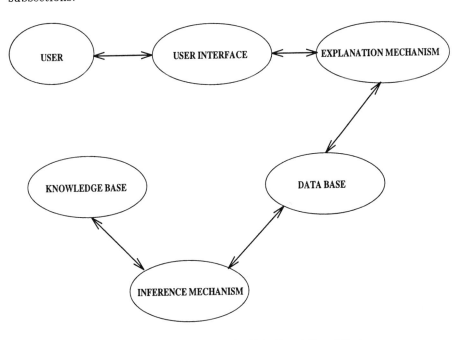

Figure 2.3.   Components of a Rule Based Expert System

2.2.2   Rule Based Architecture

The basic components of a rule based expert system are shown in Figure 2.3. The knowledge base contains a set of production rules (that is, IF . . . THEN rules). The IF part of the rule refers to the antecedent or condition part and the THEN part refers to the consequent or action part.

The database component is a repository for data required by the expert system to reach its conclusion/s based on the expertise contained in its knowledge base. That is, in certain types of expert systems, the knowledge base though endowed with expertise cannot function unless it can relate to a particular situation in the problem domain. For example, data on an applicant's present credit rating in a loan analysis expert system needs to be provided by the user to enable the expert system to transform the information contained in the knowledge base into advice. This information is stored in a data base.

The inference mechanism in a rule based system compares the data in the database with the rules in the knowledge base and decides which rules in the knowledge base apply to the data. Two types of inference mechanisms are generally used in expert systems, namely, forward chaining (data driven) and backward chaining (goal driven). In forward chaining, if a rule is found whose antecedent matches the information in the data base, the rule fires, that is, the rule's THEN part is asserted as the new fact in the database. In backward chaining on the other hand, the system forms a hypothesis that corresponds to the THEN part of a rule and then attempts to justify it by searching the data base or querying the user to establish the facts appearing in the IF part of the rule or rules. If successful, the hypothesis is established, otherwise another hypothesis is formed and the process is repeated. An important component of any expert system is the explanation mechanism. It uses a trace of rules fired to provide reasons for the decisions reached by the expert system.

The user of an expert system can be a novice user or an expert user depending upon the purpose of the expert system. That is, id the expert system is aimed at training operators in a control centre or for education, it is meant for novice users, whereas if it is aimed at improving the quality of decision making in recruitment of salespersons, it is meant for an expert user. The user characteristics determines to a large extent the type of explanation mechanism or I/O (Input/Output) interface required for a particular expert system. The input-output interface shown in Figure 2.1 permits the user to communicate with the system in a more natural way by permitting the use of simple selection menus or the use of a restricted language which is close to a natural language.

Many successful expert systems using the rule based architecture been have been built, including MYCIN, a system for diagnosing infectious diseases (Shortliffe 1976), XCON, a system developed for Digital Equipment Corp. for configuring computer systems (Kraft et al. 1984), and numerous others.

2.2.3 Rule and Frame(Object) Based Architecture

In contrast to rule base systems certain expert systems consists of both heuristic knowledge and relational knowledge. The relational knowledge describes explicit or implicit relationships among various objects/entities or events. Such knowledge is usually represented by semantic nets, frames or by objects. Figure 2.4 shows an object-oriented network for cars.

The *is-a* link is known as the inheritance or generalization-specialization link and *is-a-part-of* is known as the compositional or whole-part link. These two relational links give lot of expressive power to object-oriented representation. They help to express the hierarchies (classes) and aggregations existing in the domain. That is, *VOLKSWAGEN* and *MAZDA 131* objects are of type *SMALL*

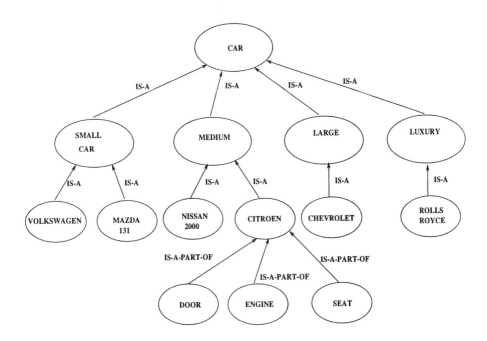

Figure 2.4.   An Object-Oriented Network

CAR. Such hybrid systems use both deductive and default (inheritance) reasoning strategies for inferencing with knowledge.

Expert systems with hybrid architectures like CENTAUR (Aitkins 1983) - a system for diagnosing lung diseases use both deductive and default reasoning for inferencing with knowledge.

2.2.4   Model Based Architecture

The previous two architectures use heuristic or shallow knowledge in the form of rules. The model based architecture on the other hand, uses an additional deep model (derived from first principles) which gives them an understanding of the complete search space over which the heuristics operate. This makes two kinds of reasoning possible: a) the application of heuristic rules in a style similar to the classical expert systems; b) the generation and examination of a deeper search space following classical search techniques, beyond the range of heuristics.

Various models can be used to describe and reason about a technical system. These include anatomical models, geometric models, functional models and causal models (Steels 1989). Anatomical models focus on the various components and their part-whole relations, whereas geometrical models focus on the general layout of the geometrical relations between components. Functional models predict the behavior of a device based on the functioning of its components. On the other hand, a causal model knows about the causal connections between various properties of the components but unlike a functional model does not know how each of the components actually works. Causal models represent an alternative, more abstract, view of a device which is particularly effective for diagnosis in cooperation with a functional model.

First generation expert systems are brittle, in the sense that as soon as situations occur which fall outside the scope of the heuristic rules, they are unable to function at all. In such a situation, second generation systems fall back on search which is not knowledge-driven and therefore potentially very inefficient. However, because these traditional search techniques can theoretically solve a wider class of problems, there is a graceful degradation of performance.

Model based reasoning strategies have been applied on various problems, e.g., XDE (Hamscher 1990), a system for diagnosing devices with heuristic structure and known component failure modes, HS-DAG (Ng 1991), a model-based system for diagnosing multiple fault cases in continuous physical devices, DIAGNOSE (Wang and Dillon 1992), a system for diagnosing power system faults, etc.

2.2.5   Blackboard Architecture

Blackboard systems are a class of systems which can include all different representational and reasoning paradigms discussed in this section. They are composed of three functional components, namely, knowledge sources component, blackboard component, and control information component respectively.

The knowledge sources component represents separate and independent sets of coded knowledge, each of which is a specialist in some limited area needed to solve a given subset of problems. The blackboard component, a globally accessible data structure, contains the current problem state and information needed by the knowledge sources (input data, partial solutions, control data, alternatives, final solutions). The knowledge sources make changes to the blackboard data that incrementally lead to a solution. The control information component may be contained within the knowledge sources , on the blackboard, or possibly in a separate module. The control knowledge monitors the changes to the blackboard and determines what the immediate focus of attention should be in solving the problem. HEARSAY-II (Erman 1980) and HEARSAY III (Balzer 1980; Erman 1981) - a speech understanding project at Stanford University is a well known example of a blackboard architecture.

2.2.6   Case-Based Reasoning Systems

In some domains (e.g. law), it is either not easy or possible to represent the knowledge using rules or objects. In these domains one may need to go back to records of individual cases that record primary user experience. Case-based reasoning is a subset of the field of artificial intelligence that deals with storing past experiences or cases and retrieving relevant ones to aid in solving a current problem. In order to facilitate retrieval of the cases relevant to the current problem a method of indexing these cases must be designed. There are two main components of a case-based reasoning system, namely, the case-base where the cases are stored and the case-based reasoner . The case-based reasoner consists of two major parts:

- mechanism for relevant case retrieval.

- mechanism for case adaptation.

Thus given a specification of the present case, the case-based reasoning system searches through the data base and retrieves cases that are closest to the current specification.

The case adaptor notes the differences between the specification of the retrieve d cases and the specification of the current case, and suggests alternatives to the retrieved cases in order to best meet the current situation.

Case based reasoners can be used in open textured domains such as the law or des ign problems. They reduce the need for intensive knowledge acquisition and try to use past experiences directly.

2.2.7   Some Limitations of Expert System Architectures

Rule-based expert systems suffer from several limitations . Among them is the limitation that they are too hard-wired to process incomplete and incorrect information. For this reason they are sometimes branded as "sudden death" systems. This limits their application especially in real-time systems where incomplete information, incorrect information, temporal reasoning, etc. are major system requirements. Further, knowledge acquisition in the form of extraction of rules from a domain expert is known to be a long and tedious process.

Model based systems overcome the major limitations of rule-based systems. However, model based systems are slow and may involve exhaustive search. The response time deteriorates further in systems which require temporal reasoning and consist of noisy data. Also, it may not be always possible to build one. Blackboard systems which combine disparate knowledge sources try to maximize the benefits of rule-based and model-based systems. The major problem, however, with these systems lies in developing an effective communication medium between disparate knowledge sources. Further, given the use of multiple knowledge sources it is not easy to keep track of the global state of the system.

Overall, these architectures have difficulty handling complex problems where the number of combinatorial possibilities are large and/or where the solution has a non-deterministic nature, and mathematical models do not exist. Artificial neural networks have been successfully used for these types of problems.

2.3 ARTIFICIAL NEURAL NETWORKS

The research in artificial neural networks[1] has been largely motivated by the studies on the function and operation of the human brain. It has assumed prominence because of the development of parallel computers and as stated in the previous chapter, the less than satisfactory performance of symbolic AI systems in pattern recognition problems like speech and vision.

The word 'neural' or 'neuron' is derived from the neural system of the brain. The goal of neural computing is that by modeling the major features of the brain and its operation, we can produce computers that can exhibit many of the useful properties of the brain. The useful properties of the brain include parallelism, high level of interconnection, self organization, learning, distributed processing, fault tolerance and graceful degradation. Neural network computational models developed to realize these properties are broadly grouped under two categories, namely, supervised learning and unsupervised learning. In both types of learning a representative training data set of the problem domain is required. In supervised learning the training data set is composed of input and target output patterns. The target output pattern acts like an external "teacher" to the network in order to evaluate its behavior and direct its subsequent modifications. On the other hand, in unsupervised learning the training

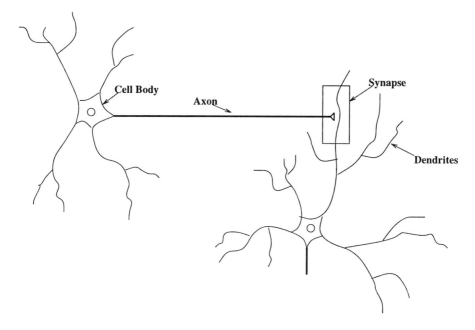

Figure 2.5.   A biological neuronBiological Neuron

data set is composed solely of input patterns. Hence, during learning no comparison with predetermined target responses can be performed to direct the network for its subsequent modifications. The network learns the underlying features of the input data and reflects them in its output. There are other categories like rote learning which are also used for categorization of neural networks.

Although, numerous neural network models have been developed in these categories, the description here is limited to the following:

- Perceptron (Supervised)

- Multilayer Perceptron (Supervised)

- Kohonen nets (Unsupervised)

- Radial Basis Function Nets (Unsupervised and Supervised)

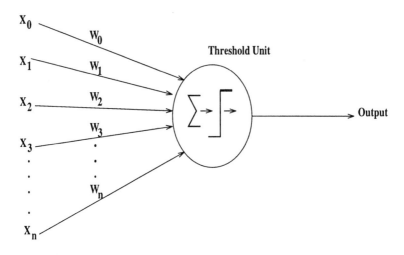

Figure 2.6.   The Perceptron

2.3.1   *Perceptron*

The perceptron was the first attempt to model the biological neuron which is shown in Figure 2.5 (Davlo et al. 1991; Beale et al. 1990). It dates back to 1943 and was developed by McCulloch and Pitts. Thus as a starting point, it is useful to understand the basic function of the biological neuron which in fact reflects the underlying mechanism of all neural models. The basic function of a biological neuron is to add up its inputs and produce an output if this sum is greater than some value, known as the threshold value. The inputs to the neuron arrive along the dendrites, which are connected to the outputs from other neurons by specialized junctions called synapses. These junctions alter the effectiveness with which the signal is transmitted. Some synapses are good junctions, and pass a large signal across, whilst others are very poor, and allow very little through. The cell body receives all these inputs, and fires if the total input exceeds the threshold value.

The perceptron shown in Figure 2.6 models the features of the biological neuron as follows:

a. The efficiency of the synapses at coupling the incoming signal into the cell body is modeled by having a weight associated with each input to the neuron. A more efficient synapse, which transmits more of the signal, has a correspondingly larger weight, whilst a weak synapse has a small weight.

b. The input to the neuron is determined by the weighted sum of its inputs
$\Sigma_{i=0}^{n} w_i x_i$
where x_i is the ith input to the neuron and w_i s its corresponding weight.

c. The output of the neuron which is on (1) or off (0) is represented by a step or heaveside function. The effect of the threshold value is achieved by biasing the neuron with an extra input x_0 which is always on (1). The equation describing the output can then be written as:
$y = f_h[\Sigma_{i=0}^{n} w_i x_i]$

The learning rule in perceptron is a variant on that proposed in 1949 by Donald Hebb, and is therefore called Hebbian learning. It can be summarized as follows:

1. set the weights and thresholds randomly.

2. present an input

3. calculate the actual output by taking the thresholded value of the weighted sum of the inputs

4. alter the weights to reinforce correct decisions and discourage incorrect decisions, i.e. reduce the error.

This is the basic perceptron learning algorithm. Modifications to this algorithm have been proposed by the well known Widrow and Hoff's (1960) delta rule. They realized that It would be best to change the weights by a lot when the weighted sum is a long way from the desired value, whilst altering them only slightly when the weighted sum is close to that required. They proposed a learning rule known as the Widrow-Hoff delta rule, which calculates the difference between the weighted sum and the required output, and calls that the error. The learning algorithm basically remains the same except for step 4 which is replaced as follows:

4. Adapt weights - Widrow-Hoff delta rule
$\Delta = d(t) - y(t)$
$w_i(t + 1) = w_i(t) + \eta \Delta x_i(t)$

$$d(t) = \begin{cases} 1, & \text{if input from class A} \\ 0, & \text{if input from class B} \end{cases}$$

where Δ is the error term, $d(t)$ is the desired response and $y(t)$ is the actual response of the system. Also $0 \leq \eta \leq$ is a positive gain function that controls the adaptation rate.

The delta rule uses the difference between the weighted sum and the required output to gradually adapt the weights for achieving the desired output value. This means that during the learning process, the output from the unit is not passed through the step function, although the actual classification is effected by the step function.

The perceptron learning rule or algorithm implemented on a single layer network, guarantees convergence to a solution whenever the problem to be solved is linearly separable. However, for the class of problems which are not linearly separable, the algorithm does not converge. This was first demonstrated by Minsky and Papert in 1969 in their influential book, *Perceptrons* using the well known XOR example. This, in fact dealt a mortal blow to the area, and sent it into hibernation for the next seventeen years till the development of multilayer perceptrons (popularly known as backpropagation) by Rumelhart et al. (1986). If the McCulloch-Pitts neuron was the father of modern neural computing, then Rumelhart's multilayer perceptron is its child prodigy.

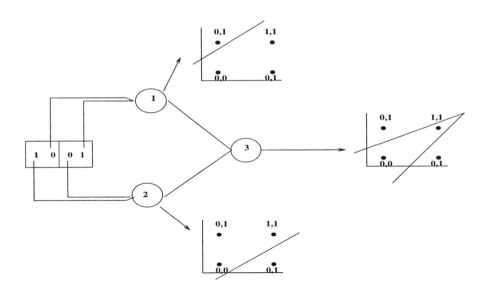

Figure 2.7.   Combination of Perceptrons to solve XOR Problem

2.3.2 Multilayer Perceptrons

An initial approach to solve problems that are not linearly separable problems was to use more than one perceptron (as shown in Figure 2.7 for the XOR problem), each set up to identify small, linearly separable sections of the outputs into another perceptron, then combining their outputs into another perceptron, which would produce a final indication of the class to which the input belongs. The problem with this approach is that the perceptrons in the second layer do not know which of the real inputs were on or not. They are only aware of the inputs from the first layer. Since learning involves strengthening the connections between active inputs and active units (Hebb 1949), it is impossible to strengthen the correct parts of the network, since the actual inputs are masked by the intermediate(first) layer. Further, the output of neuron being on (1) or off (0), gives no indication of the scale by which the weights need to be adjusted. This is also known as the credit assignment problem.

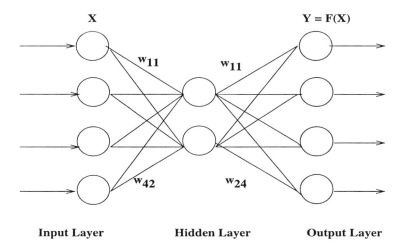

Figure 2.8.   Multilayer Perceptron

In order to overcome this problem of linear inseparability and credit assignment a new learning rule was developed by Rumelhart et al. (1986). They used a three layer multipreceptron model/network shown in Figure 2.8. The network has three layers; an input layer, an output layer, and a layer in between, the so called the hidden layer. Each unit in the hidden layer and the output layer is like a perceptron unit, except that the thresholding function is a non-linear sigmoidal function (shown in Figure 2.9),
$$f(net) = 1/(1 + e^{-knet})$$

where k is a positive constant that controls the "spread" of the function. Large values of k squash the function until as $k \to \infty$, $f(net) \to Heaviside function$.

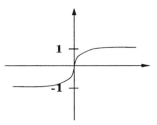

Figure 2.9.   sigmoidal functionSigmoidal Function

It is a continuously differentiable i.e. smooth everywhere and has a simple derivative. The output from the non-linear threshold sigmoid function is not 1 or 0 but lies in a range, although at the extremes it approximates the step function. The non-linear differentiable sigmoid function enables one to overcome the credit assignment problem by providing enough information about the output to the units in the earlier layers to allow them to adjust the weights in such a way as to enable convergence of the network to a desired solution state.

The learning rule which enables the multilayer perceptron to learn complex non-linear problems is called the generalized delta rule or the backpropagation rule. In order to learn successfully the value of the error function has to be continuously reduced for the actual output of the network to approach the desired output. The error surface is analogous to a valley or a deep well, and the bottom of the well corresponds to the point of minimum energy/error. This is achieved by adjusting the weights on the links between units in the direction of steepest downward descent (known as the gradient descent method). The generalized delta rule (Mclleland, et al. 1986) does this by calculating the value of the error function for a particular input, and then back-propagating (hence the name) the error from one layer to the previous one. The error term *delta* for each hidden unit is used to alter the weight linkages in the three layer network to reinforce the correct decisions and discourage incorrect decisions, i.e. reduce the error.

The learning algorithm is as follows (Beale et al. 1990):

- Initialize weights and thresholds. Set all the weights and thresholds to small random values

- Present input and target output patterns.
  Present input $X_p = x_0, x_1, x_2, ..., x_{n-1}$ and target output $T_p = t_o, t_1, ..., t_{m-1}$

where n is the number of input nodes and m is the number of output nodes. Set w_0 to be $-\theta$, the bias and x_0 to be always 1. For classification, T_p is set to zero except for one element set to 1 that corresponds to the class that X_p is in.

- Calculate actual output. Each layer calculates:
  $$y_{pj} = f[\Sigma_{i=0}^{n-1} w_{ij} x_i] = f(net_{pj}) = 1/(1 + e^{-knet_{pj}})$$

  where w_{ij} represents weight from unit i to unit j, net_{pj} is the net input to unit j for pattern p, and y_{pj} is the sigmoidal output or activation corresponding to pattern p at unit j and is passed as input to the next layer(i.e. the output layer). The output layer outputs values o_{pj}, which is the actual output at unit j of pattern p.

- Calculate the error function E_p for all the patterns to be learnt
  $$E_p = 1/2[\Sigma_j (t_{pj} - o_{pj})^2]$$
  where t_{pj} is the target output at unit j of pattern p.

- Starting from the output layer, project the error backwards to each layer and compute the error term δ for each unit.

  For output units
  $$\delta_{pj} = k o_{pj}(1 - o_{pj})(t_{pj} - o_{pj})$$
  where δ_{pj} is an error term for pattern p on output unit j

  For hidden units
  $$\delta_{pj} = k o_{pj}(1 - o_{pj})\Sigma_r \delta_{pr} w_{jr}$$
  where the sum is over the r units above unit j, and δ_{pj} is an error term for pattern p on unit j which is not an output unit.

- Adapt weights for output hidden units
  $$w_{ij}(t + 1) = w_{ij}(t) + \eta \delta_{pj} o_{pj}$$
  where w_{ij} represents the weights from unit i to unit j at time t and η is a gain term or the learning rate.

2.3.3   Radial Basis Function Net

The Radial Basis Function (RBF) inxxsupervised learning, radial basis function nets net is a 3-layer feedforward network consisting of an input layer, hidden layer and output layer as shown in Figure 2.10. The mapping of the input vectors to the hidden vectors is non-linear, whereas the mapping from hidden layer to output layer is linear. There are no weights associated with the connections from input layer to the hidden layer. In the radial basis function net, the N activation functions g_j of the hidden units correspond to a set of radial basis functions that span the input space and map. Each function $g_j(||\mathbf{x} - \mathbf{c_j})||)$ is centered about some point c_j of the input space and transforms the input vector $\mathbf{x}$ according to its Euclidean distance, denoted by $||\ ||$, from the center c_j. Therefore the function g_j has its maximum value at c_j. It has further an associated receptive field which decreases with the distance between input and center and which could overlap that of the functions of the neighboring neurons of the hidden layer.

The hidden units are fully connected to each output unit y_i with weights w_{ij}. The outputs y_i are thus linear combinations of the radial basis functions i.e.

$$y_i = \Sigma_{j=1}^{N} w_{ij} g_j(||\mathbf{x} - \mathbf{c_j}||)$$

One such set of radial basis functions that is frequently used are Gaussian Activation functions centered on the mean value c_j and with a receptive field whose size is proportional to the variances fixed for all units:

$$g_j(||\mathbf{x} - \mathbf{c_j}||) = exp(||\mathbf{x} - \mathbf{c_j}||^2/4\sigma^2)$$
where:
$$\sigma = d/\sqrt{(2N)}$$
with N number of (hidden) RBF units and d the maximum distance between the chosen centers. One could use an unsupervised learning approach to determine the centers c_j and the width σ of the receptive fields, see Haykin (1994). One could then use the delta learning rule to determine the weights between the hidden units and the output units. Since the first layer can be said to be trained using an unsupervised approach and the second using a supervised approach one could consider such a net as a hybrid net.

The RBF network is an important approximation tool because like spline functions it provides a quantifiable optimal solution to a multi-dimensional function approximation problem under certain regularization constraints concerning the smoothness of the class of approximating RBF functions.

The applications of multilayer perceptrons can be found in many areas including natural language processing (NETalk- Sejnowski and Rosenberg 1987), prediction (airlines seat booking, stock market predictions, bond rating, etc.) and fault diagnosis.

The use of neural networks is dependent upon the availability of large amounts of data. In some cases, both input and output patterns are available or known, and we can use supervised learning techniques like backpropagation, whereas in other cases the output patterns are not known and the network has to independently learn the class structure of the input data. In such cases, unsupervised

Radial Basis Function $g_j \, (\|x - c \, \|) = v_{\,j}$

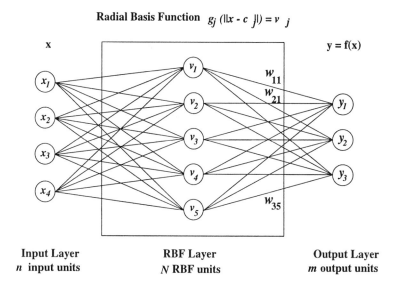

Figure 2.10.   Radial Basis Function Net

learning techniques like Kohonen nets are used. This technique is described in the following sections.

2.3.4 Kohonen Networks

Kohonen's self-organizing maps are characterized by a drive to model the self-organizing and adaptive learning features of the brain. It has been postulated that the brain uses spatial mapping to model complex data structures internally (Kohonen 1990). Much of the cerebral cortex is arranged as a two-dimensional plane of interconnected neurons but it is able to deal with concepts in much higher dimensions. The implementations of Kohonen's algorithm are also predominantly two dimensional. A typical network is shown in Figure 2.11 . The network shown is a one-layer two-dimensional Kohonen network. The neurons are arranged on a flat grid rather in layers as in a multilayer perceptron. All inputs connect to every node(neuron) in the network. Feedback is restricted to lateral interconnections to immediate neighboring nodes. Each of the nodes in the grid is itself an output node.

The learning algorithm organizes the nodes in the grid into local neighborhoods or clusters as shown in Figure 2.12 that act as feature classifiers on the input data. The biological justification for that is the cells physically close to the active cell have the strongest links. No training response is specified for

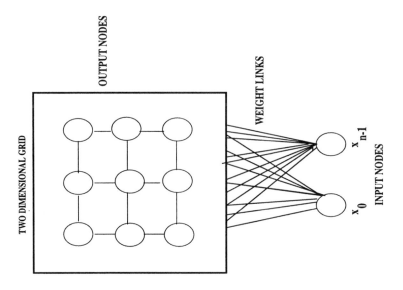

Figure 2.11.   Kohonen's Self Organizing Feature Map

any training input. In short, the learning involves finding the closest matching
node to a training input and increasing the similarity of this node, and those
in the neighboring proximity, to the input. The advantage of developing neighborhoods is that vectors that are close spatially to the training values will still
be classified correctly even though the network has not seen them before, thus
providing for generalization.

The learning algorithm is notionaly simpler than the backpropagation algorithm as it does not involve any derivatives, and is as follows (Beale et al.
1990):

1. Initialize network

    Define $w_{ij}(t)$ $(0 \leq i \leq n-1)$ to be the weight from input i to node(unit) j
    . Initialize weights from the n inputs to the nodes to s mall random values.
    Set the initial radius of the neighborhood around node j, $N_j(0)$, to be large.

2. Present input

    Present input $x_0(t), x_1(t), x_2(t), \ldots\ldots, x_{n-1}(t)$, where $x_i(t)$ is the input to node
    i at time t.

3. Calculate distances

    Compute the distance d_j between the input and each output node j, given
    by

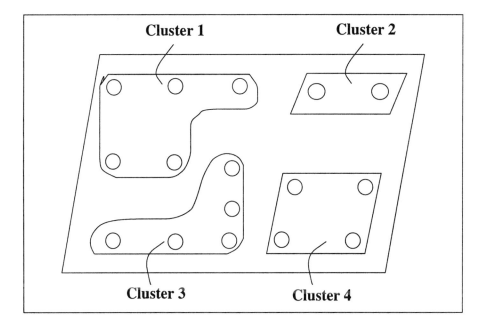

Figure 2.12.   Kohonen Network Clusters

$$d_j = \Sigma_{i=0}^{n-1}(x_i(t) - w_{ij}(t))^2$$

4. Select the minimum distance using the Euclidean distance measure

   Designate the output node with minimum d_j to be $j*$.

5. Update weights

   Update weights for node $j*$ and its neighbors, defined by the neighborhood size $N_{j*}(t)$. New Weights are:
   $w_{ij}(t+1) = w_{ij}(t) + \eta(t)(x_i(t) - w_{ij}(t))$
   For j in $N_{j*}(t)$, $0 \leq i \leq n-1$

   The term $\eta(t)$ is a gain term $(0 < \eta(t) < 1)$ that decreases in time, so slowing the weight adaptation. The neighborhood $N_{j*}(t)$ decreases in size as time goes on, thus localizing the area of maximum activity.

6. Repeat steps 2 to 5.

   The most well known application of self-organizing Kohonen networks is the Phonetic typewriter (Kohonen 1990) used for classification of phonemes in real time. Other applications include evaluating the dynamic security of power systems (Neibur et al. 1991).

2.4 FUZZY SYSTEMS

As mentioned in chapter 1, fuzzy systems provide a means of dealing with inexactness, imprecision as well as ambiguity in everyday life. In fuzzy systems relationships between imprecise concepts like "hot," "big," and "fat," "apply strong force," and "high angular velocity" are evaluated instead of mathematical equations. Although not the ultimate problem solver, fuzzy systems have been found useful in handling control or decision making problems not easily defined by practical mathematical models.

This section provides introduction to the reader on the following basic concepts used for construction of a fuzzy system:

- Fuzzy Sets

- Fuzzification of inputs

- Fuzzy Inferencing and Rule Evaluation

- Defuzzification of Fuzzy Outputs

2.4.1 Fuzzy Sets

Although fuzzy systems have become popular in the last decade, fuzzy set and fuzzy logic theories have been developed more than 25 years. In 1965 Zadeh wrote the original paper formulating fuzzy set and fuzzy logic theory. The need for fuzzy sets has emerged from the problems in the classical set theory.

In classical set theory, one can specify a set either by enumerating its elements or by specifying a function $f(x)$ such that if it is true for x, then x is an element of the set S. The latter specification is based on two-valued logic. If $f(x)$ is true, x is an element of the set, otherwise it is not. An example of such a set, S, is:

$$S = x : weight of person x > 90 kg$$

Thus all persons with weight greater than 90 kg would belong to the set S. Such sets are referred to as **crisp sets** .

Let us consider the set of "fat" persons. It is clear that it is more difficult to define a function such that if it is true, then the person belongs to the set of fat people, otherwise s/he does not. The transition between a fat and not-fat person is more gradual. The membership function m describing the relationship between weight and being fat is characterized by a function of the form given in Figure 2.13.

Such sets where the membership along the boundary is not clear cut but progressively alters are called fuzzy sets. The membership function defines the degree of membership of the set x : x fat. Note this function varies from 0 (not a member) to 1 (definitely a member).

From Figure 2.13, it can be seen that the truth value of a statement,

$person x is fat$,

varies from 0 to 1. Thus in fuzzy logic, the truth value can take any value in this range, noting that a value of 0 indicates that the statement is false and a value of 1 indicates that it is totally true. A value less than 1 but greater than 0 indicates that the statement is partially true. This contrasts with the situation in two-valued logic where the truth value can only be 0 (false) or 1 (true).

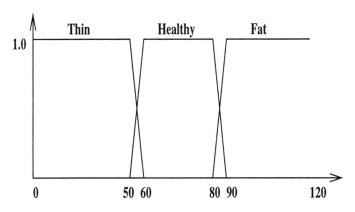

Figure 2.13.   Fuzzy Membership Function

2.4.2   Fuzzification of Inputs

Each fuzzy system is associated with a set of inputs which can be described by linguistic terms like "high," "medium" and "small" called fuzzy sets. Fuzzification is the process of determining a value to represent an input's degree of membership in each of its fuzzy sets. The two steps involved in determination of fuzzy value are:

- membership functions

- computation of fuzzy value from the membership function

Membership functions are generally determined by the system designer or domain expert based on their intuition or experience. The process of defining the membership function primarily involves:

- **defining the Universe of Discourse (UoD):** UoD covers the entire range of input values of an input variable. For example, the UoD for an input variable *Person Weight* covers the weight range of 0 to 120 Kilograms.

- **partitioning the UoD into different fuzzy sets:** A person's weight can be partitioned into three fuzzy sets and three ranges, i.e. 0-60, 50-90 and 80-120 respectively.

- **labeling fuzzy sets with linguistic terms:** The three fuzzy sets 0-60, 50-90, and 80-120 can be linguistically labeled as "thin," "healthy" and "fat" respectively.

- **allocating shape to each fuzzy set membership function:** Several different shapes are used to represent a fuzzy set. these include piecewise linear, triangle, bell shaped, trapezoidal (see Figure 2.14), and others. The shape is said to represent the fuzzy membership function of the fuzzy set.

Once the fuzzy membership function has been determined, the next step is to use it for computing the fuzzy value of a system input variable value. Figure 2.14 shows how the degree of membership or fuzzy value of a given system input variable X with value z can be computed.

2.4.3   Fuzzy Inferencing and Rule Evaluation

In order to express relationships between imprecise concepts and model the system's behavior, a fuzzy system designer develops a set of fuzzy IF-THEN rules in consultation with the domain expert. The IF part of a fuzzy rule is known as the antecedent and the THEN part is known as the consequent. The antecedent or antecedents of a fuzzy rule contain the degrees of membership (fuzzy inputs) calculated during the fuzzification of inputs process. For example consider the Fuzzy Rule 1:

IF share_price_is_decreasing AND trading_volume_is_heavy

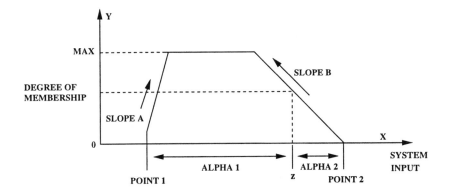

DEGREE OF MEMBERSHIP

1) COMPUTE ALPHA TERMS: ALPHA 1 = z - POINT 1, APLHA 2 = POINT 2 - z
2) IF (ALPHA 1<.0) OR (ALPHA 2 < 0) THEN DEGREE OF MEMBERSHIP = 0
   ELSE DEGREE OF MEMBERSHIP = MIN (ALPHA 1* SLOPE A, ALPHA 2* SLOPE B, MAX)

Figure 2.14.   Fuzzfication of Inputs

THEN market_order_is_sell

Here, the two antecedents *share_price_is_decreasing* and *trading_volume_is_heavy* are the rule's fuzzy antecedents. Further *share_price* and *trading_volume* are fuzzy inputs with fuzzy sets "decreasing," "stable" and "increasing," and "light," "moderate" and "heavy" respectively.

The consequent of the fuzzy rule is represented by the THEN part which in this case is *market_order_is_sell*. Generally, more than one fuzzy rule has the same fuzzy output. For example, Fuzzy Rule 2 can be :

IF share_price_is_decreasing AND trading_volume_is_moderate
THEN market_order_is_sell

In order to evaluate a fuzzy rule, rule strengths are computed based on the antecedent values and then assigned to a rule's fuzzy outputs. The antecedent values are computed based on the degree of membership of an input variable. For example, the fuzzy value of the antecedent *share_price_is_decreasing* will correspond to the degree of membership of the "decreasing" fuzzy set of *share_price* input variable. The most commonly used fuzzy inferencing method is the min-max method. In this inferencing method minimum operation is applied making the rule strength equal to the least-true or weakest antecedent value.

For example, say, for *share_price* of $ 50 , and *trading_volume* of 1000 contracts, the degree of membership values for fuzzy sets "decreasing," "heavy" and "moderate" are 0.7, 0.2, and 0.4 respectively rule strength of Fuzzy Rule 1 and 2 can be computed as follows:

Rule (firing) Strength of Fuzzy Rule $1 = min(0.7, 0.2) = 0.2$
Rule (firing) Strength of Fuzzy Rule $2 = min(0.7, 0.4) = 0.4$

The fuzzy output of selling shares is carried out to a degree reflected by the rule's strength. Since two rules apply to the same fuzzy output, the strongest rule strength is used in this case. This is done by applying the *max* operation as follows:

$market\_order\_is\_sell = max(min(0.7, 0.2), min(0.7, 0.4)) = 0.4$

2.4.4   Defuzzification of Outputs

Defuzzification is required to determine crisp values for fuzzy outputs or actions like *market_order_is_sell*. Another reason for defuzzification is to resolve conflict among competing outputs or actions. For example, competing output *market_order_is_hold* can also be trigerred based with a rule strength of say, 0.6 on the *share_price* of $ 50, and *trading_volume* of 1000 contracts with a Fuzzy Rule 3:

IF share_price_is_stable AND trading_volume_is_small
THEN market_order_is_hold

In order to resolve the conflict between competing outputs one of the common defuzzification techniques used is the center-of-gravity method . For a general case, the following steps (see Figure 2.15) are taken for defuzzification:

- determine the centroid point on the x-axis for each competing output membership function. In Figure 2.15 it is 20 and 50 for the competing outputs A and B respectively.

- compute new output membership trapezoidal areas based on rule strengths of 0.4 and 0.6 for competing outputs A and B respectively as shown in Figure.

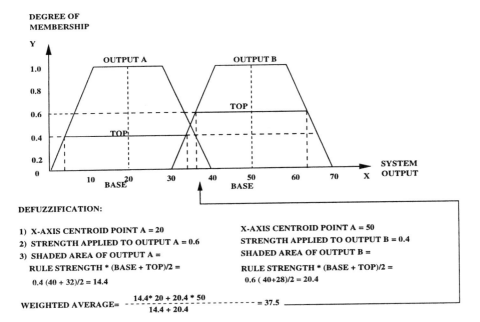

DEGREE OF
MEMBERSHIP

DEFUZZIFICATION:

1) X-AXIS CENTROID POINT A = 20   X-AXIS CENTROID POINT A = 50
2) STRENGTH APPLIED TO OUTPUT A = 0.6   STRENGTH APPLIED TO OUTPUT B = 0.4
3) SHADED AREA OF OUTPUT A =   SHADED AREA OF OUTPUT B =
   RULE STRENGTH * (BASE + TOP)/2 =   RULE STRENGTH * (BASE + TOP)/2 =
   0.4 (40 + 32)/2 = 14.4   0.6 (40+28)/2 = 20.4

$$\text{WEIGHTED AVERAGE} = \frac{14.4 * 20 + 20.4 * 50}{14.4 + 20.4} = 37.5$$

Figure 2.15.   Defuzzfication of Outputs

- compute the defuzzified output by taking a weighted average of x-axis centroid points and the output membership areas (for rule strengths 0.4 and 0.6 respectively).

Fuzzy systems have been applied in a number of industrial control applications under the following conditions:

- When the system is highly nonlinear and mathematical models do not exist or are difficult to build.

- When parameters of the system change.

- Where sensor accuracy is a problem.

- where the system may receive conflicting or uncertain input but must still make correct control decisions.

The satisfaction of these conditions provides number of advantages to systems modeled on fuzzy logic. These include increased efficiency of control cycles which may result in reduced power consumption or higher stability, requirement of less expensive and less accurate sensors because of the ability to model imprecision in sensor data, and ability to extract knowledge from an expert using everyday language. Inspite of a number of advantages fuzzy systems also suffer from some disadvantages. These include problems with learning of control process, tuning/optimization of membership functions, and optimization of rules. Genetic algorithms which are good at optimization problems are described in the next section.

2.5   GENETIC ALGORITHMS

Genetic algorithms refer to a class of adaptive search procedures based on principles derived from the dynamics of natural population genetics (Grefenstette 1990). These search procedures exploit the mechanics of natural selection and natural genetics. Survival of the fittest is thus combined with mechanisms for generation of new candidates to form search algorithms.

John Holland (1975) first developed genetic algorithms. His aim was to abstract and rigorously explain the adaptive process of natural systems. From this, software systems that capture the evolutionary mechanisms of natural systems have been developed. These genetic algorithms provide robust search procedures for complex spaces.

Some of the general characteristics of genetic algorithms include (Goldberg 1989):

- working with a coding of the parameter set

- searching from a population of points, rather than a single point

- using a pay-off function

- using probabilistic transition rules.

We will now discuss these characteristics. First, the parameter set used in a problem is coded as a finite length string. To explain this, consider the following:

- Problem : determine the best days for making interstate phone calls.

- In this case, the parameters are the seven days of the week. Therefore, to code the parameter set as a finite length string, a string of length 7 with each element representing a different day of the week is chosen.

In general, the parameter set is coded into:

$$A = a_1 \ a_2...a_n,$$

where $a_1\ a_2...a_n$ represents the individual parameters in the set of size n. The individual parameters are binary, with values of either 1 or 0. This encoding is required by the genetic algorithm for several reasons, including the flexibility to apply the three operators of reproduction, crossover, and mutation on the strings, as discussed below. New strings are created by applying these three string operators. Genetic algorithms actually work by processing and manipulating a group of coded strings. As these coded strings are used in determining the solution, genetic algorithms are regarded as working with a group of points simultaneously. To search effectively through the strings, some form of pay-off value is required for each string. These pay-off values are objective function values used to guide the search through the strings (Goldberg 1989).

The coding of the parameter set is usually over an extended alphabet and not just over 0 and 1. This extended alphabet includes a hash symbol #, representing a don't care or wild card symbol (Booker et al. 1990; Goldberg 1989). Thus, the # matches either a 0 or 1, as illustrated below:

- Suppose the following strings are in the population being considered:

  0110011   1111111
  0010010   0111010

- The string ##1##1#, represented in the extended alphabet, matches each string in the population above. To match this string, a string must have a 1 in both positions 3 and 6; the other elements are not relevant.

The flexibility of this notation is apparent, as large groups of patterns can be stated in a concise form. Thus, the notation greatly simplifies the analysis of genetic algorithms because it explicitly recognizes all the possible similarities in a population of strings.

There are three important operators commonly used in genetic algorithms; reproduction, crossover, and mutation. The first operator, reproduction, is a process in which individual strings are copied according to their objective function values (Goldberg 1989). These objective function values represent a measure of goodness that needs to be maximized. Copying strings with higher fitness values gives those strings a higher probability of contributing one or more of their offspring to the next generation. In the natural world, this is equivalent to natural selection, resulting in the survival of the fittest. Crossover occurs when members of the newly reproduced strings are mated at random. This is the second operator used by genetic algorithms. Goldberg (1989) argues that much of the power of genetic algorithms derives from the combination of reproduction and crossover which provides structured, though randomized, information exchange. Mutation, the third operator, usually plays a less significant role

To illustrate these operators, let string 1 be 11011 with an objective function value of 0.25, and string 2 be 11101 with an objective function value of 0.50.

- Applying the reproduction operator to the two strings results in string 2 being reproduced because it has a higher objective function value.

- Applying the crossover operator, assuming the crossover point in the string is the third position, results in the following:
  original strings: 11011 gives new strings: 11111
  and   11101   11001

- Applying the mutation operator to the two strings may result in the following (mutation being the random inverting of a bit in the string):
  original strings: 11011 gives new strings: 11010
  and   11101   11001

Comprehension of a genetic algorithm's manipulation of building blocks is crucial (Booker et al. 1990). Building blocks are high performance schemata that are combined to form strings with expected higher performance. The schemata or similarity templates are strings over the extended alphabet 0,1,# (Booker et al. 1990; Goldberg 1989). According to Booker et al. (1990):

A genetic algorithm rapidly explores the space of schemas, a very large space, implicitly rating and exploiting the schemas according to the strengths of the rules employing them.

Two useful schemata properties are (Goldberg 1989):

- order - the number of fixed positions present in each schema

- length - the defining length of each schema; that is, the distance between the first and last specific string positions.

The generating procedure samples the schemata with above average instances more regularly. This results in further confirmation of the useful schemata and degradation of the less useful. The overall average strength is thus increased. This leads to an ever-increasing criterion that a schema must meet to be above average. Also the generating procedure uses the distribution of instances rather than the most recent, best schemata. This helps the system to be more robust and to overcome local minima which misdirect development. The better than average building blocks are rapidly accumulated by the generating procedure. The strengths are determined by the regularities and interactions in the environment (Booker et al. 1990).

2.6 KNOWLEDGE DISCOVERY AND DATA MINING

Thanks to the computer revolution, organizations today have gigabytes of data in their computer systems and do not know what to do with it. Knowledge discovery in databases area has become significant in the light of the fact that organizations need to extract meaningful and novel knowledge from the gigabytes of stored data in order to become more productive and competitive. Knowledge discovery in data bases involves mining of data using intelligent methodologies (sometimes in conjunction with data abstraction methodologies like data warehouse (Inmon and Kelly 1993) and object-oriented data modeling), and meaningful interpretation of the mined data.

Data Mining using intelligent methodologies essentially consists of examining a database of stored examples to extract any underlying patterns in the data. In contrast to a Data Base Management System (DBMS) which will retrieve one or more records that satisfy a specification, the data mining approach looks for underlying patterns. If neural nets are used for data mining, these underlying patterns will be coded into the weights. If symbolic patterns such as production rules need to be extracted, then approaches as those based on deci sion trees, neural networks and Bayesian networks have been employed for extraction of symbolic rules (Fayyad et al. 1996). Chapter 5 in this book describes in more detail the motivations for knowledge discovery in databases and various hybrid applications which involve use of intelligent methodologies, genetic , and data warehouse and object-oriented methodologies.

2.7 OBJECT-ORIENTED METHODOLOGY

The recent research in object-oriented software engineering, artificial intelligence, and databases has shown objects provide a powerful and comprehensive mechanism for capturing the relationships between concepts in terms of their structure (Cox 1986; Booch 1986; Kim et al. 1988; Myer 1988; Unland et al. 1989; Coad et al. 1990, 92; Rumbaugh 1990; Dillon and Tan 1993, and others). Computationally, they offer more powerful encapsulation than other knowledge representation formalisms like frames (Dillon and Tan 1993). They also have special features like message passing and polymorphism which make them more attractive computationally than other symbolic knowledge representation formalisms semantic networks and frames.

As a result of research in these two communities and in artificial intelligence, a common set of characteristics which define the general object-oriented model have emerged. These characteristics are: unique object identifier, data[2] and operations, encapsulation, inheritance, composition, message passing and polymorphism.

Object-oriented methodology has been used in this book in the context of its knowledge modeling features like inheritance and composability, and software implementation features like encapsulation, message passing, polymorphism and reusability. These features are now briefly described in the following subsections.

2.7.1   Inheritance and Composability

The structured representation of real world concepts includes various kinds of relationships.   Inheritance and composition are two constructs which have given expressive power to knowledge representation formalisms for relating real world concepts in a meaningful fashion. These two relational constructs are also important features of the object-oriented methodology.

Inheritance allows real-world objects to be organized into classes so that objects of the same class can have similar properties, and more specific classes or subclasses may inherit properties of the more general classes. In fact inheritance is also used as a mechanism for default reasoning in computer applications where knowledge is organized hierarchically. Put another way inheritance makes extensive use of abstraction. The classes at the higher levels in the hierarchy are expressed as generalizations of the lower level classes (which are their specializations). The relationship is expressed through an *IS-A* link. The two basic modes in which inheritance can be realized are single and multiple inheritance. In single inheritance, there is a single *IS-A* or *INSTANCE-OF* link from a class or instance at a lower level to class at a higher level, whereas in multiple inheritance there can be multiple *IS-A* links from a class at a lower hierarchical level and classes at higher level/s.

Aggregation or composition is another common form of relationship observed in hierarchical structures. It represents the whole-part concept used in our everyday life (Britannica 1986) and is incorporated in the object-oriented model. A composite object consists of a collection of two or more heterogeneous, related objects referred to as component objects. The component objects have a *PART-OF* relationship to the composite object. Each component object may, in turn, be a composite object, thus resulting in a "component-of" hierarchy. There are other forms of non-hierarchical relationships like *ASSOCIATION* (Dillon and Tan 1993) and others obtained from entity-relationship models (Coad and Yourdon 1990; Hawryszkiewyz 1991) which are also modeled as extensions to the object-oriented methodology.

2.7.2   Encapsulation

Encapsulation is a property of object-oriented models by which all the information (i.e. data and behavior) of a object is captured under one name, that is the object name. For example a real world object like *Chair* encapsulates attribute values that define the *Chair*, methods that are applied to change the attributes of *Chair*, and other related information. This notion of encapsulating information related to a particular concept does not distinguish between the type of attributes or methods used to define that concept. In other words, it is a useful software implementation methodology for realizing heterogeneous architectures involving more than one intelligent methodology.

2.7.3   Message Passing

When integrating two fundamentally different paradigms it becomes important to ascertain the communication mechanism between the two from a computational point of view. The object-oriented model provides a uniform communication mechanism between objects. In order to communicate, objects pass messages. A message defines the interface between the object and its environment. Essentially, a message consists of the name of the receiving object, a message name or method selector and arguments of the selected method.

2.7.4   Polymorphism

In large scale domains, genericity is an important element to promote comprehensibility and intelligibility of the domain . Polymorphism (Pressman 1992; Dillon and Tan 1993) is another feature of the object-oriented models which brings about the genericity in terms of the behavior of different objects or concepts in the domain. It is one of the key features of object-oriented programming(Blair et. al. 1989; Pressman, 1992; Dillon and Tan 1993). It allows object-oriented systems to separate a generic function from its implementation. These generic functions or virtual functions (as they are called sometimes) provide the ability to carry out function overloading (Berry 1988; Dillon and Tan 1993).

2.7.5   Reusability

Reusability is an important characteristic of a high quality software. That is, software should be designed and implemented so that it can be reused in many different programs. Reuse in object-oriented software engineering occurs in principle through inheritance. If two classes have a similar set of attributes then then the similar attributes can be used to create an abstract class. Then these two original classes inherit from the abstract class (which may not be always meaningful in itself (Jacobson et al. 1995)). Another way is to reuse application classes by creating a class library. For example, a customer class modeled in a bank ATM (Automatic Teller Machine) software application can be reused in a loan approval software application which also models the customer class.

2.8 AGENTS AND AGENT ARCHITECTURES

Intelligent agents and multi-agent systems are one of the most important emerging technologies in computer science today. A dictionary definition of the term "agent" is: *An entity authorized to act on another's behalf*. The definitional term "another's behalf" in this book refers to a problem solver or a user. The definitional term "entity" refers to the agent which maps percepts (e.g. inputs) to actions for achieving a set of tasks or goals assigned to it in a largely non-deterministic environment. The four ingredients Percept, Action, Goal and Environment (PAGE) of an agent have been derived from Russell and Norvig (1995). Table 2.1 provides a PAGE description of systems modeled as agent types. In fact Russell and Norvig (1995) define an agent as consisting of an architecture and a program (*agent = architecture + program*). As a software program it maps percepts to actions and in the process exhibits the following characteristics:

Table 2.1.   PAGE Description of Agent Types

Agent Type	Percepts	Actions	Goals	Environment
Fruit Storage Control System	Temprature, Humidity Readings	Control Fruit Weight Loss Control Fruit Disease	Retain Fruit Freshness	Controlled Storage
Oil Dewaxing System	Oil Type, Oil Inflow Rate, Tank Oil Level	Adjust Oil Inflow Rate Valve, Adjust Oil Outflow Rate Valve	High Quality Dewaxed Lubricant Oil	Petroleum Plant
Inventory Mgt. System	Sales Forecast Exisiting Stock	Stockpile, Liquidate, Replenish	Minimize Storage Cost	Inventory and Sales Databases, User

- **Autonomy:**An agent should be able to excise a degree of in its operations. It should be able to take initiative and exercise a non-trivial degree of control over its own actions.

- **Collaboration:** An agent should have the ability to collaborate and exchange information with other agents in the environment to assist other agents in improving their quality of decision making as well as its own.

- **Flexibility and Versatility:** An agent should be able to dynamically choose which actions to invoke, and in what sequence, in response to the state of its external environment. Besides, an agent should have a suite of problem solving methods from which it can formulate its actions and action sequences. This facility provides versatility as well as more flexibility to respond to new situations and new contexts.

- **Temporal History:** An agent should be able to keep a record of its beliefs and internal state and other information about the state of its continuously changing environment. The record of its internal state helps it to achieve its goals as well as revise its previous decisions in the light of new data from the environment.

- **Adaptation and Learning:** An agent should have the capability to adapt to new situations in its environment. This includes the capability to learn from new situations and not repeat its mistakes.

- **Knowledge Representation:** In order to support its actions and goals with an agent should have the capabilities and constructs to properly model structural and relational aspects of the problem domain and its environment.

- **Communication:** An agent should be able to engage in complex communication with other agents, including human agents, in order to obtain information or request for their help in accomplishing its goals.

- **Distributed and Continuous Operation:**An agent should be capable of distributed and continuous operation (even without human intervention) in one machine as well as across different machine for accomplishing its goals.

An agent program with above characteristics can be a single agent system or a multi-agent system. Multi-agent systems are concerned with coordinating problem solving behavior amongst a collection of agents. Each agent in a multi-agent system represents a specific set of problem solving skills and experience (or a specific PAGE description). The intention is to coordinate the skills, knowledge, plans and experience of different agents to pursue a common high level system goal.

An agent program describes the behavior of an agent in the sense that for a given set of percepts or inputs a particular action is performed. A number of agent programs can be found to assist an user in e-mail filtering, on line news management, and in various other manufacturing and business areas (Maes et al. 1994; Dinh 1995). Agent applications can also be found in the areas of air-traffic control, network resource allocation, and user-interface design. On the other hand, *an agent architecture outlines how the job of generating actions from percepts to actions is organized* (Russell and Norvig 1995). A more elaborate definition has been provided by Maes (1994). Maes defines an agent architecture as *a particular methodology for building agents. It specifies how the agent can be decomposed into the construction of a set of component modules and how these modules should be made to interact. The total set of modules and their interactions has to provide an answer to the question of how the sensor data and the current internal stste of the agent determine its actions and future internal state of the agent. An architecture encompasses techniques and algorithms that support this methodology.*

This organization or methodology as has been described in this book can be outlined at a task structure level and at a computational level. The task and computational level facilitate the the PAGE description of an agent at a higher level which to a large extent is domain independent and programming language independent.

The software Various agents At the computational level the agent can be seen as a software program (with a set of characteristics described in the preceding section) which is run on a hardware computing device with traditional von Neumann architecture or other special purpose hardware architecture. In addition to the hardware architecture on which it is run an agent can also have a software architecture in which it is programmed

2.9 SUMMARY

A number of methodologies have been used in this book. These include expert systems, case-based reasoning, artificial neural networks, fuzzy systems, genetic algorithms, knowledge discovery and data mining, object-oriented software engineering, and agents and agent architectures. Expert systems, artificial neural networks, fuzzy systems and genetic algorithms the four most widely used intelligent methodologies for developing intelligent systems. Expert systems can be broadly grouped under four architectures, namely, rule based, rule and frame (object) based, model based, and blackboard architectures. These architectures employ a number of representation formalisms including, predicate calculus, production rules, semantic networks, frames and objects. Case-based reasoning methods are used in domains like law where it is not possible to represent the knowledge using rules or objects However, these architectures suffer from number of limitations including combinatorial explosion of rules, inability to to handle problems of non-deterministic or fuzzy nature, and others. Artificial neural networks which are used for problems with random and non-deterministic character can be grouped under supervised and unsupervised learning. Although there are number of supervised and unsupervised neural network based learning algorithms, the most commonly used ones are backpropagation and self organizing Kohonen maps. Radial basis function networks which incorporate both unsupervised and supervised characteristics are also used. Fuzzy systems which like artificial neural networks involve approximate reasoning have been widely used in various industrial applications. Fuzzy system construction involves determination of fuzzy sets and fuzzy membership functions, fuzzification of inputs, fuzzy inferencing and rule evaluation, and defuzzification of outputs.

Genetic algorithms refer to a class of adaptive search procedures based on princ iples derived from the dynamics of natural population genetics. The general characteristics of genetic algorithms include working with a coding of the parameter set, searching from a population of points, rather than a single point, using a pay-off or fitness function, and using probabilistic transition rules.

Knowledge discovery in databases and data mining methodology uses one or more intelligent methodology described in the preceding paragraphs to extract meaningful knowledge from gigabytes of data stored organizational databases. It is motivated by the need of organizations to become more competitive and productive.

Object-oriented software engineering methodology today is one of the pre mium software engineering methodologies for building software systems. Besides its ability to structure data through inheritance and composability relationships and other non-hierarchical relationships, its encapsulation, and polymorphic properties make it more attractive from a software implementation viewpoint. Agents and agent architectures are one of the most important emerging technologies in computer science today. They provide a means to map percepts to actions and in the process incorporate very sophisticated characteristics in a software program, namely, autonomy, collaboration, flexibility and versatility, adaptation and learning, complex communication, and others.

Notes

1. This term is used interchangeably with connectionist networks, connectionist models, connectionist systems or simply neural networks/nets throughout the book.

2. Data is used here interchangeably with the term " attribute".

References

Aitkins, J. (1983), "Prototypical Knowledge for Expert Systems," *Artificial Intelligence*, vol. 20, pp. 163-210.

Balzer, R., Erman, L. D., London, P. E. and Williams, C. (1980), "Hearsay-III: A Domain-Independent Framework for Expert Systems," *First National Conference on Artificial Intelligence (AAAI)*, pp. 108-110.

Beale, R., and Jackson, T. (1990), *Neural Computing: An Introduction*, Bristol: Hilger, U.K.

Berry, J. T. (1988), *C++ Programming*, Howard W. Sams and company, Indianapolis, Indiana, USA.

Booch, G. (1986), "Object-Oriented Development," *IEEE Transactions Software Engineering*, vol. SE-12, no. 2, pp. 212-222.

Booker, L. B., Goldberg, D. E., and Holland, J. H. (1990), "Classifier Systems and Genetic Algorithms," *Machine Learning: Paradigms and Methods*, MIT Press, Cambridge, MA, pp. 235-82.

Encyclopedia Britannica, (1986), Articles on "Behavior, Animal," "Classification Theory," and "Mood," *Encyclopedia Britannica, Inc.*

Chandrasekaran, B. (1990), "What Kind of Information Processing is Intelligence," *The Foundations of AI: A Sourcebook*, Cambridge, UK: Cambridge University Press, pp. 14-46.

Coad, P. and Yourdon, E. (1990), *Object-Oriented Analysis*, Prentice Hall, Englewood Cliffs, NJ, USA.

Coad, P. and Yourdon, E. (1992), *Object-Oriented Design*, Prentice Hall, Englewood Cliffs, NJ, USA.

Cox, B. J. (1986), *Object-Oriented Programming*, Addison-Wesley.

Davlo, E., and Naim, P. (1991), *Neural Networks*, Macmillan Computer Science Series, U.K.

Dillon, T. and Tan, P. L. (1993), *Object-Oriented Conceptual Modeling*, Prentice Hall, Sydney, Australia.

Dinh, S. (1995), "Intelligent E-Mail Agents," *Honors Thesis*, School of Computer Science and Computer Engineering, La Trobe University, Melbourne, Victoria - 3083, Australia.

Erman, L. D., Hayes-Roth, F., Lesser, V. R. and Reddy, D. R. (1980), "The Hearsay-II Speech-Understanding System: Integrating Knowledge to Resolve Uncertainty," *ACM Computing Surveys*, vol. 12, no. 2, June, pp. 213-53.

Erman, L. D., London, P. E. and Fickas, S. F. (1981), "The Design and an Example Use of Hearsay-III," *Seventh International Joint Conference on Artificial Intelligence*, pp. 409-15.

Fayyad, U., Piatetsky-Shaprio, G., and Smyth, P. (1996b), "From Data Mining to Knowledge Discovery in Databases," *AI Magazine*, AAAI Press, vol. 17, no. 3, pp. 37-54.

Grefenstette, J. J. (1990), "Genetic Algorithms and their Applications," *Encyclopaedia of Computer Science and Technology*, vol. 21, eds. A Kent and J.

G. William, AIC-90-006, Naval Research Laboratory, Washington DC, pp. 139-52.

Goldberg, D. E. (1989), *Genetic Algorithms in Search, Optimization and Machine Learning*, Addision-Wesley, Reading, MA, pp. 217-307.

Hamscher, W. (1990), "XDE: Diagnosing Devices with Hierarchic Structure and Known Failure Modes," *Sixth Conference of Artificial Intelligence Applications*, California, pp. 48-54.

Hawryszkiewyz, I. T. (1991), *Introduction to System Analysis and Design*, Prentice Hall, Sydney, Australia.

Hayes-Roth, F., Waterman, D. A. and Lenat, D. B. (1983), *Building Expert Systems*, Addison-Wesley.

Haykin, S. (1994), *Neural Networks, A Comprehensive Foundation*, IEEE Press, Macmillan College Publishing Company, Englewood Cliffs, NJ, USA.

Hebb, D. (1949), *The Organization of Behavior*, Wiley, New York.

Holland, J. (1975), *Adaptation in Natural and Artificial Systems*, University of Michigan Press, Ann Arbor, Michigan, USA.

Inmon, W.H., and Kelley, C., (1993), *Rdb/VMS, Developing the Data Warehouse*, QED, Publication Group, Boston, USA.

Jacobson, I., Christerson, M., Jonsson, P., and Overgaard, G. (1995), *Object-Oriented Software Engineering*, Addison-Wesley, USA.

Kim, Ballou, Chou, Garza and Woelk (1988), "Integrating an Object-Oriented Programming System with a Database System," *ACM OOPSLA Proceedings*, October.

Kohonen, T. (1990), *Self Organization and Associative Memory*, Springer-Verlag.

Kolodner, J. L. (1984), "Towards an Understanding of the Role of Experience in the Evolution from Novice to Expert," *Developments in Expert Systems*, London: Academic Press.

Kraft, A. (1984), "XCON: An Expert Configuration System at Digital Equipment Corporation," *The AI Business: Commercial Uses of Artificial Intelligence*, Cambridge, MA: MIT Press.

Maes, P.(1994), "Agents that Reduce Work and Information Overload," in *Communications of the ACM*, pp. 31-40.

McClelland, J. L., Rumelhart, D. E. and Hinton, G.E. (1986), "The Appeal of Parallel Distributed Processing," *Parallel Distributed Processing*, vol. 1, Cambridge, MA: The MIT Press, pp. 3-40,

Minsky, M. and Papert, S. (1969), *Perceptrons*, MIT press.

Minsky, M. (1981) "A Framework for representing Knowledge," *Mind Design*, Cambridge, MA: the MIT Press, pp. 95-128.

Myer, B. (1988), *Object-Oriented Software Construction*, Prentice Hall.

Neibur, D. and Germond, A. J. (1992) "Power System Static Security Assessment Using The Kohonen Neural Network Classifier," *IEEE Transactions on Power Systems*, May, vol. 7, no. 2, pp. 865-72.

Newell, A. (1977), "On Analysis of Human Problem Solving," *Thinking: Readings in Cognitive Science*, Cambridge UK: Cambridge University Press.

Ng, H.T., 1991, "Model-Based , Multiple-Fault Diagnosis of Dynamic, Continuous Physical Devices," *IEEE Expert*, pp. 38-43.

Pressman, R. S. (1992), *Software Engineering: A Practioner's Approach*, McGraw-Hill International, Singapore.

Quillian, M. R. (1968), "Semantic Memory," *Semantic Information Processing*, Cambridge, MA: The MIT Press, pp. 227-270.

Rumbaugh, J. et al. (1990), *Object-Oriented Modeling and Design*, Prentice Hall, Englewood Cliffs, NJ, USA.

Rumelhart, D. E., Hinton, G. E. and Williams, R. J. (1986), "Learning Internal Representations by Error Propagation," *Parallel Distributed Processing*, vol. 1, Cambridge, MA:The MIT Press, pp. 318-362.

Russell, S., and Norvig, P. (1995), *Artificial Intelligence - A Modern Approach*, Prentice Hall, New Jersey, USA, pp. 788-790.

Schank, R. C. (1972), "Conceptual Dependency," *Cognitive Psychology*, vol. 3, pp. 552-631.

Schank, R. C. and Abelson, R. P. (1977), *Scripts, Plans, Goals and Understanding*, Hillsdale, NJ: Lawrence Erlbaum.

Sejnowski, T.J., and Rosenberg, C.R. (1987), "Parallel Networks that Learn to Pronounce English Text," *Complex Systems*, pp, 145-168.

Shortliffe, E.H. (1976), *Computer-based Medical Consultation: MYCIN*, New York: American Elsevier.

Steels, L. (1989), "Artificial Intelligence and Complex Dynamics," *Concepts and Characteristics of Knowledge Based Systems*, Eds., M. Tokoro, et al., North Holland, pp. 369-404.

Unland, R, and Schlageter, G. (1989), "An Object-Oriented Programming Environment for Advanced Database Applications," *Journal of Object-Oriented Programming*, May/June.

Wang, X. and Dillon, T. S. (1992), "A Second Generation Expert System for Fault Diagnosis," *Journal of Electrical Power and Energy Systems*, April/June, 14 (2/3), pp. 212-16.

Widrow, B., and Hoff, M.E. (1960), "Adaptive Switching Circuits," *IRE WESCON Convention Record*, Part 4, pp. 96-104.

Zadeh, L.A. (1965), "Fuzzy sets," *Information and Control*, vol. 8, pp. 338-353.

3   INTELLIGENT FUSION AND TRANSFORMATION SYSTEMS

3.1   INTRODUCTION

In recent times, research efforts have been directed towards hybridization of various intelligent methodologies in order to solve complex industrial problems. At the same time these research efforts have been undertaken to develop a better understanding of the human information processing system. Today, one can find a number of applications involving hybridization of intelligent methodologies like knowledge based systems, fuzzy systems, genetic algorithms, and case based reasoning with artificial neural networks. The central methodology of many hybrid systems has been artificial neural networks. In fact, 2/3rd of the applications involving intelligent hybrid systems use neural networks. Broadly, in these applications neural networks are either used as the primary problem solving entity, or are used in conjunction with other intelligent methodology/ies which have a distinct and separate role to play in the problem solving process. The former are categorized as fusion and/or transformation based approaches, and the later are categorized as combination approaches. However, fusion and/or transformation based approaches center around not only neural networks but also center around the other other methodology, namely, genetic algorithms. Similarly, the combination approaches can involve intelligent methodologies other than neural networks also. The goal of this chapter is to provide an overview to the reader about fusion and transformation based synergies with neural network and genetic algorithms as the primary problem solving entities. In this direction, this chapter looks at the neuro-symbolic systems, neuro-fuzzy systems, genetic-neuro systems, and genetic-fuzzy systems.

3.2 FUSION AND TRANSFORMATION

The difference between fusion and transformation can be understood based on the tasks and constraints associated with these concepts. Fusion occurs in tasks primarily where the task complexity is high, and adaptation, performance and optimization are important to the solution of the problem. Transformation Fusion implies that it is possible to fuse the knowledge and/or reasoning from one intelligent methodology to another. The fusion process generally involves hard wiring representation of one intelligent methodology into another. That is, the transition from one representation to another does not occur naturally. In fusion based systems the task and constraints (e.g. adaptation, optimization and performance) are generally satisfied within one representation or methodology (e.g. artificial neural network). In fusion based systems the solution to a task or problem is realized in the framework of one intelligent methodology (e.g. artificial neural network). (e.g. fusion of fuzzy knowledge and inference in a neural network).

Transformation on the other hand, occurs in situations where knowledge required to accomplish the task is not available and one intelligent methodology depends upon another intelligent methodology for its reasoning or processing (e.g extraction of rules from a neural networks so that the rule-based system can reason with the extracted rules). Transformation systems also generally use numerical data as a starting point for the solving the problem, whereas fusion systems generally use symbolic or fuzzfied data to start with for fusion purposes. In transformation systems the task constraints are satisfied through two representations or two methodologies (e.g. one representation using artificial neural network used for learning the rules and the other representation using extracted rules for reasoning).

Transformation can also take place in situations where transformation occurs in the form of optimization. It requires transformation of an optimized representation from one methodology to another (e.g. rule refinement - extraction of optimized rules from an optimized neuro-symbolic network). Such systems are called fusion and transformation systems where unoptimized representation A is fused into representation B and then optimized representation A is extracted from optimized representation B.

The intelligent fusion and transformation systems based on the above distinction are described in this chapter. The transformation systems are also described in chapter 4 in the context of knowledge discovery and data mining.

3.3 NEURAL NETWORK BASED NEURO-SYMBOLIC FUSION AND TRANSFORMATION SYSTEMS

All intelligent methodologies are inspired by one or more aspects of human cognition. In that light, artificial neural networks and symbolic systems are many times said to be inspired by the automated/perceptual and analytical processes respectively of human cognition (Chandrasekaran 1990; Lallement et al. 1995). A good proportion of neural network research is based on two premises. The first premise is that using symbols as a starting point it is possible to develop a more flexible architecture for problem solving in a neural network framework by exploiting the inherent strengths of neural networks. The second premise is that it is possible to cover the full range of human information processing (automated and analytical) using neural networks alone without any help from symbolic models.

The neural network applications which are based on the the first premise can be categorized as top down neuro-symbolic systems, whereas those based on the second premise can be categorized as bottom up neuro-symbolic systems.

The top down approach centers around creation of neural network structure to represent a symbolic function. Unlike the neurobiological motivations of the bottom up approach, one motivation behind the top down strategy is that a number of limitations of the symbolic AI formalisms (e.g. brittleness of rules, lack of graceful degradation, etc.) can be overcome through the neural network representation. The top down systems start with symbolic problem description as shown in Figure 3.1.

The bottom up approach is based on the premise that the emergent symbolic behavior has neuronal underpinnings and thus works its way up from sensor motor responses to concept formation, language processing, and high level symbolic functions. The bottom up systems generally use numerical or continuous data as a medium for problem description as shown in Figure 3.1. In the rest of this section, top down neuro-symbolic fusion systems and bottom up neuro-symbolic transformation systems are described.

3.3.1 Top Down Neuro-Symbolic Fusion Systems

The top-down neuro-symbolic systems can be broadly put into three categories:

- Modeling symbolic knowledge structures using neural networks.

- Modeling symbolic reasoning using neural networks.

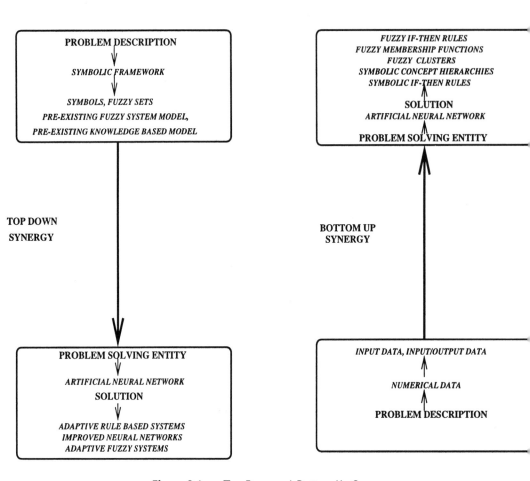

Figure 3.1.   Top Down and Bottom Up Synergy

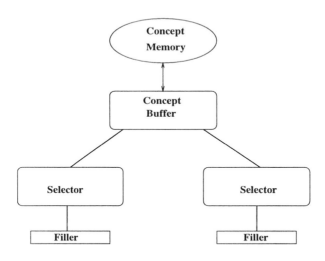

Figure 3.2.   The Architecture of DUCS © 1987 IEEE

■ Modeling decision trees using neural networks.

3.3.1.1   Modeling Symbolic Knowledge Structures using Neural Networks.   There have been many approaches to embody symbolic knowledge structures in artificial neural networks[1] (Touretzky and Hinton 1988; Shastri 1988; Hinton 1990; Plate 1995). Some of them are distributed, like Touretzky's (1988), in which each concept like frame is represented by a pattern of features. Others are localist, such as Shastri's in which single nodes are associated with concepts, and the causal relationship (IS-A) between concepts is represented by the strength of connections between them.

The work in these approaches is closely related to that in natural language understanding. In Touretzky's (1987) scheme called DUCS (Dynamically Updatable Conceptual Structures), both slots and slots fillers are represented as bit vectors. The architecture of DUCS as shown in Figure 3.2 consists of a concept memory at the top level where different concepts are held at the same time. This is followed by a concept buffer which holds multiple slot-filler relationships as patterns of activity of different concepts. Individual concepts are retrieved from the concept memory by activating several concept's slots at

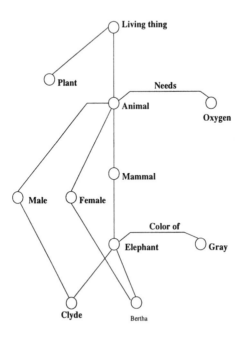

Figure 3.3.   NETL Example © 1987 IEEE

the same time so as to uniquely identify the concept. The next level consists of multiple selectors which are attached to the concept buffer to pull different slot-filler relationships from it. The selectors store the association between filler vector **v** and slot name vector **a**, as a pattern in the selector units. The selectors also allow to remove additional slot-filler associations. All the bits in the selector are copied to the concept buffer in the same places. The concept buffer is the same size as the selector. The concept buffer's contents are then used as input to the concept memory. The concept buffer thus serves as a connector between between the concept memory and the selectors.

DUCS gets around the problems of making copies of the entire network to retrieve several slot-filler relationships with traditional localist approaches to concept representation by using the concept buffer. However, performance degrades as slots are added and the available units become saturated. Further, DUCS model does not deal with an important property like inheritance and default values associated with frames. The inheritance property has been incorporated by Fahlman and Hinton (1987) in the NETL system designed and implemented by them. The NETL system (Figure 3.3) is a neural network implementation of a semantic network in which nodes represent noun-like concepts and the links represent relationships between these concepts. Fahlman and Hinton, use a marker passing scheme to retrieve information for from the system. If the problem is to find color of Clyde the elephant, a marker 1 is set on the Clyde node. Then the system repeatedly orders any is-a link with marker 1 set on the node below it to pass it on the node above. Once the marker 1 is set on all the nodes from whom Clyde has inherited the properties, a marker 2 is set on all the color-of links of all the superior nodes including the color-of link of the Clyde node. These are then retrieved by the system. Recognition tasks of a limited kind can also be performed by NETL because of its capacity to do set intersections in parallel.

As stated by the authors themselves, on the positive side, NETL examines the descriptions in memory at once in constant time, not depending on the heuristics that might prematurely rule out the right answer. On the negative side, it can be used only in clean domains because every feature is an atomic entity which is either present or absent. Additionally, as with most localistic representation, there is always a chance of node explosion and losing total information about a node if it dies down.

Shastri (1988) has developed a connectionist model for knowledge representation which has adopted ideas from semantic nets, evidential reasoning, and inheritance hierarchies. He constrains the possible set of inferences that the network can form to those inferences that people seem to make automatically. These inferences include inheritance of class properties to instances of those classes, and recognition of class membership. He also incorporates knowledge about how things are correlated in his model. This he does by developing statistical correlations based on experience between structural links knowledge (is-a, is-a-part-of, and occurs-during) of different concepts and the property knowledge(like, color,dimensions,etc). These statistical correlations are then used to calculate probabilistic evidence of a structural link, a property and its value. This also helps him to deal with exceptions where there are conflicting ideas about a certain concept. He uses a "principle of relevance" for inheritance of default values. For example, consider following structural links between three concepts, *Ford Car, American Car*, and *Car*.

Ford is an American Car and an American Car is a Car

The most immediate inheritance link of *Ford Car* in the above description is *American Car*. Now the conflict arises when the *Ford Car* concept has to inherit the default value of the property color. The concept *Car* has a color property value "25% of all cars are white" associated with it whereas, the concept *American Car* has a color property value "35% of all American Cars are white". Since *Ford* is an American car it is more intuitive and relevant to inherit the color property value "35% of all American Cars are white" from its most immediate structural link *American Car*.

The connectionist network built by him has six kinds of nodes: (1-3) nodes representing concepts, properties and values, (4) binder nodes that bind concept, property, and value nodes together, (5) enable nodes (which allow the network to distinguish between recognition and inheritance queries, the two main types of queries that the system is designed to deal with), and (6) relay-nodes, which implement links between subset and superset nodes. Inheritance is implemented in the following manner: If B is a superset of A that has property P, and there is no superset of A that is more specific that also has property P, then there is a connection between the binder node for the B-P relationship and the binder node for the A-P relationship, with a weight equal to the ratio of the incidence of the properties in the two nodes. He has also applied his network formalism to problems involving inheritance from multiple categories.

Since Shastri's weights are based on relative frequency of occurrence, the learning rule is roughly compatible to the Hebbian rule. The work is interesting as it captures the important properties of a concept like frames. The model has not only a capacity to identify a concept directly linked to a pattern of inputs but also infer or elicit other related concepts.

3.3.1.2   Modeling Symbolic Reasoning Using Neural Networks.

Neural network based top down symbolic reasoning models center around connectionist production system (Tourtezky and Hinton, 1988; Shastri 1988; Shastri and Ajjanagadde 1989; Ajjanagadde and Shastri 1991; Shastri and Ajjanagadde 1993; Park et al. 1995; and others), Commonsense reasoning (Sun 1991 94; Kokinov 1995) and neurally embedded decision trees (Sethi 1990, 91).

Connectionist Production System. The problem considered to be critical in operation of an AI production system in a connectionist paradigm is pattern matching and variable binding. Tourtezky and Hinton, 1988 have addressed this problem by developing a distributed connectionist production system (DCPS), consisting of binary threshold units. One of the restrictive assumptions in DCPS is that the working memory consists of symbols which are organized in triplets, such as (FAA). They use two types of production rules. The first type, (F A A) (F B B) $\rightarrow$ +(G A B) -(F A A) - (F B B) can be interpreted as follows: replace the triples (F A A) and (F B B), if they are present in the working memory, with the triple (G A B). In the first type left-hand side

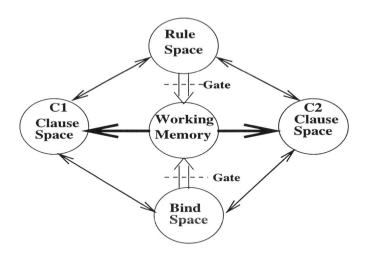

Figure 3.4.   Block Architecture of DCPS

consists of only two clauses that specify triples whereas the right-hand side actions that modify working memory by adding or deleting triples can be any number. In the second type, (=x A B) (=x C D) → +(G =x P) -(=x R =x) they add variables to the first position of a triple if it appears on the left hand side and anywhere if it appears on the right hand side.

DCPS is comprised of five groups of cells as shown in Figure 3.4. The main one of these is the working memory. There are two "clause spaces" called C1 and C2 each holding a single triple. There is a space representing the production rules and another space representing the variable bindings. Cells in the clause space, production rules space, binding space and working memory are interconnected with each other. The system behaves like a standard production system, which repeats a recognize-act cycle over and over; recognizing the left side of a rule, and taking the action specified on the right side of that rule. In the recognize phase, cells in working memory influence cells in the C1 and C2 clause spaces, and the rule and binding spaces. Relaxation takes place until a state that corresponds to a match is achieved.

Other researchers in this area have further extended this work to incorporate variable and dynamic binding associated with production system (Shastri 1988;

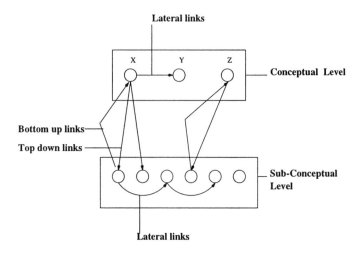

Figure 3.5.   CONSYDERR Architecture

Shastri and Ajjanagadde 1989; Ajjanagadde and Shastri 1991; Shastri and Ajjanagadde 1993; Park 1995). The architectures employed by these researchers
use localist representations. Additionally applications based on connectionist
production systems using fuzzy logic have also been built (Kasabov, 1994, 95).

Commonsense Reasoning. Sun, (1991,94) has developed a connectionist
model for robust reasoning (CONSYDERR) shown in Figure 3.5 to account
for the brittleness problem in rules to certain extent in reasoning patterns of
commonsense knowledge. The model uses both local and distributed representations. Local representation reflects the conceptual knowledge and the
distributed representation reflects the sub-conceptual knowledge. The synergy
between the two representations, and rule-based reasoning and similarity-based
reasoning helps to deal with problems such as partial information, no exact
match, property inheritance, rule interaction, etc.. CONSYDERR operates
in three phases, namely, Top-down phase, Activation and Settling phase, and
Bottom-up phase. The top-down phase activates the distributed representations based of the concept activated at the conceptual level. This will result

in the partial activation of the distributed representations of other concept/s based on a similarity measure. In the settling phase, the top-down links between the conceptual and sub-conceptual level take effect and certain other distributed representations are partially activated based on the similarity of concepts on the conceptual level and the lateral links (representing similarity between distributed representations) at the sub-conceptual level. Finally, in the bottom-up phase the partially activated distributed representations activate a conceptual node in the conceptual level which leads to inference of a new concept.

The architecture does solve the brittleness problem by taking advantage of similarity based reasoning.

Neurally Embedded Decision Trees. The inherent strengths like learning, generalization, and noise tolerance of neural networks are also used effectively by embedding a decision tree in a neural network structure. Sethi (1990, 91) maps a decision tree into a neural network. The number of neurons in the first layer (called the portioning layer) of the layered network equals the number of internal nodes of the decision tree. Each of these neurons implements one of the decision functions of the internal nodes. All leaf nodes have a corresponding neuron in the second hidden layer (called the ANDing layer) where the ANDing is implemented. The number of neurons in the output layer equals the number of distinct classes or actions. This tree-to-network architecture is called an Entropy Net (Sethi 1990, 1991). The Entropy Net is not a fully connected neural network since the decision tree nodes are hard wired into the neural network structure.

The hard wiring of the decision tree nodes into the neural network, helps to overcome the credit assignment problem and account for each neuron in the Entropy Net. When an example given to the network is known to be an instance of class C, then only one neuron in the ANDing layer produces an output of 1. The other neurons produce an output of 0. From the learning point of view, the activation of the neuron producing the highest output is further enhanced, whereas, the activation of the other neurons is further suppressed (i.e. a winner-take-all strategy). The sigmoid function is associated with activation of every neuron, thus converting a binary signal into an analog one. This helps to enhance the generalization of the net, noise in the input data, and overcomes the rigidity associated with the decision tree classifiers.

3.3.2 Top Down Neuro-Symbolic Fusion and Transformation Systems

As outlined in section 3.2 neuro-symbolic systems which include fusion and transformation primarily involve optimization. Unlike in fusion based systems where optimization is realized in a single configuration (e.g. neural network),

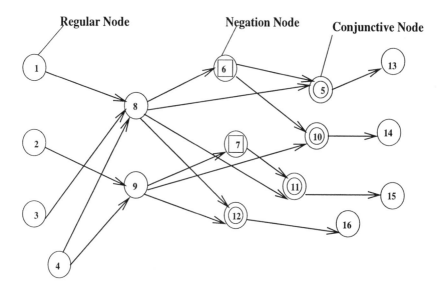

Figure 3.6. Expert Network for 4-2-4 Encoder/Decoder Expert System © 1992 IEEE

optimization in fusion and transformation systems is realized in two configuration (e.g. an optimized neural network, and extracted optimized rules).

Lacher et al. (1992) have developed what they call "Expert Networks" (shown in Figure 3.6) that are useful for both inference and optimization. Expert networks are defined as event-driven, acyclic networks of neural objects (artificial neurons) derived from expert systems. The initial network topology is determined by the knowledge base and the network dynamics is determined by the inference engine. The two distinguishing features in their approach is introduction of back-propagation learning and automation of the knowledge acquisition task for certainty factors. Back-propagation algorithms are applied on layered feed-forward networks. They replace the layered structure with an event-driven model and the feed-forward requirement with an acyclic network. The role of time which is implicit in the layered structure is not so in the event-driven structure where activation is controlled by the flow of the data. The back-propagation algorithm is reformulated in this new setting.

The artificial neural object (also called node) is defined as an artificial neuron with its attendant states, I/O, and processes: internal state, processing state, incoming connections and connection strengths, outgoing connections, combining function, and output function, together with a communications facility to signal changes in the processing state (waiting, processing, ready) to adjacent nodes in the network.

In order to fuse the knowledge base of a MYCIN type expert system Lacher, et al. use three types of nodes, namely, regular nodes, conjunction nodes and negation nodes. To test the learning of certainty factors they fuse an incorrectly functioning (with certainty factors randomly changed or replaced) expert system into an expert network. Back-propagation learning algorithm is applied on the expert network. After learning the expert network is transformed into the expert system. The new expert system is then compared to original expert system with correct certainty factors. It is observed that comparison is favorable if a) the learning has been successful and b) the solution to the inference problem is unique within a given rule structure. This can be seen as an example of fusion and transformation where the declarative and reasoning aspects of an incorrectly defined expert system are fused into a neural network like representation and then transformed into a correct or optimized expert system. As stated by the Lacher, et al. themselves the use of different types of nodes makes a provision for greater learning, but results in rule proliferation.

Fu and Fu (1990) also describe a neuro-symbolic system which optimizes or refines an existing rule base system.

3.3.3   Bottom Up Neuro-Symbolic Transformation Systems

The bottom up approach encompasses two research strategies:

- Neuronal Symbol Processing based on biological neurons

- Extraction of symbolic structures from Artificial Neural Networks (ANN)

3.3.3.1   Neuronal Symbol Processing. This research strategy models the emergent symbolic human behavior in biological reality. Its goal is to model brain's high level functions using properties of biological neural networks. One of strongest proponents of this strategy, Edelman (1992) has outlined a Theory of Neuronal Group Selection (TNGS), also called neuronal darwinism which involves a process of natural and somatic selection. TNGS consists of three tenets, developmental selection, experimental selection, and reentrant mapping. The three tenets are related to the proposed multilevel human development and refinement process starting from genes to proteins, from cells to orderly development, from electrical activity to neurotransmitter release, from sensory sheets to maps, from shape to function and behavior, and from social communication back to any and all the levels.

The first tenet, Development selection is based on topobiological competition, suggesting preliminary self organization of the human brain, where population of variant groups of neurons through a process of somatic selection form neural networks representing the anatomy or different functions of the brain.

This preliminary process is is called primary repertoire and forms primary maps.

The second tenet, Experiential selection relates to selective strengthening or weakening of the neural networks as a result of behavior. So, although the anatomical pattern of the brain is not altered, it is further refined through behavior to form multiple secondary maps which are linked to primary maps. This process is called the secondary repertoire.

The third tenet, reentrant mapping deals with connecting psychology (i.e response to external stimuli) to physiology (described by formation of primary and secondary maps in the brain). It explains how the brain maps interact and coordinate with each other through a process called reentry to do perceptual categorization. The latest artifact of these three tenets Darwin III developed by Edelman is able to demonstrate higher order brain functions like perceptual categorization, memory and learning. However, it will be some time before real world applications can be contemplated on the TNGS theory.

3.3.3.2   Extraction of Symbolic Structures from ANNs.

The extraction of symbolic structures from ANNs is a weak bottom up approach as the ANNs used only bear a general similarity to the biological neurons. This research strategy is based on the premise that non-predetermined and new symbolic structures can be extracted from trained ANNs. It uses discrete and/or continuous data to train the ANNs. The trained ANNs are then used with help of special algorithms for extraction of rules and concept hierarchies.

Gallant (1988) has developed a connectionist expert system which deals with a specialized domain in medical diagnosis, acute theoretical diseases of the sarcophagus. There are nodes in three layers corresponding to each symptom (layer 1), disease (layer 2), and treatment (layer 3) that is relevant to this domain; the nodes are 3-valued, with 1 denoting the activation of the disease, symptom, or treatment, -1, the deactivation-activation and 0, the absence of knowledge. Gallant uses a learning algorithm that he calls the pocket algorithm. The basic idea behind this algorithm is that for each vector of weights impinging on an internal or output node, there is another vector in your "pocket" as it were. Over the course of training, if vector of weights results in a longer string of correct classifications than the vector of weights in your pocket, then you replace the pocket weights with this vector. The pocket algorithm does monotonically improve performance by selecting a better set of weights, but is not guaranteed to converge with good performance. Gallant's system also explains its behavior by extracting rules from the network. For instance, node $u5$(state 1) might receive input from nodes $u1, u2, u3$, and $u4$, with states 1, -1, 1, -1 and weights -3, 6, -5, and 4 respectively. Nodes $u2$ and $u3$ would be chosen to participate in the rule because their absolute weights are largest and sum of their absolute weights (11) exceeds the sum of all the remaining weights (7). This results in the formation of the rule: if ((not $u2$) and $u3$). In a similar way, may other rules can be formed.

Gallant (1988) in his system called MACIE (Matrix Controlled Inference Engine) does a connectionist implementation of the backward chaining technique normally used in expert systems. As in all expert systems, it is useful to know what additional data to gather in order to prove or disprove a hypothesis.

The configuration of MACIE is same as the connectionist system described above with symptom, disease, and treatment layers. In MACIE these hypotheses are seen as output variables of a connectionist network. If there are a collection of output variables, none of which has been verified or falsified (possibly representing diseases or treatments), the system selects a variable u among these with the highest associated confidence. This confidence level is defined as the weighted sum of the confidences of all the cells impinging upon a given unknown cell (computed one level at a time, from bottom up - i.e. from the input layer), normalized by the total weight attached to unknown cells. It then repeatedly searches backward in the system through the layers, starting with u, starting with u, at each layer choosing the unknown variable with the largest weight connection to the variable in the layer above. Eventually, in this fashion, an input variable is reached and the data can be gathered on the value of this variable.

Saito and Nakano (1988) apply a connectionist network to problem of muscle contraction headache in medical diagnosis. They use a three layer feed-forward network with 216 input units corresponding to 216 symptoms, 72 hidden units and 23 output units corresponding to 23 diseases. Backpropagation is used to train the network. They use a Relation factor (RF) which relates the symptom to the disease. RF is determined by two methods. In the first method, the effect of a single symptom on the presence of a given disease is determined ignoring the effects of other symptoms. In the second method, the estimate is given for an actual set of patients. RFs relate only single symptoms with single diseases. For relating more than one symptom to a disease, they consider a set of symptoms, setting all the symptoms in such a set "on" in neural network, and then seeing if a given disease is activated. If the given disease is activated an affirmative rule is generated. Similarly , negative rules are generated by considering symptoms for which a given disease is de-activated. These rules are used to confirm patients' symptoms, extract symptoms that the patients are partially conscious of and reject symptoms that are as a result of errors made by the patient in course of answering questions thus improving the quality of diagnosis.

Many papers (Hudson et al. 1991; Towell and Shavlik 1992; Sestito and Dillon 1991, 94) have since been written for including ones for learning control heuristics for theorem provers (Sperduti et al. 1995) .

Hudson, et al. (1991) uses a three layered feed-forward neural network for extracting rules and making yes or no surgery decision for patients suffering from carcinoma of the lung. The input nodes, represent data values for signs, symptoms, and test results, which may be continuous or discrete. Each node above the input level computes an output which is a weighted sum of its inputs. The weighting factors are determined by a learning algorithm . The learning algorithm uses data of known classification to determine the weighting factors. The rules extracted through neural network are incorporated with approximate reasoning, i.e. a degree of importance coefficient is associated with each antecedent of the rule. These rules are thus able to model realistic situations to a greater degree.

Sestito and Dillon (1994) describe their system BRAINNE (Building Representations for AI using Neural NEtworks) for extraction of production rules from domains with both symbolic discrete attributes and continuous numeric attributes. The five step method employed by Sestito and Dillon (1994) for extraction of conjunctive rules is as follows

- Train a modified, multi-layered neural network with one hidden layer on the input-output data set using backpropagation. The modified net has an extended input set consisting of the original inputs and desired outputs as additional inputs.

- Determine the contribution of an original input to an output using the Sum of Squares Error (SSE) measurement:
  $SSE_{ab} = \sigma_j(W_{bj} - W_{aj})$
  where a is an input attribute, b is an additional input (desired output), and W_{aj} and W_{bj} are the weight between a and hidden unit j, respectively. The smaller the SSE value, the greater the contribution of the original attribute to the output. This is based on the premise that there is a strong association between an input and an additional input (desired output) if their weight links to each hidden unit are similar.

- Negate the original inputs and use these as inputs to a single-layered neural network with the desired outputs as its outputs. Train the net using Hebb's Rule to obtain a measure of irrelevance of the original inputs to the outputs. The smaller the weight value, the greater the contribution of the original attribute to the output.

- Calculate the product of the inhibitory weight values and the SSE measurements for all combinations of the inputs a and the outputs b. The smaller the product value, the greater the contribution of the original attribute to the output.

- For each input, sort the product list as the contributory attributes in the antecedent of the corresponding conjunctive rule. A clear cut-off occurs if one of the two consecutive products is 2 or 3 times larger than the other.

The extraction of rules through use of artificial neural networks is a useful mechanism which can be employed by knowledge engineers for knowledge acquisition. It can appreciably reduce the time for extracting rules from the domain expert. However, it is applicable in domains where lot of domain data is easily available. Extraction of rules can get complex if there are too many variables and overlap between input patterns is high.

3.4 NEURAL NETWORK BASED NEURO-FUZZY FUSION AND TRANSFORMATION SYSTEMS

Artificial neural networks and fuzzy systems are dynamic model free function estimators. The synergy between the two estimators is based on exploiting the inherent strengths of the neural networks, such as generalization, graceful degradation, retrieval from partial information, learning from well defined patterns, and properties of fuzzy systems, such as abstract or high level reasoning and human-like responses in cases involving imprecise data.

The interest in neuro-fuzzy synergy has been heightened by the their complementary nature and also because of the enormous success of fuzzy logic based applications in the industry.

The neuro-fuzzy fusion systems like the neuro-symbolic fuzzy systems can be broadly categorized as top down and bottom up neuro-fuzzy systems as shown in Figure 3.1 As can be seen in Figure 3.1, the top-down neuro-fuzzy systems solve the problem using the symbolic medium to describe the problem. (e.g. fuzzy sets, pre-existing fuzzy system model) On the other hand, the bottom-up neuro-fuzzy systems solve the problem using the continuous medium (e.g. input-output data sets) to describe the problem.

3.4.1 Top Down Neuro-Fuzzy Fusion Systems

Top-down fuzzy-neuro systems can be categorized based on the motivation behind the fusion of the two technologies:

- Fuzzification of Input Data.

- Optimization of Fuzzy Systems.

- Implementation of Fuzzy Logic Operations in ANNs.

3.4.1.1   Fuzzification of input-output data. ANNs are used as pattern classifiers in various domains including classification, diagnosis, prediction and control. The performance of the ANNs in these domains is to a large extent dependent on the quality of the input data set. A number of input variables to the ANN in these domains are imprecise in their natural form. These imprecise or linguistic variables are denoted by words like "large," "medium," "small," or "hot," "mild," and "cold". Zadeh (1965) introduced fuzzy sets as means for dealing with these imprecise variables. He proposed the use of membership functions which will properly reflect the imprecision in these variables. For example, one of the characteristics which distinguishes different cats (e.g. *Wild-Cat, Puma, Leopard, Tiger,* and *Lion*) is their size. The size characteristic is described in the form of three linguistic variables, namely Large, Medium, and Small. Large (150-180cm), Medium(100-150cm) and Small(0-100cm) represent length range of different cats. If these three linguistic variables are part of an input pattern to a neural network used for classification of cats, then the values of these linguistic variables can be represented in two ways. The linguistic variables can have discrete value (1 or 0) depending upon their presence or absence in a pattern or the linguistic variables can be represented as fuzzy membership functions shown in Figure 3.7 reflecting their imprecise nature. The latter case provides a more accurate description of the problem than the former, and to that extent improves the quality of the input data set and performance of the ANN. The input layer of an ANN in such an arrangement represents the linguistic variables for fuzzification of input data. Thus such an arrangement represents fusion of fuzzified input data with the neural network.

3.4.1.2   Optimization of Fuzzy Systems. In the previous section, it has been shown how fusion between neural networks and fuzzy systems can lead to the improvement in the performance of neural networks. In this section, it is shown how the synergistic characteristics can be usefully employed to optimize/improve the performance of the fuzzy systems.

A number of complex control applications are difficult to model mathematically because of their nonlinear and time varying behaviors. For example, robot motion control application may require a robot to be robust in order to recover from failures, adapt to UN-modeled changes in the environment, and so on. A number of neural, and fuzzy controllers have been built to model such control applications. The main disadvantage of the neural controllers is that the control operation cannot be stated explicitly. On the other hand, the main disadvantage of the fuzzy controllers is that, although they are robust like the neural controllers, but unlike the neural controllers, they are not adaptive (Lee 1990). More so, optimization of fuzzy controllers (in terms of the fuzzy membership functions and rules) based on fuzzy logic techniques can be difficult. This is

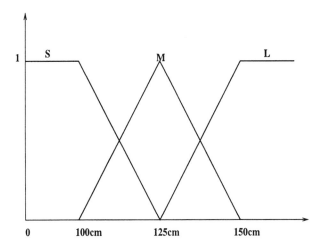

Figure 3.7.   Fuzzy Representation of the Size of Cats

because little knowledge is available on the exact influence of the the fuzzy parameters on the behavior of the controller. Thus the adaptive properties of neural networks can be used to optimize the performance of a fuzzy rule based controller or any fuzzy rule based system for that matter. This can be done by fusion of fuzzy rules and membership functions into a neural network framework and employing the supervised or reinforcement learning algorithms to optimize the performance. The supervised learning algorithm like backpropagation fine tunes the membership functions through adaptation of weights (Hung 1993) or through changing weight links (Sulzberger 1993) resulting in change of shape of a membership function. For example, consider the following fuzzy if-then control rule extracted from a domain expert to control an inverted pendulum's dynamical system

IF angle error is positive small
AND angle rate error is negative small
THEN motor force is positive small

The above rule can be represented in a supervised multilayer neural network as shown in Figure 3.8. The supervised multilayer neural network can now be trained to optimize the shape of membership functions *Angle Error* and *Angle Error Rate*. Once the optimization is accomplished the optimized multilayer neural network can be used as it is or if required the optimized fuzzy if-then rule base can be extracted using rule extraction techniques (Sulzberger 1993).

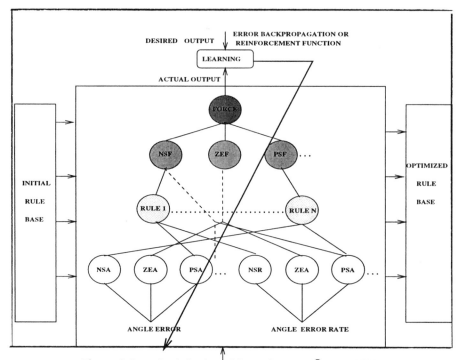

Figure 3.8.   Optimization of Fuzzy Systems © 1993 IEEE

In fact, the multilayer neural network in Figure 3.8 and the weight links between different layers can also be seen as a model of a fuzzy inference system. The neural network in Figure 3.8 is a neuro-fuzzy inference system which implements a fuzzy inference system in a functional sense. Fuzzy inference can be considered as a general concept which is implemented in a neural network framework as a by-product of other motivations, namely optimization, adaptation, building neuro-fuzzy controllers (Bulsari 1992; Nauck 1993), using neural network as a basis for analyzing and improving different fuzzy representation and inference methods (Ishibuchi 1994; Lee 1994; Keller et al. 1992b).

Fuzzy systems also suffer from their inability to deal with noisy or missing data. Artificial neural networks with their partial immunity to noise, generalization and parallel processing properties can help to alleviate these problems. Keller and Tahani (1992a) in their paper demonstrate how through such a mapping a neuro-fuzzy system can be made insensitive to noisy input distributions, and even handle rule conflict resolution. Besides taking advantage of the inherent structural properties of neural networks through such a fusion, the learning properties of the backpropagation algorithm have also been effectively used to fine tune input membership functions of the fuzzy inputs.

3.4.1.3   Implementation of Fuzzy Logic Operations in Neural Networks .

In conventional, supervised backpropagation neural network the nodes usually employ the weighted-sum function to compute the activation of an output node. In certain neuro-fuzzy systems the weighted-sum function is replaced by fuzzy logic based min-max operations (Watnabe et al. 1990; Hayashi et al. 1992; Pedrycz et al. 1993; Kwan et al. 1994). The multiplication part of the weighted-sum function is replaced by the min operation and the addition part of the weighted-sum function is replaced by the max operation (Watnabe et al. 1990). The structure of such a logical neuron is shown in Figure 3.9. The activation of the neuron is determined as follows:

$$y = max((min(w_1, x_1), (min(w_2, x_2)), ...(min(w_n, x_m)))$$
$z = y - th$, where th is the threshold value for activation of the neuron

The desired output value d is used to compute the error to be propagated back into the network.

The use of min-max logical neurons can result in enhancing the speed of operation (i.e reduced computation time) of the neural network and robustness in separability of the input patterns (Watnabe 1990). Maeda et al. (1993) have developed a Fuzzy Logic Inference Procedure network (FLIP-net) for a road tunnel ventilation control system. The backpropagation learning technique is used for tuning the shapes of membership functions. The logical relationships between fuzzy rules and membership functions, however remain the same. This feature facilitates adding and modification of fuzzy rules and membership functions by the developer.

A point to be noted here is that the use of logical neurons can also result in convergence problems in a gradient descent based learning method because of discontinuous derivatives of some logical functions (Pedrycz 1991).

3.4.2   Bottom Up Neuro-Fuzzy Fusion Systems

As outlined by Takagi et al. (1992a, 92b) the motivation behind the bottom up fusion of neural networks and fuzzy systems is to take advantage of the learning properties of artificial neural networks. As shown in Figure 3.1, bottom-up neuro-fuzzy fusion systems use the continuous medium, i.e., numerical data in the form of input-output data pairs or simply input data to describe the problem.

Bottom up neuro-fuzzy fusion systems can be categorized into systems which:

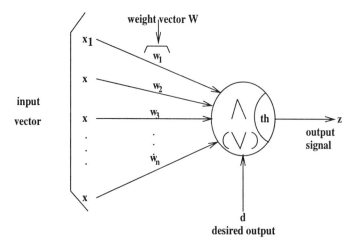

Figure 3.9.   Neural Network with Fuzzy Neurons (Watnabe et al. 1990)

- learn fuzzy clusters.

- learn fuzzy membership functions.

- learn fuzzy if-then rules.

3.4.2.1   Learn Fuzzy Clusters.   Recently, Kohonen self organizing maps and other competitive learning algorithms have been used for pattern classification (Kohonen 1990; Carpenter et al. 1994). In these neural network algorithms, the neurons compete according to some sort of distance metric, usually the Euclidean one, to learn the current input pattern. Based on the Euclidean distance only one neuron will win and learn the current input pattern. The concept "win" in this setting is a crisp set (i.e. has a value of 1 (win), or 0 (not win)). The shortcoming of this conventional approach is that information concerning the closeness of input patterns and overlaps between learnt clusters is not properly accounted for in the winner-take-all training process. Besides, the neurons with their weight vectors far away from the winning neuron may never learn and result in underutilization of the network's resources. These problems can be overcome by considering "win" as a fuzzy set, where every neuron to a certain degree wins and consequently learns according to its "win" membership. This would result in generation of membership values of input data to each output class or cluster.

The benefits of such a approach are well illustrated by (Bezdek 1993). Consider, two pattern classifiers representing two output classes *Peach*, and *Plum*.

Table 3.1.   Crisp and Fuzzy Sets

Object	Crisp Classification			Fuzzy Classification		
	X_1	X_2	X_3	X_1	X_2	X_3
Peaches	1	0	0	0.9	0.2	0.4
Plums	0	1	1	0.1	0.8	0.6

The first classifier represents the two output classes as two crisp subsets, whereas the second represents them as two fuzzy sets.

Now, the two classifiers are presented with three input vectors, x_1, x_2, and x_3 representing the classes *Peach*, *Plum*, and *Nectarine* respectively. Figure 3.1 shows the crisp and fuzzy classification of the three input vectors. In the crisp/hard classification x_3 is labeled *Plum*, which is incorrect as *Nectarine* is a hybrid of *Peach* and *Plum*. In the fuzzy classification, the input vector x_3 is classified with 0.6 membership of the *Plum* class, and 0.4 membership of the *Peach* class which is a more accurate classification given the fact that the fuzzy classifier was trained only on two classes.

A number of real world problems involve overlapping or fuzzy class boundaries. The above rationale for fuzzy partitioning of input space has been incorporated in the Kohonen self-organizing networks by integrating a third layer called the membership layer as shown in Figure 3.10 (Huntsberger 1990). In fact, most of the fuzzy-neuro fusion systems applied to clustering and classification are versions of previous existing neural networks, which have adopted the concept of fuzzy membership functions.

In Figure 3.10, the feedback path between the third layer and the input nodes is used to incorporate fuzzy membership values in the weight updates of the output nodes in the competitive/distance layer.

The above membership value determination is based on the iterative algorithm developed by (Bezdek 1981).

The algorithm designed by Huntsberger (1990) has been further extended and improved by Bezdek (1992) with a new family of algorithms called Fuzzy Kohonen Clustering network algorithms. Besides, a similar concept of learning fuzzy clusters has been incorporated in the other competitive learning algorithm, namely ART1 (Carpenter et al. 1992, 94).

In the preceding paragraphs it has been shown how fuzzy clusters can be learnt using unsupervised neural networks. It is also possible to incorporate fuzzy classification using the conventional multilayer backpropagation network.

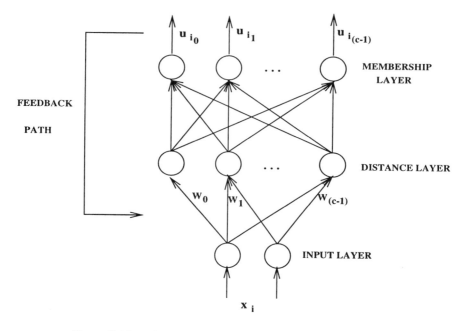

Figure 3.10.   Learning Fuzzy Clusters (Huntsberger et al. 1990)

Pal and Mitra (1992) use the weighted distance of a training pattern criterion to compute its degree of membership of a particular class. In fact, Keller (1985) is one of the pioneers in incorporating fuzzy membership functions in the perceptron algorithm.

3.4.2.2   Learn Fuzzy Membership Functions.   One of the central issue in building any fuzzy system is how do we design the membership functions (Takagi and Hayashi 1992b)? The fuzzy system builder generally makes an assumption on the shape of the membership functions. The most common assumptions are triangular or trapezoidal membership functions. If these membership functions are not found to be suitable, then heuristic adjustments are made or simulations of the linguistic variables (if possible) need to be performed to determine the shape. Thus there is no definite method to determine the membership functions. This problem is accentuated by another possibility as highlighted by (Takagi and Hayashi 1992b). Generally, the membership function for each linguistic variable is determined separately/independently from one another. However, this may not be possible in some situations. Consider the fuzzy rule:

If the temperature is slightly higher and the humidity is lower,
Then the power has to be lowered slightly.

In the above rule, temperature and humidity are not independent of each other. The membership function exists in three dimensional space consisting of the membership value axis and the temperature-humidity plane. Thus it may be not be possible to use experience to determine the membership function in multidimensional space as the number of linguistic variables increase.

The artificial neural networks (of backpropagation variety) because of their learning properties and ability to handle multiple constraints can be usefully employed to learn fuzzy membership functions in multidimensional space. This concept has been well exploited by (Takagi and Hayashi 1991, 1992a, 92b) to not only learn fuzzy membership functions but also to learn and refine fuzzy if-then rules. The neural network architecture used for learning fuzzy membership functions is shown in Figure 3.11. There are basically four phases in the neuro-fuzzy system developed by (Takagi and Hayashi 1992a, 92b). The first phase deals with determining the number of rules and the domain (cluster) of each rule using an unsupervised learning algorithm (e.g. Kohonen Networks). That is, each cluster will represent a set of input patterns or rules. In the second phase, supervised multilayer (backpropagation) neural network is used to identify the constitution of each IF part (represented by an input pattern) in terms of its membership to the domain/cluster of each rule. The first and the second phase are represented by the **NNmem** box in Figure 3.11. For example, if there are three clusters, C_1, C_2, and C_3, then the output of the supervised neural network will define the membership function of an input pattern $(x_1, x_2, ..., x_n)$ to each of these clusters. The third phase relates to identification of each THEN part. NN_1 in Figure 3.11 is used to identify labels in classes for each cluster C_1. For example, if cluster C_1 represents input patterns 1 $(x_{11}^c, x_{21}^c, ..., x_{n1}^c)$ to 5, then the neural network NN_1 in Figure 3.11 is trained in the supervised mode based on input-output data pairs for these five patterns. The final output y^* from the neuro-fuzzy system for a given input pattern is obtained by weighting the outputs from THEN part NNs with membership ship vector $\mathbf{W}$ (w^1, w^2, w^r) from the **NNmem**.

The neuro-fuzzy system developed by (Takagi and Hayashi 1991, 1992a, 92b) is a comprehensive fuzzy reasoning system which exploits the inherent strengths of neural networks fully. Takagi and Hayashi (1992a, 92b) also use the architecture in Figure 3.11 to refine the fuzzy if-then rules.

3.4.2.3 Learning Fuzzy If-Then Rules.
The development of conventional fuzzy rule based systems involves use of domain experts to extract the fuzzy if-then rules. This knowledge acquisition process, as in case of knowledge

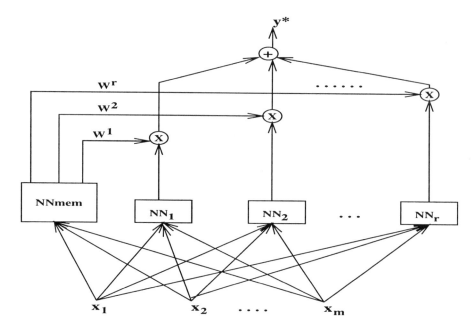

Figure 3.11.   NN Architecture for Fuzzy Reasoning
Reprinted with permission of the publisher from
("NN-driven fuzzy reasoning," by Takagi & Hayashi),
International Journal of Approximate Reasoning,
Vol. 5, pp. 191-212, © 1991, by Elsevier Science Inc.

based systems, can be tedious, and at times difficult. The neuro-fuzzy system developed by (Takagi and Hayashi 1992a, 92b) and described in the preceding section not only learns the fuzzy membership functions but also fuzzy if-then rules. Another method of learning fuzzy if-then rules is to use unsupervised (Kohonen Self-Organizing Map (SOM)) and supervised Kohonen Learning Vector Quantization (LVQ) in conjunction with one another to extract the principle fuzzy rules from the input-output data set (Hung 1993). In this arrangement the unsupervised method is used initially to cluster the input space. In order to determine clear demarcation of the class boundaries supervised Kohonen LVQ method is employed. The quantization vectors with clear class boundaries then represent the principle fuzzy rules.

Hung (1993) has applied above two stage training process for learning principle fuzzy rules to control the inverted pendulum's dynamical system. The control system consists of two input signals, namely error of angle, and error of angle rate and one one output signal, namely motor force to the system. It is shown that the operation of the control system based on principle fuzzy rules from the two stage learning process is more efficient and accurate as compared to the operation based on the control rules extracted from the human expert.

3.5 RECAPITULATION

In section 3.3, and 3.4, neuro-symbolic fusion and transformation systems and neuro-fuzzy fusion and transformation systems have been described. In these systems neural network has been employed as the primary problem solving entity. Neuro-symbolic systems have been categorized as top down and bottom up neuro-symbolic fusion systems. Top down neuro-symbolic fusion systems use the symbolic medium as the starting point and exploit the inherent strengths of neural networks. Top down neuro-symbolic fusion systems are subdivided into three categories, namely modeling symbolic knowledge structures using neural networks, modeling symbolic reasoning using neural networks, and modeling decision trees using neural networks. Top down neuro-symbolic systems involving fusion and transformation have also been outlined.

The bottom up neuro-symbolic transformation systems unlike the top down use numerical data in the form of input-output data pairs or simply input data as the starting point and exploit the learning properties of neural networks. In this arrangement applications have been developed based on the neuronal symbol processing approach grounded in biological neurons and extracting symbolic structures (e.g. concept hierarchies, rules) from artificial neural networks.

Neuro-fuzzy systems like the neuro-symbolic systems are also categorized as top down and bottom up neuro-fuzzy systems. Top down neuro-fuzzy fusion systems involve fuzzification of input data, optimization of fuzzy systems, modeling fuzzy inferencing in artificial neural networks and implementation of fuzzy logic operations in artificial neural networks. On the other hand, bottom up neuro-fuzzy transformation systems involve learning fuzzy clusters, learning fuzzy membership functions, and learning fuzzy if-then rules.

3.6 GENETIC ALGORITHMS BASED FUSION AND TRANSFORMATION SYSTEMS

In the last four years, there has a been substantial interest in genetic algorithms as an alternative to neural networks for solving optimization tasks. Some of the advantages of which have made genetic algorithms (Nobre 1995; Lee and Takagi 1993) as an attractive alternative are as follows:

- GA's are largely unconstrained by the limitations of the classical search methods (continuity, derivative existence (e..g back propagation neural networks), and others);

- GA's work with a coding of the process parameters on strings, not directly on themselves;

- GA's move in parallel search from a population of points, then the probability for locating false peaks in multimodal spaces is reduced. They include random elements like crossover and mutation which help them to avoid local minima;

- GA's only require a objective function for evaluation of the optimality of each solution.

Keeping in view the main theme of this chapter, in this section applications where genetic algorithms are the primary problem solving entity are described. These applications fall in the following categories:

- Genetic-Fuzzy Fusion and Transformation Systems

- Genetic-Neuro Fusion and Transformation Systems

A primary application of genetic-fuzzy fusion and transformation is automation of design of fuzzy systems. On the other hand, genetic-neuro fusion and transformation systems involve optimization of connection weights, neural network structure and input data set. These fusion and transformation systems are described in the rest of this section.

3.6.1 Automating Design of Fuzzy Systems

Fuzzy systems design consists of three stages, namely, definition of fuzzy membership functions, determining the number of useful rules and rule-consequent parameters (Lee and Takagi 1993).

Most of the neural network based approaches for optimizing fuzzy systems described in section 3.4.1.2 have centered around tuning or learning membership functions, and/or learning/optimizing number of fuzzy rules.

A number of applications using genetic algorithms have been developed recently for optimizing one or two of the three fuzzy system design stages (Nobre 1995; Ishibuchi et al. 1993; Karr 1985, 91, 92). One of the simple methods of converting a fuzzy membership function into a genetic bit string representation is shown in Figure 3.12 (Nobre, 1995). The 1's in the bit string correspond to the value 1 in the fuzzy sets *VL (Very Low)*, *L (Low)*, *M (Medium)* , *H (High)*, and *VH (Very High)* respectively. The values in (b) are the real values of temperature corresponding to the value 1 in the various fuzzy sets. After crossover, mutation, reproduction, and the best fit or tuned fuzzy membership function, the corresponding bit string, and real value representations are shown in Figure 3.13.

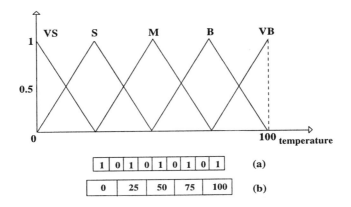

Figure 3.12.   Fuzzy Membership Functions © 1995 IEEE

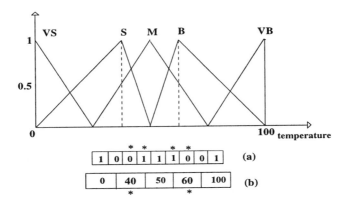

Figure 3.13.   Genetically Tuned Fuzzy Membership Functions © 1995 IEEE

However, as highlighted by Lee and Takagi (1993) there are very few approaches involving genetic algorithms which attempt to optimize all three aspects of the fuzzy system design. The fact of the matter is that all three stages of fuzzy system design are interdependent. Thus by optimizing one or two of them one can at best hope to achieve a sub-optimal solution.

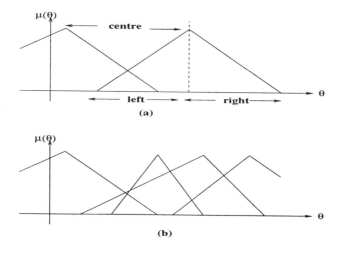

Figure 3.14.   TSK Model © 1993 IEEE

Lee and Takagi (1993) have developed a novel approach which uses genetic algorithms for simultaneously determining the membership functions, the number of fuzzy rules, and the rule-consequent parameters at the same time. They use the Takagi-Sugeno-Kang (TSK) parameterized fuzzy model (Takagi and Sugeno, 1985) to demonstrate their approach. In the TSK model the membership functions are parameterized as shown in Figure 3.14. As shown in Figure 3.14 each membership function has a triangular shape with three parameters, namely center, left base, and right base. The center is encoded as the distance from the previous center, and the base values are encoded as the distance from the corresponding center point. The values of these three parameters are used in conjunction with the boundary conditions of an application to eliminate the the unnecessary fuzzy sets. For example, in the inverted pendulum application shown in Figure 3.15 (and the corresponding control system diagram shown in Figure 3.16), the θ fuzzy membership sets (representing the vertical angle as shown in Figure 3.15) with center positions greater than 90 degrees can be discarded. The fuzzy rules in the TSK model are represented as:

IF X is A and Y is B THEN $C = w_1 X + w_2 Y + w_3$

where w_1, w_2, and w_3 are the rule-consequent parameters. As can be seen from Figure 3.14, each membership function requires three operand the consequent of each fuzzy rule also requires three parameters.

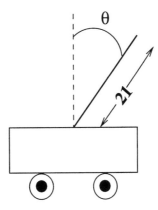

Figure 3.15.   Inverted Pendulum Control Application ©️ 1993 IEEE

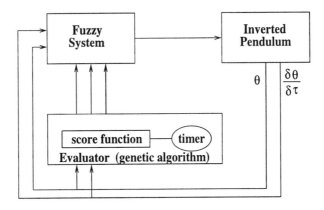

Figure 3.16.   Inverted Pendulum Control System Diagram ©️ 1993 IEEE

A two dimensional input space with p and q fuzzy sets respectively will have $p+q$ membership functions, $p * q$ number of fuzzy rules, and $3*(p+q)$ number of system parameters. The genetic/chromosome representations of the membership functions and rule-consequent parameters are shown in Figure 3.17. The gene map based on a maximum limit of ten fuzzy sets or fuzzy partitions for input variables θ, and $\delta\theta/\delta t$ (representing the angle rate) respectively is shown in Figure 3.18. This arrangement leads to $3*(10+10) = 360$ parameters, 100 rules, and 2880 bits.

The goal of the inverted pendulum control application is to balance the pole in the shortest time possible for a wide range of initial conditions. The fitness or the scoring function used for determining the best solution generated by

centre	left base	right base
101001100	1001100	0101100

membership function chromosome (MFC)

w_1	w_2	w_3
101001100	1001100	0101100

rule-consequent parameters chromosome (RPC)

Figure 3.17.   Chromosomes Representation Inverted Pendulum Control Problem
© 1993 IEEE

the genetic algorithm is derived from the three possible termination conditions (i.e., pole balanced, time limit expired before pole could be balanced, and the pole fell over or θ is greater than 90 degrees) shown in Figure 3.19. Based on the terminating conditions each trial is scored as follows:

$$score_{raw}(t_{end}) = a_1(t_{max} - t_{end}) + a_2 reward - (polebalanced)$$
$$score_{raw}(t_{end}) = reward - (timeexpiredt_{max} = t_{end})$$
$$score_{raw}(t_{end}) = b.t_{end} - (polefellover)$$

$where a_1, a_2, and reward are constants.$

In addition to the raw scoring function, a penalty strategy is also employed, wherein, the solution with minimum number of rules are scored higher than others.

The fitness score used is a composite score accumulated over eight trials with a wide range of initial conditions.

Genetic algorithm with a population size of 10, two point crossover and mutation operators has been employed. The member with the highest fitness value advanced to the next generation. The automated fuzzy system design generated through this arrangement had four fuzzy control rules with parametric structure of each rule being

If θ is A_i and $\delta\theta/\delta t$ is B_i THEN $y = w_{1i}\theta + w_{2i}\delta\theta/\delta t + w_{3i}$

The corresponding values of the membership functions A_i and B_i, and rule-consequent parameters w_{1i}, w_{2i}, and w_{3i} for each rule was also generated by the genetic algorithm. The four fuzzy control rules took abut 0.4 seconds for balancing the pole for various initial conditions. Although, Lee and Takagi (1993) employed this design strategy for a two dimensional control problem,

fuzzy variable			fuzzy variable		rule-consequent parameters		
MFC_1	....	MFC_{10}	MFC_1	....	MFC_{10}	RPC_1	....	RCP_{10}

Figure 3.18.   Gene Map of Inverted Pendulum Control Problem © 1993 IEEE

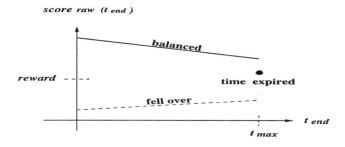

Figure 3.19.   Three Terminating Conditions and Scoring Function © 1993 IEEE

the concept can be usefully employed for problems with more than two dimensions.

The intelligent genetic-fuzzy system described in this section involves fusion and transformation. The membership functions and rule-consequent parameters are fused in a chromosome or a genetic representation. The fitness score is used as a means for optimizing the genetic representation. Once the optimum genetic representation is found it is transformed into fuzzy control rules.

3.6.2   Genetic Algorithm Based Optimization of Neural Network Weights

As has been described in chapter 2, neural networks (especially back propagation) are a versatile methodology which have been used in solving numerous non-linear problems. However, in order to utilize neural networks like back propagation a number of choices have to be made. These include, number of hidden layers, number of hidden units, the type of learning rule and transfer function, parameters like learning rate and momentum, and quality of input data set for improving performance.

These choices are made in order to develop the most effective or optimal neural network topology to solve a given problem. However, making these choices can be very time consuming and frustrating. Besides, there is another problem, and that is, local minima which inflicts the most widely used back propagation neural networks. Genetic Algorithms (GA) which are useful for optimization problems have been used for training of neural networks and opti-

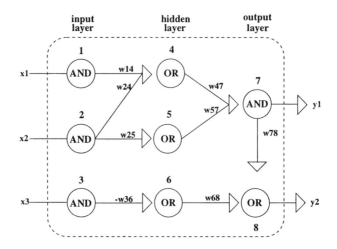

Figure 3.20.   Neuro-Fuzzy Network © 1995 IEEE

mization of neural network topology (Whitley et al. 1990; Fukuda et. al. 1993; Romamiuk 1994; Williams et al. 1994). One of the main reasons for using genetic algorithms is because of their capacity to converge to an optimum local or global after locating the region containing the optimum (Nobre 1995). In order to train neural networks or learn the connection weights, the weights are encoded as binary strings and tuned through use of reproduction, crossover, and mutation operators. The fitness function used is the squared error cost function (Nobre 1995).

3.6.3 Genetic Algorithms for Optimizing Neural Network Structure

Nobre (1995) outlines how the neural network structure can be optimized using genetic algorithms. In section 3.4.1.3 it has been shown how fuzzy logic neurons can be realized in a neural network structure. Nobre (1995) employs genetic algorithms to optimize the structure of a neuro-fuzzy network developed by Pedrycz (1993) and shown in Figure 3.20. The logical neurons in Figure 3.20 are represented as binary vectors. An AND neuron is represented as zero, and an OR neuron is represented as one. The connections between neurons in input, hidden, and output layer are represented in $n * n$ binary matrix . Given that, there are eight neurons in three layers, the connections are represented in a $8 * 8$ binary matrix as shown in Figure 3.21. There are four representative states, 00 - no connection; 01 - negative or inhibitory connection; 10 - positive or excitory

	1	2	3	4	5	6	7	8
1	11	11	11	11	11	11	11	11
2	00	00	00	10	10	10	10	10
3	00	00	00	00	00	10	00	00
4	00	00	00	00	00	00	01	00
5	00	00	00	00	00	00	00	10
6	00	00	00	00	00	00	00	10
7	00	00	00	00	00	00	00	00
8	00	00	00	00	00	00	10	10

Figure 3.21.   Binary Matrix Representation of Neuro-Fuzzy Net Topology © 1995 IEEE

connection; 11 - unused connection. The direction of the connection can also be represented separately using the binary notation, where 1 indicates direct connection or forward direction and 0 indicates reverse connection or reverse direction.

After reproduction, crossover, and mutation the binary matrix of the the optimum neuro-fuzzy network topology is shown in Figure 3.22. The regenerated neuro-fuzzy network transformed from the optimized binary matrix is shown in Figure 3.23. It may be noted that in the regenerated neuro-fuzzy net, not only the connections have been changed but also the fuzzy neurons have changed from OR to AND and AND to OR in the hidden and output layers respectively.

This section completes the description of neural network based and genetic algorithm based fusion and transformation systems. The next chapter looks at intelligent hybrid applications based on combination of two or more intelligent methodologies.

3.6.4   Genetic Algorithms for Optimizing Input Data

One of main problems associated with the use of artificial neural networks for bank note recognition is their large size. One of the ways to reduce the size of artificial neural networks is to reduce the dimensionality of the input vectors through preprocessing. Previously Fast Fourier Transforms (FFT) have been proposed for preprocessing the input image of bank notes before being fed for recognition to the artificial neural network. However, because of their complexity, FFT's are not a preferred method for implantation in current bank note recognition machines. Takeda and Omatu (1994) have successfully employed a

$$
\begin{array}{cccccccc}
 & 1 & 2 & 3 & 4 & 5 & 6 & 7 & 8 \\
\end{array}
$$

$$
\begin{array}{c}
1 \\ 2 \\ 3 \\ 4 \\ 5 \\ 6 \\ 7 \\ 8
\end{array}
\left[
\begin{array}{cccccccc}
00 & 00 & 00 & 10 & 00 & 00 & 00 & 00 \\
00 & 00 & 00 & 10 & 10 & 00 & 00 & 00 \\
00 & 00 & 00 & 00 & 00 & 01 & 00 & 00 \\
00 & 00 & 00 & 00 & 00 & 00 & 10 & 00 \\
00 & 00 & 00 & 00 & 00 & 00 & 10 & 00 \\
00 & 00 & 00 & 00 & 00 & 00 & 00 & 10 \\
00 & 00 & 00 & 00 & 00 & 00 & 00 & 10 \\
00 & 00 & 00 & 00 & 00 & 00 & 00 & 00
\end{array}
\right]
$$

Figure 3.22.   Binary Matrix Corresponding to Optimum Neuro-Fuzzy Net Topology
© 1995 IEEE

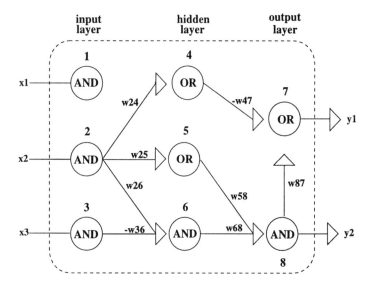

Figure 3.23.   Regenerated Neuro-Fuzzy Network © 1995 IEEE

GA-neuro approach for recognition of seven kinds of US dollar notes, namely, $1, $2, $5, $10, $20, $50, and $100.

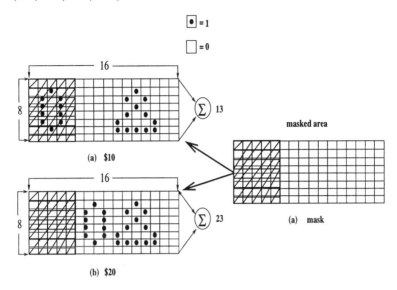

Figure 3.24.   Mask and Slab Values © 1994 Springer-Verlag

Takeda and Omatu (1994) use a slab architecture which amounts to covering or masking some parts of the two-dimensional image of the bank notes as shown in Figure 3.24. The masking is done to ensure different denomination bank notes have different slab values. This helps the neural networks to distinguish between different bank notes. The sum of the unmasked pixels is then called a slab value. For example, consider the slab or input input of the $10 and $20 bank notes shown in Figure 3.24. It can be seen that without the mask the number of pixels or slab value of both the bank notes is 23. However, as shown in Figure 3.24 the slab value with the mask is 13 and 23 for $10 and $20 bank notes respectively. Sixteen such slab values (called a mask set) are generated by randomly masking the input image. Eack slab value represents a unique mask and the sixteen values form one input vector to the artificial neural network. The problem then is how to determine the most effective mask set which achieves the desired or target recognition ability. GA's have been precisely used for this purpose, i.e. developing the optimum mask set for recognition of bank notes. Figure 3.25 shows the flowchart of optimizing the mask set by GA. The GA processes of crossover, mutation, and selection are employed for optimization. The crossover is brought about by sampling unique masks from two mask sets (say, A and B). Each mask set is represented as a

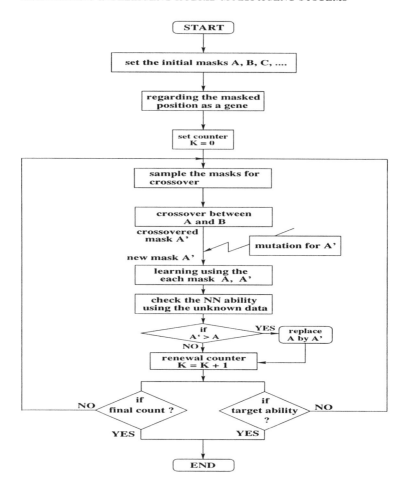

Figure 3.25.   Flow Chart for Optimizing Mask Set by GA © 1994 Springer-Verlag

set of bit string as shown in Figure 3.26. In the bit strings used, 1 represents a masked part, and 0 represents the unmasked part. In all 10 mask sets are used. Two bit values are reversed randomly in the crossovered mask during mutation. A standard back propagation artificial neural network configuration is used with (see Figure 3.27) 16 input units, 16 hidden units, and 7 output units. If the generalization (target ability) of the artificial neural network with crossovered mask set (say, A') is better than the generalization with the parental one (say , A), then A is replaced with A'. Otherwise the whole process is repeated till the

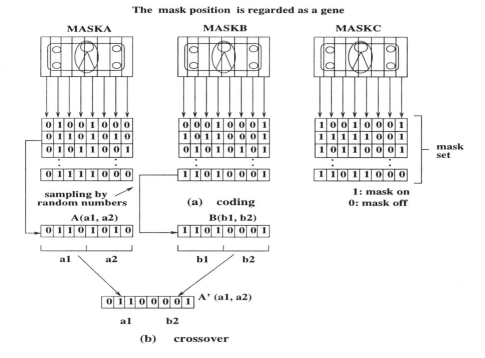

Figure 3.26.   Neural Network Structure for Currency Recognition © 1994 Springer-Verlag

desired generalization or target ability is achieved. If the desired generalization is not obtained then the mask set with the best target ability is retained.

In their GA-neuro hybrid money recognition system, Takeda and Omatu realize two objectives. Firstly, they reduce the size of the Neural networks by employing the slab architecture (Widrow et. al. 1992) and secondly, reduce the learning time of the neural networks by using GA's as preprocessors for optimizing the quality of input vectors.

3.7 SUMMARY

Fusion and transformation approaches towards building intelligent hybrid systems primarily involve neural networks and genetic algorithms. Neural networks based fusion and transformation systems. A number of intelligent Neurosymbolic and Neuro-fuzzy systems have been developed based on the fusion and transformation concepts. Neuro-symbolic and neuro-fuzzy fusion and transformation systems employ a top down or a bottom up knowledge engineering

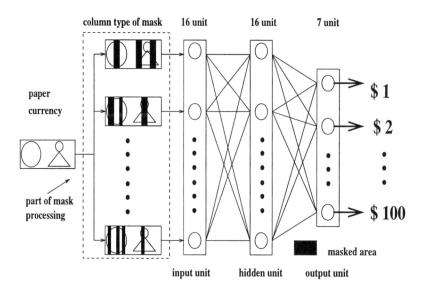

Figure 3.27.   Neural Network Structure for Currency Recognition © 1994 Spr inger-Verlag

strategy. Top down neuro-symbolic and neuro-fuzzy fusion systems use the symbolic medium as the starting point and exploit the inherent strengths (e.g., adaptation, optimization and generalization) of neural networks. The bottom up neuro-symbolic and neuro-fuzzy transformation systems unlike the top down use numerical data in the form of input-output data pairs or simply input data as the starting point and exploit the learning properties of neural networks. Also the bottom-up hybrid systems are mostly transformational systems as they operate on numerical data whereas top down are mostly fusion systems because they operate on symbolic or fuzzified data. Top down neuro-symbolic fusion systems are subdivided into three categories, namely modeling symbolic knowledge structures using neural networks, modeling symbolic reasoning using neural networks, and modeling decision trees using neural networks. Top down neuro-symbolic systems are also used for optimization of symbolic rule based systems through fusion and transformation.

In the bottom up arrangement arrangement applications have been developed based on the neuronal symbol proces sing approach grounded in biological neurons and extracting symbolic structures (e.g. concept hierarchies, rules) from artificial neural networks.

Top down neuro-fuzzy fusion systems involve fuzzification of input data, optimization of fuzzy systems, modeling fuzzy inferencing in artificial neural networks and implementation of fuzzy logic ope rations in artificial neural networks. On the other hand, bottom up neuro-fuzzy transformation systems involve learning fuzzy clusters, learning fuzzy membership functions, and learning fuzzy if-then rules.

Genetic algorithms can be seen as an alternative optimizing technique to neural networks. Genetic algorithms based fusion and transformation systems are largely used for optimization problems. Intelligent genetic-fuzzy systems are used for automating and optimizing the design of fuzzy systems. Genetic-neuro or genetic-neuro-fuzzy system as described in this chapter are used for optimizing the connection weights, neural network structure or topology, and input data set to a neural network. In the next chapter combination approaches are discussed which involve all the four intelligent methodologies are discussed.

Notes

1. The terms artificial neural networks and neural networks have been used interchangeably throughout the book.

References

Ajjanagadde, V. and Shastri, L. (1991), "Rules and Variables in Neural nets," *Neural Computation*, vol. 3, pp. 121-134.

Bezdek, J.C. (1981), *Pattern Recognition with Fuzzy Objective Function Algorithms*, Advanced Applications in Pattern Recognition, Plenum Press.

Bezdek, J.C., Tsao, E.C., and Pal, N.R. (1992), "Fuzzy Kohonen Clustering Networks," *IEEE International Conference on Fuzzy Systems* , San Diego, pp. 1035-1043.

Bezdek, J.C. (1993), "A Review of Probabilistic, Fuzzy, and Neural Models for Pattern Recognition," *Journal of Intelligent and Fuzzy Systems*, vol. 1 no. 1, pp. 1-25.

Bulsari, A., Saxen, H., and Kraslawski, A. (1992), "Fuzzy Simulation by an Artificial Neural Network," *Engineering Applications of Artificial Intelligence*, vol. 4, no. 5, pp. 404-406.

Carpenter, G. A., Grossberg, S., Markuzon, N., Reynolds, J.H., and Rosen, D. B. (1992), "Fuzzy ARTMAP: A Neural Network Architecture for Incremental Supervised Learning of Analog Multidimensional Maps," *IEEE Transactions on Neural Networks*, vol. 3, no. 5, pp. 698-713.

Carpenter, G. A., and Gjaja, M. N. (1994), "Fuzzy ART Choice Functions," *Proceedings of the World Congress on Neural Networks*, vol. 1, no. 5, pp 713-722

Chandrasekaran, B., "What kind of Information Processing is Intelligence," *Foundations of Artificial Intelligence*, Cambridge University Press, pp14-45, 1990.

Edelman, G. (1992), *Bright Air, Brilliant Fire: On the Matter of the Mind*, Raven Press, New York, USA.

Fahlman, S. E. and Hinton, G. E. (1987), "Connectionist Architectures for Artificial Intelligence," *IEEE Computer*, January, pp. 100-109.

Fu, L. M., and Fu, L. C. (1990), "Mapping Rule-Based Systems into Neural Architecture," *Knowledge-Based Systems*, vol. 3, no 1, pp.48-56

Fukuda, T., Ishigami, H. et. al. (1993), "Structure Optimization of Fuzzy Gaussian Neural Network using Genetic Algorithm," *Proceedings of Intelligent Computer Aided Manufacturing*.

Gallant, S. (1988), "Connectionist Expert Systems," *Communications of the ACM*, February pp. 152-169.

Hayashi, Y., Czogala, E., and Buckley, J.J. (1992), "Fuzzy Neural Controller," *Proceedings of 1992 IEEE International Conference on Fuzzy Systems*, San Diego, USA, pp. 197-202.

Hinton, G. E. (1990), "Mapping Part-whole Hierarchies into Connectionist Networks," *Artificial Intelligence*, vol. 46, no. 1-2, pp. 47-76.

Hudson, D. L., Cohen, M. E. and Anderson, M. F. (1991), "Use of Neural Network Techniques in a Medical Expert System," *International Journal of Intelligent Systems*, Vol. 6, pp213-223.

Hung, C.C. (1993), "Building a Neuro-Fuzzy Learning Control System," *AI Expert*, vol. 8, no. 11, pp. 40-49.

Huntsberger, T.L., and Ajjimarangsee, P. (1990), "Parallel Self-Organising Feature Maps for Unsupervised Pattern Recognition," International Journal of General Systems, vol. 16, no.4, pp. 357-372.

Ishibuchi, H., Tanaka, H., and Okada, H. (1994), "Interpolation of Fuzzy if-then Rules by Neural Networks," *International Journal of Approximate Reasoning*, January, vol. 10, no. 1, pp. 3-27.

Ishibuchi, H., Nozaki, K., Yamamoto, N., and Tanaka, H. (1993), "Genetic Operations for Rule Selection in Fuzzy Classification Systems," *Fifth IFSA World Congress*, pp. 769-772.

Lacher, R. C. (1992), "Back-Propagation Learning in Expert Networks," *IEEE Transactions on Neural Networks*, vol. 3, no 1, pp. 62-71.

Lee, C.C. (1990), "Fuzzy Logic in Control Systems: Fuzzy Logic Controller - Part I-II," *IEEE Transactions on Systems, Man and Cybernetics*, vol. 20, no 2, pp. 404-435.

Lee, M., Lee, S., and Park, C.H. (1994), "A New Neuro-Fuzzy Identification Model of Nonlinear Systems," *International Journal of Approximate Reasoning*, January, vol. 10, no. 1, pp. 29-44.

Karr, C., Freeman, L., and Meredith, D. (1989), "Improved Fuzzy Process Control of Spacecraft Autonomous Rendezvous using a Genetic Algorithm,"

Proc. of the SPIE Conference on Intelligent Control Adaptive Systems, pp. 274-283.

Karr, C. (1992), "Design of an Adaptive Fuzzy Logic ," *Proc. of International Conference on Genetic Algorithms*, pp. 450-457.

Kasabov, N. K. (1994), "Hybrid Connectionist Fuzzy Systems for Speech Recognition and the use of Connectionist Production Systems," *Advances in Fuzzy Logic, Neural networks and Genetic Algorithms* , pp. 19-33.

Kasabov, N. K. (1995), "Hybrid Fuzzy Connectionist Rule-based Systems and the Role of Fuzz Rules Extraction," *1995 IEEE International Conference on Fuzzy Systems* , vol. 1, pp. 49-56.

Kokinov, B. (1995), "Micro-Level Hybridization in the Cognitive Architecture DUAL," *Working Notes on Connectionist-Symbolic Integration: From Unified to Hybrid Approaches, IJCAI95*, Montreal, Canada, pp. 37-43.

Keller, J.M., and Hunt, D.J. (1985), "Incorporating Fuzzy Membership Functions into the Perceptron Algorithms," *IEEE Transactions Pattern Analysis and Machine Intelligence*, PAMII, pp. 693-699.

Keller, J.M., and Tahani, H. (1992a),"Implementation of Conjunctive and Disjunctive Fuzzy Logic Rules with Neural Networks," *Fuzzy Models for Pattern Recognition*, edited by Bezdek and Pal, IEEE Press, pp. 513-528.

Keller, J. M., Krishnapuram, R., and Rhee, F.C. (1992b), "Evidence Aggregation networks for Fuzzy Logic Inference," *IEEE Transactions on Neural Networks*, September, vol. 2, No.5, pp. 769-769.

Kohonen, T. (1990), *Self Organization and Associative Memory*, Springer-Verlag.

Kwan, H.K., and Cai, Y. (1994), "A Fuzzy Neural Network and its Application to Pattern Recognition," *IEEE Transactions on Fuzzy Systems*, vol. 2, no 3, pp. 185-193.

Lallement, Y., and Alexandre, F. (1995), "Cognitive Aspects of Neurosymbolic Integration," *Working Notes of IJCAI95 Workshop on Connectionist-Symbolic Integration: From Unified to Hybrid Approaches*, Montreal, Canada, pp. 7-11.

Maeda, A., Ichimori, T., and Funabashi, M. (1993), "FLIP-net: A Network Representation of Fuzzy Inference Procedure and its Application to Fuzzy Rule Structure Analysis," *Proceedings of Second IEEE International Conference on Fuzzy Systems*, pp. 391-395.

Nauck, D., and Kruse, R. (1993), "A Fuzzy Neural Network Learning Fuzzy Control Rules and Membership Functions by Fuzzy Error Backpropagation," *Proceedings of the IEEE International Conference on Neural Networks*, San Francisco, vol. 2, pp. 1022-1027.

Nobre, F. S. M. (1995), "Genetic-Neuro-Fuzzy Systems: A Promising Fusion," *1995 IEEE International Conference on Fuzzy Systems*, vol. 1, pp. 259-266.

Pal, S. K., Mitra, A. (1992), "Multilayer Perceptron, Fuzzy Sets, and Classification," *IEEE Transactions on Neural Networks*, vol. 3, no. 5, pp. 683-697

Pedrycz, W. (1991), "Neurocomputations in Relational Systems," *IEEE Transactions on Pattern Analysis and Machine Intelligence*, vol. 13 no 3, pp. 289-297.

Park, N. S., and Robertson, D. (1995), "A Localist Network Architecture for Logical Inference Based on Temporal Synchrony Approach to Dynamic Variable Binding," *Working Notes on Connectionist-Symbolic Integration: From Unified to Hybrid Approaches, IJCAI95*, Montreal, Canada, pp. 63-68.

Pedrycz, W. (1992), "Fuzzy Neural Networks with Reference Neurons as Pattern Classifiers," *IEEE Transactions on Neural Networks*, vol. 3 no 5, pp.770-775.

Pedrycz, W., and Rocha, A. (1993), "Fuzzy-Set based Models of Neurons and Knowledge based Neurons," *IEEE Transactions on Fuzzy Systems*, vol. 1 no 4, pp.254-266.

Plate, T. A. (1995), "A Distributed Representation for Nested Compositional Structure," *Working Notes on Connectionist-Symbolic Integration: From Unified to Hybrid Approaches, IJCAI95*, Montreal, Canada, pp. 69-74.

Romaniuk, S. (1994), "Applying Crossover Operators to Automatic Neural Network Construction," *Proceedings of the First IEEE Conference on Evolutionary Computation*, vol. 2 pp. 750-753.

Saito, K. and Nakano, R. (1988), "Medical Diagnostic Expert System Based on PDP Model," *Proceedings of the IEEE International Conference on Neural Networks*, San Diego, USA.

Sestito, S. and Dillon, T. (1991), "The use of Sub-Symbolic Methods for the Automation of Knowledge Acquisition for Expert Systems," *Eleventh International Conference on Expert Systems and their Applications*, France, pp. 317-28.

Sestito, S. and Dillon, T. (1994), *Automated Knowledge Acquisition*, Prentice Hall, Sydney, Australia.

Sethi, I.K. (1990), "Entropy Nets: From Decision Trees to Neural Networks," *Proceedings of the IEEE*, vol. 78, no 10, pp. 1605-13.

Sethi, I.K. (1991), "Decision Tree Performance Enhancement using an Artificial Neural Network Implementation," *Artificial Neural Networks and Statistical Pattern Recognition: Old and New Connections*, eds I.K. Sethi & A.K. Jain, Elssevier Science Publishers, Amsterdam, pp. 71-86.

Shastri, L. (1988), "A Connectionist Approach to Knowledge Representation and Limited Inference," *Cognitive Science*, vol. 12, pp. 331-392.

Shastri, L. and Ajjanagadde, V. (1989), "A Connectionist System for Rule Based Reasoning with Multi-Place Predicates and Variables ," *Technical Re-*

port *MS-CIS-8906*, Philadelphia, PA: University of Pennysylvania Computer and Information Science Department.

Shastri, L. and Ajjanagadde, V. (1993), "From Simple Associations to Systematic Reasoning: A Connectionist Representation of Rules, Variables, and Dynamic Bindings using Temporal Synchrony, " *Behavioral and Brain Sciences*, vol. 16, no. 3.

Sun, R. 1991, "Integrating Rules and Connectionism for Robust Reasoning," in *Technical Report No. CS-90-154*, Waltham, MA: Brandeis University. Computer Science Dept.

Sun, R. (1994), "CONSYDERR: A Two Level Hybrid Architecture for Structuring Knowledge for Commonsense Reasoning," *Proceedings of the First International Symposium on Integrating Knowledge and Neural Heuristics*, Florida, USA, pp. 32-39.

Sulzberger, S.N., Tschichold-Giirman, N., and Vestli, S.J. (1993), "FUN: Optimization of Fuzzy Rule Based Systems using Neural Networks," *Proceedings of the IEEE International Conference on Neural Networks*, San Francisco, vol. 1, pp. 312-316.

Takagi, H., and Sugeno, M. (1985),"Fuzzy Identification of Systems and its Applications to Modeling and Control," *IEEE Transactions on Systems Man and Cybernetics*, vol. 15, no. 1 pp. 116-132.

Takagi, H., and Hayashi, I. (1991),"NN-driven Fuzzy Reasoning," *International Journal of Approximate Reasoning*, vol. 5, pp. 191-212,

Takagi, H., and Hayashi, I. (1992a),"NN-driven Fuzzy Reasoning," *Fuzzy Models of Pattern Recognition*, edited by Bezdek and Pal, IEEE Press, pp. 496-512.

Takagi, H., and Hayashi, I. (1992b),"Neural Networks Designed on Approximate Reasoning Architecture and their Applications," *IEEE Transactions on Neural Networks*, September, vol. 2, no. 5, pp.752-760.

Lee, M. A., and Takagi, H. (1993), "Integrating Design Stages of Fuzzy System using Genetic Algorithms," *1993 IEEE International Conference on Fuzzy Systems using Genetic Algorithms* vol. 1, pp. 612-617. bibitemTakeda94 Takeda, F., and Omatu, S. (1994), "A Neuro-Money Recognition Using Optimized Masks by GA," *Advances in Fuzzy Logic, Neural Networks and Genetic Algorithms*, pp. 190-202.

Towell, G. and Shavlik, J. (1992), "The Extraction of Refined Rules from Knowledge-based Neural Networks," *Machine Learning Research Group Working Paper 91-4*, University of Wisconsin, Madison, USA.

Touretzky, D. S. (1987), "Representing Conceptual Structures in a Neural Network," *Proceedings of the First IEEE Conference on Neural Networks*, San Diego, USA.

Touretzky, D. S., and Hinton, G. E. (1988), "A Distributed Connectionist Production System," *Cognitive Science*, vol. 12, pp. 423-466.

Watnabe,T., Matsumoto, M, and Hasegawn, T. (1990), "A Layered Neural Network Model using Logic Neurons," *Proceedings of the International Conference on Fuzzy Logic & Neural Networks*, Iizuka, Japan, pp. 675-678.

Whitley, D., Strakweather, T., and Bogart, C. (1990), "Genetic Algorithms and Neural Networks: Optimizing Connection and Connectivity," *Parallel Computing*, vol. 14 pp. 347-361.

Widrow, B., Winter, R. G., and Baxter, R. A. (1988), "Layered Neural Nets for Pattern Recognition," *IEEE Transactions on Acoustics, Speech and Signal Processing*, vol. 36, no. 7, pp. 1109-1118.

Williams, B., Bostock, R., Bounds, D., and Harget, A. (1994), "Improving Classification Performance in the Bumptree Network by Optimizing Topology with a Genetic Algorithm," *Proceedings of the First IEEE Conference on Evolutionary Computation*, vol. 1 pp. 490-495.

Zadeh, L.A. (1965), "Fuzzy Sets," *Information and Control*, vol. 8, pp. 338-353.

4 INTELLIGENT COMBINATION SYSTEMS

4.1 INTRODUCTION

In the last decade intelligent methodologies like artificial neural networks, fuzzy systems, genetic algorithms, and knowledge based systems (or expert systems) have been applied in a number of industrial applications. However, the approach of using these technologies by themselves alone has exposed some limitations which have discouraged their use in some industrial applications. For example, knowledge based systems are also sometimes known as "sudden death systems" because of their brittleness, and lack of graceful degradation. Artificial neural networks on the other hand, have problems generating explanations for their results (although, to some extent this problem has been addressed by the rule extraction techniques described in the last chapter). Genetic algorithms and fuzzy systems are known to be computationally expensive. Besides, as indicated in the last chapter, fuzzy systems have problems in determining the membership functions.

These limitations are symptomatic manifestations of complexities associated with real world problems. It is many times difficult to model effectively solution to complex real world problems (especially real time) using any one intelligent methodology. In the last chapter it has been shown how artificial neural networks on their own have been able to address a number of limitations of knowledge based systems and fuzzy systems through fusion and transformation. Also it has been shown how genetic algorithms can alleviate some of the problems associated with the design of neural networks and fuzzy systems through fusion and transformation. In this chapter, combination approaches involving synergistic combination of two or more intelligent methodologies are described. The two or more intelligent methodologies play a distinct role in the problem solving process. This hybrid approach addresses complex real world problems based on the premise that it is beyond the reach of a single intelligent methodology to cover the entire spectrum of human cognition and information processing. In other words, various intelligent methodologies reflect different aspects of human cognition and by combining them together one can develop a better and more complete understanding of the human information processing system, and at the same time develop more powerful problem solving strategies.

4.2 INTELLIGENT COMBINATION APPROACHES

Intelligent hybrid systems based on the combination approach are described under the following categories:

- Neuro-Symbolic Combination Systems

- Symbolic-Genetic Combination Scheduling System

- Neuro-Fuzzy Combination Systems

- Neuro-Fuzzy-Case-Based Combination Systems

- Combination Approaches in Intelligent Hybrid Control Systems

4.3 NEURO-SYMBOLIC COMBINATION SYSTEMS

In order to broaden the scope of application of neural networks and symbolic knowledge based systems a number of neuro-symbolic combination systems have been developed. Although numerous applications have been built using this combination approach, the ones covered here have been put into two categories, namely, neuro-semantic network combination systems, and neuro-symbolic classification and diagnostic systems.

4.3.1 Early Work Neuro-Symbolic Combination Systems

Some of the early work in neuro-symbolic combination systems has been done by Hendler (1989), Hiri (1990), and Gutknecht et al. (1990). Hendler (1989) describes an experimental system which combines symbolic semantic networks with a back propagation neural network (BPNN) for performing reasoning about perceptual similarities. The paper uses this approach to identify a correct solution to a problem in a natural language planning system. It uses a marker-passing algorithm with some heuristics for converting symbolic activation at the leaves of the semantic network to numeric activation of the input nodes of a BPNN. In this neuro-semantic network application, neural network and semantic networks have been used as coprocessors.

In another hybrid approach (Klusch 1993) a combination of back propagation neural network and semantic networks is used for classification of stars. In this application the hybrid arrangement functions in a top down or top return mode and bottom up mode. In the top down mode the semantic network and neural network function in a manner similar to that proposed by Hendler (1989). In the bottom up mode neural network is used for classification and semantic network is used for evaluation of the classification results. Here the neural network and semantic network are used as coprocessors in the top down arrangement and semantic network. Hiri (1990) proposes a combination (with help of two small examples) where he uses neural networks to classify subsymbolic sensory data knowledge in the symbolic form. The neural net is triggered from the conditional part of a rule in the rule base. The motivation for such a combination is to avoid loss of information in transformation of numeric data into symbolic form. Gutknecht et al. (1990) describe a hybrid approach for solving a high-school physics problem. They use neural network to predict the subgoal at the lowest level for solving a particular problem. They support the concept of step wise or incremental training rather than one step training.

Hilario (1995) has outlined the processing role of symbolic systems and neural networks in neuro-symbolic systems. The processing roles identified are chain processing, subprocessing , coprocessing, and metaprocessing . In chainprocessing one component functions as a main processor, whereas the other undertakes pre or postprocessing role. In subprocessing, one of the two components is embedded in and subordinated to the other. On the other hand, in metaprocessing one component functions as a base problem solver and the other plays a metarole such as monitoring and control. Coprocessing (as has already been described) implies both components are equal partners in the problem solving process. Although such a distinction is informative, as stated by Hilario (1995) *it has little bearing on central issues of neurosymbolic integration.*

4.3.2   Neuro-Symbolic Combination Systems for Diagnostic Applications

Becraft et al. (1991) combine neural networks with expert systems for fault diagnosis in a chemical process plant based on sensor measurements. Two levels of back propagation neural networks are used to implement the diagnostic strategy. A model based expert system is used for initializing the neural networks and confirming or rejecting the diagnosis of the neural networks. The expert system also handles novel situations where the neural networks give incorrect diagnosis. In this combinatory configuration the neural networks and expert systems function as black boxes in a complementary manner.

On similar lines, a hybrid performance monitoring and fault diagnosis system in a telecommunication network has been developed by Senjen et al. (1993) for diagnosing network faults. In the hybrid configuration, the back propagation neural network are used for generating the possible faults for various network components. The inputs to the network are symptoms which are derived by continuously monitoring the network performance data. The symptoms are localized for each network component. The possible faults generated by the neural networks are then fed into an expert system which determines the cause of these faults by performing detailed fault diagnosis. Many times possible faults may have a common cause because all network are related/connected in the telecommunication network. The expert system produces the active list of prioritized faults which can then be pursued by the network manager for remedial action.

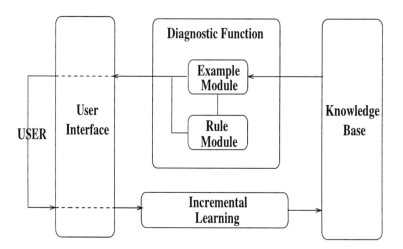

Figure 4.1.   Neuro-Symbolic Combination System for Fault Diagnosis (Lim et al. 1991)

A framework for combining fault diagnosis and incremental knowledge acquisition in connectionist expert systems has been proposed by Lim et al. (1991). The framework consists of an example module(EM) and rule module(RM) for diagnosis, and a knowledge base for learning past and new cases as shown in Figure 4.1. The knowledge base consists of a three layered (symptom,hidden,fault) neural network and is connected to the EM. The initial symptoms are fed into EM which inferences a list of possible faults using the nearest neighbor algorithm. In case it cannot configure the fault/s, Rule Module(RM) is activated to pinpoint the fault and the new example is learnt by the neural network by invoking an incremental learning algorithm based on Euclidean distance measure. The list of possible faults are verified by the user by replacing the faulty component suggested by the system. In case the fault still persists, the EM uses backward chains through the neural network using the information heuristic scheme to solicit the values of the relevant unknown symptoms which are useful in discriminating the possible faults. The system has been developed for Singapore Airline to assist technicians in diagnosing avionic equipment.

4.4 SYMBOLIC-GENETIC COMBINATION SCHEDULING SYSTEM

Expert systems have been used for scheduling systems in air traffic management, maintenance scheduling in power systems and in various other areas. Hamada et al. (1995) report on application of an expert system in conjunction with genetic algorithms for an automatic scheduling system for Nippon Steel. Figure 4.2 shows the block diagram of the major processes involved in the production of stainless steel for which the scheduling system has been developed.

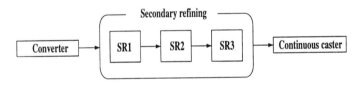

Figure 4.2.   Major Processes in Stainless Steel Production © 1995 IEEE

The converter oxidizes iron and produces molten steel which is fed into secondary refinement processes, SR1, SR2, and SR3 respectively. The secondary refinement processes alloy different types of steel and produces pots or charges of molten steel. The continuous caster casts the molten steel which is then cut into several slabs and sent to the rolling mills to produce the finished product. Three categories (A, B, and C) of stainless steel are produced. The continuous caster must be fed with multiple charges in order to work efficiently. The goal

scheduling system is to develop a production schedule which minimizes the total time to process all orders and satisfy the 16 operational constraints shown
in Table 4.1. The time minimization problem is aggravated by the fact that
the time taken by the continuous caster to cast one charge (T_c) is less than the
time taken by the secondary refinement process to produce a charge (T_s). Since
multiple charges must be fed into the continuous caster, there is waiting time
involved until other charges (i.e 2 or 3 additional charges needed for continuous
casting) are ready. If a charge of molten steel has to wait for more than 120
minutes for other charges from the secondary refinement process then it has to
be reheated (thus increasing the heating costs).

The operational constraints shown in Table 4.1 are divided into two groups,
namely, Charge-group Forming (CCF), and Charge-group Ordering (CCO).
Charge groups have to be as large as possible (subject to the satisfaction of the
CCF constraints) in order to increase the efficiency of the continuous caster.
On the other hand, the charge groups have to be scheduled in a proper sequence
which satisfies the CEO constraints.

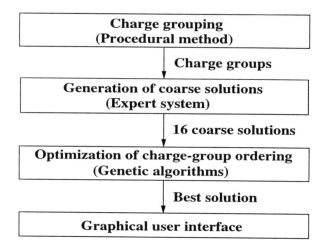

Figure 4.3.   Flow Chart of Scheduling System © 1995 IEEE

The flow chart of the scheduling system is shown in Figure 4.3. The pro-
cedural method in Figure 4.3 is used to form charge groups through charge
ordering (Figure 4.4a), charge grouping (Figure 4.4b), and charge-group di-
viding (Figure 4.4c). The charge ordering, grouping and group-dividing is done
in a manner so as to satisfy the CCO and CCF constraints. For example, the
charge groups as shown in Figure 4.4b are sorted out from wide to narrow in
order to satisfy constraints CCF-1 and CCF-4. In order to satisfy constraint

Table 4.1. Operational Constraints in Production Scheduling of Stainless Steel © 1995 IEEE

GROUP	DESCRIPTION	PRIORITY
Constraints on charge-group forming (CCF)	1. Identical type only for multiple charge continuous casting	1
	2. Multiple-charge continuous casting permissible up to four continuous runs.	1
	3. Maximum range of width change for multiple-charge continuous casting is 300 mm	1
	4. Width change in multiple continuous casting made from larger one	1
	5. Single-charge casting permissible up to three continuous runs	1
Constraints on charge-group ordering (CCF)	1. Production to be made in the order Category A $\Rightarrow$ Category B $\Rightarrow$ Category C	1
	2. To start from Category A with less content of element a	1
	3. To start from Category A in single-charge casting	1
	4. To start from a type with higher content of element b	3
	5. Type A5 to be placed in first 10 charges	1
	6. Type B1 to be started from one with less content of element c	1
	7. Possible to process type B1 up to three charges a day	2
	8. Pot used for a charge of type B1 with higher content of element d s not usable for the nexy charge with less content of element d	1
	9. Pot used for a charge of type with higher content of element e is not used for the next charge of type B2 containing element e 6	1
	10. Type B3 not to be ploaced in last four charges	1
	11. Possible to process type C1 up to seven charges a day	2
	12. Types C2 and C3 to be processed before type C4	1
	13. Types C4 and C5 to be processed before type C6	1
	14. For type C7, pot used for Category A is not usable	1
	15. Type C7 not to be placed in first four charges	1
	16. Delivery time	2

CCF-3, the charges with less than 300 mm difference are grouped in one group as shown in Figure 4.4b. Finally, the charge groups in Figure 4.4b are further sub-divided into into groups A-G1, A-G2, A-G3, and so on in order to satisfy constraint CCF-2.

An expert system with 153 rules is built to satisfy all the 16 constraints. One constraint is represented in several rules that modify the candidate solution in order to satisfy the constraints. The schedule produced by the expert system is not necessarily the best or quasi-optimal one as the rules for what is a "good" schedule cannot be extracted from the human operator.

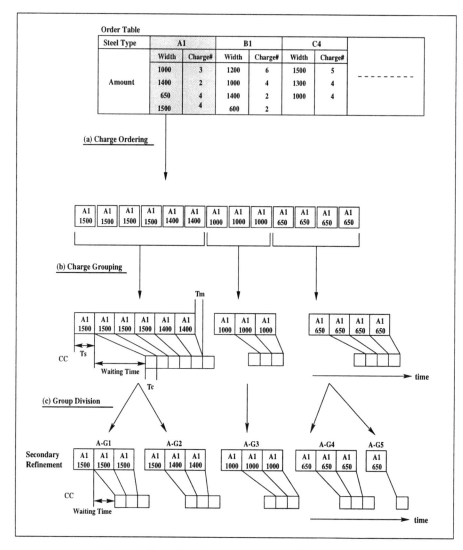

Figure 4.4.   Forming Charge Groups © 1995 IEEE

Thus, genetic algorithms are used to search the quasi-optimal sequence of charge groups. In order to build the initial population, the expert system is used to generate 16 coarse sequences, each of which is generated by rules corresponding to one of the 16 constraints. A gene map or chromosome representing a sequence is shown in Figure 4.5. The population size, probability of crossover, and mutation were decided through trial and error as 250, 50%, and 40% respectively. The objective of the GA is to minimize the fitness function:

$$e = TotalTime + \sigma w_i * c_i$$

where, c_i is the penalty that is positive (1) when the ith constraint is not satisfied, and 0 otherwise; w_i is the positive weight for c_i and is determined by importance of the constraint. The schedule generated by hybrid arrange-

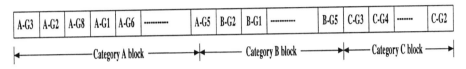

Figure 4.5.   Chromosome for the Scheduling System © 1995 IEEE

ment was 38 minutes shorter in processing time than corresponding man-made schedule, and increased productivity by 10%.

4.5 NEURO-FUZZY COMBINATION SYSTEMS

The previous chapter looked at how neural networks could be used for optimization of fuzzy systems. However, there are a number of applications where the division of labor between fuzzy systems and neural networks is distinct and independent of each other. These neuro-fuzzy combination systems can be categorized as:

- Neuro-Fuzzy Load Forecasting System with Fuzzy Preprocessing

- Neuro-Fuzzy Number Recognition System with Modular Nets, and

- Neuro-Fuzzy Condition Based Maintenance Coprocessing System

4.5.1 Neuro-Fuzzy Load Forecasting System with Fuzzy Preprocessing

As front end preprocessors fuzzy logic is not only used for fuzzification of input data but also for building a fuzzy knowledge base (fuzzy expert system). The fuzzification of input data and fuzzy knowledge bases are used to enhance the accuracy (in terms of prediction) of the neural networks and model common-sense knowledge used by the human experts in a problem domain respectively.

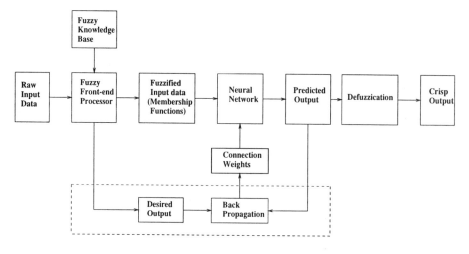

Figure 4.6.   Overall Neuro-fuzzy Architecture © 1994 IEE

Neural networks in this hybrid approach are used for their learning, generalization, retrieval from partial information, graceful degradation, and noise handling properties.

A hybrid neuro-fuzzy system developed by Srinivasan et al. (1994) for predicting future load values with a lead time of 24 hours has been shown to outperform the existing autoregressive models, and other stand alone neural network and fuzzy logic approaches. The overall architecture of the hybrid load forecasting system is shown in Figure 4.6. A more detailed version outlining the processing flow is shown in Figure 4.7.

4.5.1.1   Role of Fuzzy Frontend Processor.   As can be seen from Figure 4.6 and 4.7 respectively, the fuzzy frontend processor performs two functions, namely fuzzification of input data, and application of fuzzy knowledge base or fuzzy expert system (built by human experts) on the input data. Input data like maximum temperature (*max_t), minimum temperature (*min_t), maximum humidity (*max_h), proximity to holiday (*hol_d), seasonal trend (*trend), and average temperature of previous day (*prev_av) are fuzzified using predefined membership functions. The fuzzy expert system consists of commonsense rules in the domain. These commonsense rules are related to change in load forecast based on the weather forecast. For example, if the weather forecast is "*maximum temperature 33 degrees centigrade and heavy rains in late afternoon,*" then the following rule will apply:

IF max_t is high (0.8)
AND max_h is high (0.9)
AND heavy rains in the afternoon *(1.0)*
THEN change_in_load in the afternoon *is PS (Positive Small)*

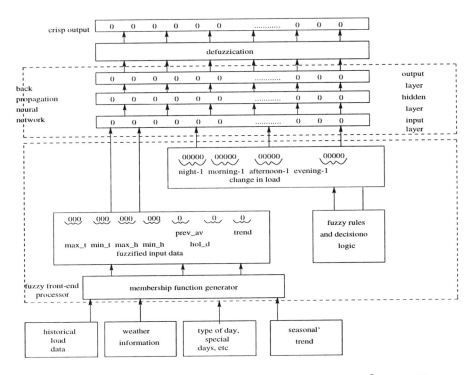

Figure 4.7.   Processing Flow of Neuro-Fuzzy Load Forecaster © 1994 IEE

Positive Small is a predefined membership function. Similarly other commonsense rules related to *change_in_load* forecast based on special events (e.g. sporting event), and particular day of the week (e.g. Friday) are also used. After the application of a fuzzy rule, the *change_in_load* is used as fuzzy input variable along with other fuzzified input variables as shown in Figure 4.7.

4.5.1.2   Role of Neural Networks. In the fuzzy-neuro load forecaster, neural networks are used for their learning, generalization, noise handling, retrieval from partial information, and parallel processing properties. A standard three layer neural network trained with backpropagation is used for predicting the 24 hour load profile.

4.5.2 Neuro-Fuzzy Number Recognition System with Modular Nets

The approximation capabilities of fuzzy systems are also usefully employed for monitoring the performance of modular neural networks neural networks. In this hybrid neuro-fuzzy approach, the fuzzy system is used as a postprocessor. A generic fuzzy rule for monitoring the performance of a neural network can be:

IF a neural net N is a kind of feedforward neural net
AND its classification results are UNSATISFACTORY
THEN decrease its reliability SLIGHTLY

Fuzzy logic can also be used for postprocessing where modular neural networks or neural network committees are employed for improving the classification accuracy (Cho 1994). The idea behind modular neural networks shown in Figure 4.8 is to use a set of networks as possible generalizers and achieve better generalization through a winner-takes-all strategy or by combining the outputs by simple averaging. In this arrangement all the neural networks are trained on the same task but with different points of view or different sets of training data. However, the problem with this arrangement is that simple averaging or winner-takes-all strategy is not good enough if we have networks with different classification accuracy.

Cho (1994) has shown that by using fuzzy integral introduced by Sugeno (1977) as postprocessors, the neuro-fuzzy hybrid system performs better than simple averaging, winner-take-all or even weighted average. The fuzzy integral e can be computed by:

$$e = max_{i=1}^{n}[min(h(y_i), g(A_i))]$$
where
$A_i = y_1, y_2, ..., y_i,$ $g(A_1) = g(y_1) = g^1,$ *and*
$g(A_i) = g^i + g(A_{i-1}) + \lambda g^i g(A_{i-1}),$ *for* $1 < i \leq n,$ $\lambda \in (-1, +\infty),$ *and* $\lambda \neq 0.$

The term $h(y_i)$ is an indication of how certain we are in the classification of object A to be in class w_k using the neural network y_i. This $(h(y_i))$ is in fact the output activation of the output unit representing class w_k in neural network y_i. g^i is the fuzzy density of neural network y_i depending upon how good this neural network has performed on the validation data. It represents the degree of importance of the neural network y_i. The *min* operation indicates the degree to which class w_k satisfies both the criteria $h(y_i)$, and $g(A_i)$. The *max*

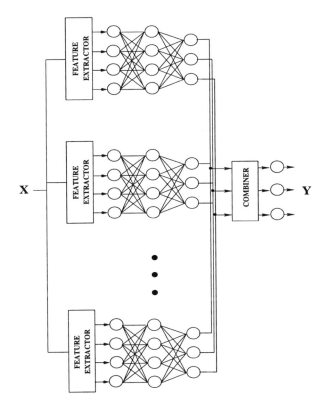

Figure 4.8.   A Modular Neural Network Scheme © 1994 Springer-Verlag

operation takes the biggest of all these terms. Thus the fuzzy integral computes the maximal grade of agreement between between the objective evidence and expectation. More details on the properties of fuzzy integral used for this application are described in Cho (1994).

For example, consider the modular neural network arrangement shown in Figure 4.9 for a two class problem. The modular networks are used to determine whether the input image represents the digit 6 (class 1) or 4 (class 2). Figure 4.9 also shows the corresponding fuzzy density values g^1 (0.34), g^2 (0.32), and g^3 (0.33) of three neural networks NN^{f1}, NN^{f2}, and NN^{f3} respectively. The output values as shown in Figure 4.9 for class 1 the output values are $h(y_1)$ = 0.6, $h(y_2)$ = 0.7, and $h(y_3)$ = 0.1. For class 1, the output values are $h(y_1)$ = 0.6, $h(y_2)$ = 0.3, and $h(y_3)$ = 0.4. Table 4.2 shows how the fuzzy integral is applied to come to a consensus decision (where $H(E) = min(h(y_i), g(A_i))$ that the input image belongs to class 1. Cho (1994) shows that neuro-fuzzy

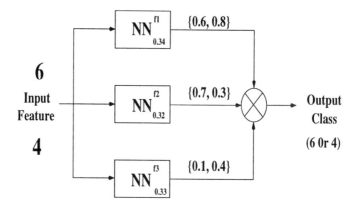

Figure 4.9. Modular Neural Networks for a Two Class Problem © 1994 Springer-Verlag

Table 4.2. Fuzzy Integral Based Consensus Decision © 1994 Springer-Verlag

Class	$h(y_i)$	$g(A_i)$	H(E)	max (H (E))
1	0.7	$g(\{y_2\}) = g^2 = 0.32$	0.32	0.6
	0.6	$g(\{y_2, y_1\}) = g^2 + g^4 + \lambda g^2 g^4 = 0.66$	0.6	
	0.1	$g(\{y_3, y_2, y_1\}) = 1.0$	0.1	
2	0.8	$g(\{y_1\}) = g = 0.34$	0.34	0.4
	0.4	$g(\{y_1, y_3\}) = g^4 + g^3 + \lambda g^1 g^3 = 0.67$	0.4	
	0.3	$g(\{y_1, y_3, y_2\}) = 1.0$	0.3	

hybrid system for recognition of handwritten characters in arabic outperforms the modular neural networks using the averaging and voting method in terms of recognition accuracy.

4.5.3 Neuro-Fuzzy Condition Based Maintenance Coprocessing System

A hybrid approach for implicit and explicit automated reasoning for condition-based maintenance (CBM) has been developed by Garga (1996). The development of the new approach is motivated by describing the relevance of automated reasoning to CBM, and to the smart ship project, and by reviewing the challenges in designing an automated reasoning system. The concept of the new approach is then described and illustrated with an example. In an illustrative example, an intelligent oil analysis system, a set of explicit rules is built, for ascertaining whether the oil filter is clogged based on pressure, temperature, and flow measurements. Then the rules are used to train a layered feedforward neural network to capture the explicit knowledge. Finally, a much smaller rule set is derived from the neural network which represents the same explicit knowledge that was obtained from a domain expert. The neural network can be retrained on actual sensor data to refine the knowledge base and to incorporate implicit system knowledge. Rules can also be extracted from the resulting network. The notable features of the novel approach are its ability to encapsulate explicit and implicit knowledge and its flexibility in allowing updates of the knowledge base and extraction of a parsimonious and consistent set of rules from the network. The approach is truly hybrid as it combines fuzzy logic and rule-based systems, to capture the expert's explicit knowledge, and neural networks, to provide a parsimonious representation of the knowledge base and for their adaptability. The approach can benefit shipboard CBM efforts in automating intelligent condition monitoring and by providing an efficient implementation mechanism for distributed diagnostics.

4.6   NEURO-FUZZY-CASE COMBINATION SYSTEM FOR FASHION SHOE DESIGN

Fashions change from season to season, and even mid-season. However fashion trends have a cyclic behavior. A fashion 'look' will often go 'out' and a number of years later make another appearance. An additional factor in women's fashion is the diversity. Even within a single trend, there are a multitude of variations. People usually want something that is 'in', but a little different, or a little better than the other products on the market. A company that can create new designs to satisfy this demand increases its chance of success by giving itself an edge in the marketplace. In a company that produces women's fashion shoes there are hundreds of designs produced each year and a designer forgets all but a few of the main ones in a relatively short span of time. Thus, when a fashion trend makes a second (or third) appearance, the previous designs have been forgotten and the new shoes are re-designed from scratch. This forgetting and starting over procedure is a problem that if overcome could lead to a faster design stage (and hence decrease response time to markets) and also to the input of a wider variety of ideas than the designer has access to at any given time.

4.6.1   Case Based Reasoning for Fashion Shoe Design

In order to develop a fashion shoe design system (Main, Dillon and Khosla 1996) a Case-Based Reasoning (CBR) system is chosen as there are no given rules that could be implemented directly in a Rule-Based system and, addi-

SPECIFICATION

STYLE NAME:- **GIGI**	
LAST:- **RAVEL**	
PATTERNS:- "GRETEL" COMPLETE	
GEORGIE" BACK STRIP SIZES:-	
INSOLE:- **RAVEL - CODE 054**	

SOCK:- **FULL SOCK**	
SOCK STAMP:- **VARIOUS**	
SOLE UNIT:- **765 PU - CODE 118**	VAMP:- Nº/ 1, 2 & 3 FRONTS U/LAY SKIVE
	EDGE SKIVE BACKS

SIZE RANGE

SIZE:-	5 -	6 -	7 -	8 -	9 -	10 -	11	QTRS:-
UNIT SIZE:-	36	37	38	39	40			

STIFFENER:- NONE	OTHER COMPONENTS:-
TOE PUFF:-	
BOX:-	SPLITTING
SHANK:-	
SPECIAL COMMENTS	PERFORATION:-
*SIZE RANGE OF SOLES REDUCED TO FIT RAVEL LAST	ORNAMENTS
	ELASTIC:-
	THREAD:- N º/25
	STRIPPING/BINDING:-
	SUNDRIES:-

Figure 4.10.   Shoe Design Specification Sheet

tionally, CBR allows one to store past cases and reason with them. Case-based reasoning consists of: storing and indexing cases so that the appropriate ones are retrieved; retrieval of the appropriate case; and then adapting the retrieved case to help solve the current problem case. Problems remain with the case-based approach both in deriving efficient characteristics for indexing and with the retrieval methods used for recovering the past case(s). There are several main types of retrieval that have been traditionally used in case-based systems. The simplest way of retrieving cases is using a nearest neighbor search. A nearest neighbor search looks for those cases stored in memory that have the greatest number of characteristics that are the same as the current problem. The weighting of characteristics used has to be determined by the implementer. It is frequently difficult to decide which characteristics of a case are more important than the others, in that they contribute more to the successful outcome of a particular case. Inductive approaches to case-retrieval use statistical algorithms to determine which characteristics of cases are most important in the retrieval of those cases. Inductive approaches are often better at achieving correct weightings than the weightings used in nearest neighbor search, but inductive indexing methods have their own drawbacks. One problem is that a large number of cases are required for deriving satisfactory weightings for the attributes of the cases. Another is that not all cases are determined by the same order of importance of their characteristics. For example, one case may be primarily determined by three particular characteristics and those characteristics may have little or no importance in determining the outcome of another case. Knowledge based approaches to the retrieval of cases use domain knowledge to determine the characteristics that are important for the retrieval of a particular case, for each individual case. Unfortunately, to implement this type of indexing system, a large amount of domain knowledge must be available, acquired and applied to the cases.

In order to address the problem, a Neuro-Fuzzy-Case Combination System has been designed that stores the past shoe designs and retrieves a number of designs that are close to the design specification that the designer inputs. The output from the system is a set of previous designs that most closely approximate the desired design. These designs are used to develop adaptations that provide the desired design. For developing the intelligent combination system a collection of previously existing designs (about 400) from Main Jenkin Pty. Ltd. (a Melbourne, Australia based footwear manufacturer), has been used as a case base. Figure 4.10 shows a sample shoe design specification sheet.

The following two sections describe the role of neural networks, and fuzzy systems in the fashion shoe design system.

4.6.1.1   Role of Neural Networks in Fashion Shoe Design.   Neural networks are used for indexing and retrieving the case which fits most closely to the feature description provided by the input pattern. In an attempt to deal with the complexity of case retrieval associated with 400 design cases hierarchical neural networks have been used. The output classes associated with the neural networks at different levels are shown in Figure 4.11 . Another reason behind the neural network arrangement shown in Figure 4.11 is to enhance the indexation accuracy. Supervised neural networks using backpropagation algorithm are employed for training. A set of input features used at the level 1 and 2 neural networks are shown in Table 4.3. As can be seen from Table 4.3, the input features used to describe the cases are boolean, continuous, multivalued and fuzzy.

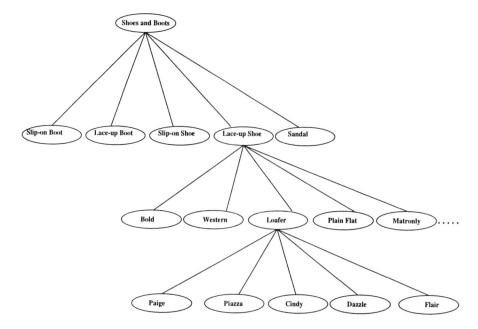

Figure 4.11.   Output Classes of Neural networks in Fashion Shoe Design

4.6.1.2   Role of Fuzzy Logic in Fashion Shoe Design. Fuzzy logic has been used for fuzzification of input features. For example, the *inside-back height* feature of a shoe can be described as high (a boot), medium (a shoe) or low (a type of sandal), but some shoes will not fit into the crisp separations of categories. If a shoe has a height of 1.9", it would be nice to say that it is partly a shoe and partly a boot. This can be represented for input to the neural network by using 3 attributes: 'low', 'medium' and 'high'. Each shoe would have a value for each of these three variables. A low shoe (e.g. a mule) would have attributes with these values: low:1 medium:0 high:0. A boot or shoe with a large back-height would have the following: low:0 medium:0 high:1. A traditional shoe (of 1.5") would have the following values: low:0 medium:1 high:0. This is very straight forward for the instances that fall exactly into one of the main categories. However, for those cases that do not fit so nicely (e.g. the 1.9" case), a fuzzy membership function shown in Figure 4.12 that determines the value for in-between cases has been used. For shoe height of 1.9", the corresponding values of low, medium, and high fuzzy sets are 0, 0.2, and 0.8 respectively.

Table 4.3.   Some Input Features in Fashion Shoe Design System

Attribute	Type	Description	Range
apron-front	boolean	Whether the shoe has an apron front	
open-heel	boolean	Whether the shoe has no covering at the heel	
open-toe	boolean	Whether the shoe has a peep toe	
sling-back	boolean	Whether the shoe has a strap at the back	
spur-straps	boolean	Whether the shoe has straps around the ankle	
toe-seam	boolean	Whether the shoe has a seam across the toe	
percentage-cut-out	continuous	The amount cut out or perforated on the shoe	0-100
fastening	multivalued	The method by which the shoe is held on	laces, elastic, buckle, zip, none
sole type	multivalued	The type of sole on a shoe	cleat, resin, saw-tooth, compound, layered, covered, bertie-type, wedge
toe shape	multivalued	The shape of the toe of the shoe	round, pointed, chiselled, bump
heel-height	fuzzy	The height of the shoe's heel	low, medium, heigh
inside-back-height	fuzzy	The attribute by which the height of the shoe is measured	low, medium, high
sole-thickness	fuzzy	The thickness of the shoe's sole	flat, medium, thick
boot-height	fuzzy	Whether a boot is a low boot, mid-boot or high boot	low, medium, high

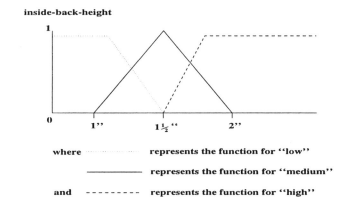

Figure 4.12.   Membership Function of Input Feature "inside-back height"

4.7   COMBINATION APPROACH BASED INTELLIGENT HYBRID CONTROL APPLICATIONS

Historically, the motivation for development of hybrid fuzzy control systems is somewhat similar to that of hybrid knowledge based systems. Some of the pio-

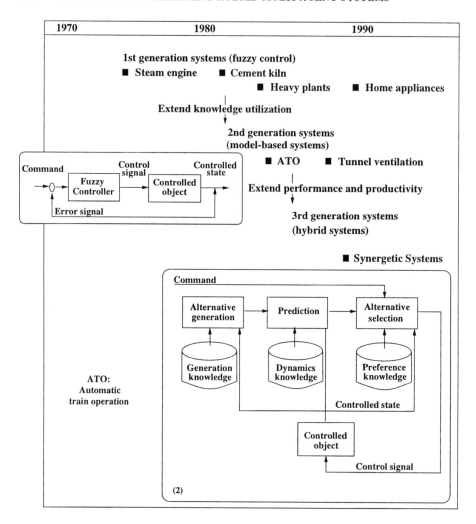

Figure 4.13.   History of Fuzzy Application Systems © 1995 IEEE
(1) Fuzzy Control Scheme as Nonlinear Feedback Logic
(2) Predictive Fuzzy Control Scheme (Model Based Systems

neering work in fuzzy control systems (as an alternative to conventional control systems) was done by Mamdani (1974), Ostergaard (1977), and Holmblad and Ostergaard (1982) when they developed fuzzy control systems for steam engines and cement kiln processes. The fuzzy control systems in this generation

(1970-80) were based on shallow reasoning or on rules acquired hybrid control systems from the experts. In the second generation, fuzzy control systems with model based architectures were developed. In the model based control systems an additional deep model (derived from first principles) of the system which gives an understanding of the complete search space over which the heuristics operate (Steels 1986). This makes two kinds of reasoning possible: a) the application of heuristic rules in a style similar to the rule based fuzzy expert systems; b) examination of a deeper search space, beyond the range of heuristics. Some fuzzy control systems in this generation include Automatic Train Operation (ATO) fuzzy control system (Yasunobu and Miyamoto 1983), and Tunnel Ventilation Control (TVC) system (Funabashi et al. 1991). The conceptual scheme of the ATO system shown in Figure 4.13 involved generation knowledge (to generate possible control actions), dynamics knowledge (to predict outcome of a generated control action), and preference knowledge (to evaluate the control outcomes for selecting the best control action). The dynamic and preferences knowledge in this application is primarily model based and does not involve linguistic variables. However, as the complexity of the control systems grew, it became more difficult to develop mathematical models. Further aspects related to learning and optimization became important factors in building fuzzy control systems. This has lead to the third generation of fuzzy control system, namely, hybrid fuzzy-control systems. The present hybrid fuzzy-control systems involve combination of two or more intelligent methodologies like fuzzy logic, neural networks, and genetic algorithms. For example, in the ATO application in Figure 4.13, genetic algorithms can be used generating optimum control actions, and neural networks or fuzzy-neural networks can be used for predicting the control outcomes. The applications which are described in the following sections belong to the third generation of fuzzy control systems.

4.7.1 Fuzzy-Neuro Hybrid Control for Pole Balancing

Real world process control applications are generally complex, nonlinear, time invariant and multi-objective. These constraints make their mathematical modeling difficult. Fuzzy logic and neural networks have been employed to model these real processes which cannot be modeled mathematically. More recently, combined hybrid neuro-fuzzy systems have been developed in the process control area. Fuzzy control can be applied to invoke or activate neural networks. A fuzzy rule can use the degree to which its condition (antecedent) clause is satisfied to determine/adjust its action. Thus depending upon the partial matching result, a fuzzy rule may or may not activate a neural network module. On the other hand, neural networks can be used to modify or adapt the fuzzy control rules in order to get the desired control action. Lee (1991) describes a self-learning cart-pole balancing fuzzy-neuro controller shown in Figure 4.14. The basic idea is to use the learning properties of neural networks to modify the control output of the fuzzy control rules in a manner which limits the vertical angular displacement of the pole to maximum of 12 degrees. The fuzzy-neuro controller consists of three components, a fuzzy rule based controller module, a neural network module, and a cart-pole system.

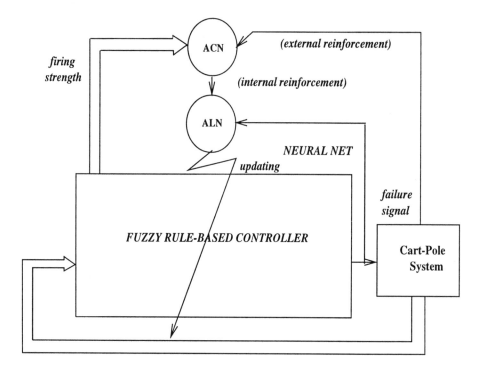

Figure 4.14.   Intelligent Hybrid Controller for Pole-Balancing (Based on Lee 1991

The fuzzy rule based controller consists of fuzzy control rules like:

IF the angle of the pole is positive large
AND angular velocity is positive large
THEN the force is positive large

The purpose of the fuzzy rule base module is to sense the system inputs like angle of the pole, and angular velocity and and fire the fuzzy control rules in parallel. The *n* defuzzified consequents of the fired fuzzy control rules along with the recommended control action are fed as inputs to the neural network module.

The neural net module consists of two neuron-like elements, namely, an Associative Critic Neuron (ACN) and an Associative Learning Neuron(ALN). The ACN receives inputs from the fuzzy rule based controller, and an external reinforcement input from the cart-pole system and produces an internal reinforcement or critic output . The external reinforcement output is basically a feedback signal from the cart-pole system with a value -1 if the pole is not balanced and zero if the pole is balanced based on the previous control action. The ALN receives three types of inputs, namely, internal reinforcement input from the ACN, and n consequents of the fired fuzzy control rules along with the control action from the fuzzy rule based controller. Based on these inputs ALN produces n associative weights as outputs which are then used to modify the consequents of the fuzzy control rules to produce the desired a control action for pole balancing. It may be noted that the ACN and ALN are self-trained with little *a priori* knowledge (in terms of reinforcement from the cart-pole system) for determining the desired control action.

The cart-pole control system described in this section is a classical fuzzy control application. In the next section a neuro-fuzzy control application in practical use in a petroleum plant is described.

4.7.2   Neuro-Fuzzy Hybrid Control in Petroleum Plant

The neuro-fuzzy application described in this application employs fuzzy logic rules to model expert knowledge and neural networks for modeling non linear time variant control processes. Tani et al. (1996) describe a neuro-fuzzy hybrid control system to control the tank level of a solvent dewaxing plant shown in Figure 4.15 which consists of non-linear processes like refrigeration (shown as crystallizer in Figure 4.15) and filtering. The neuro-fuzzy system developed by them has been in practical use for three years and saves at least 1000 hours per person per year.

The purpose of the solvent dewaxing plant is to produce lubricant oils by removing the waxing component of the feed oil fed into the dewaxing plant. The waxing component can otherwise, cause the the lubricant oil to freeze at low temperatures. The dewaxing plant is used for processing different kinds of feed oils. As shown in Figure 4.15 the dewaxing process involves five steps. In the first step primary solvent is added to the feed oil fed into the dewaxing plant. After adding the primary solvent the feed oil is refrigerated in a crystallizer and a secondary solvent is added. The third step involves removal of the frozen wax from the oil by using filters and a filter solvent. A vacuumed rotating drum which forms part of the filter is used to separate the frozen wax from the

oil. Once the oil has been separated from the wax it is filled or buffered into
a tank. Finally, the dewaxed oil is sent to a heater for evaporating the solvent
and producing low fluid point lubricant oil.

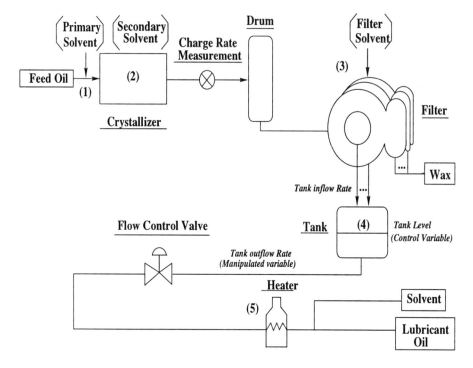

Figure 4.15.   The Process Flow of a Dewaxing Plant © 1996 IEEE

The tank level needs to be controlled in a manner so as to prevent rapid
changes in the tank outflow rate which in turn can cause changes in the temperature of the heater and affect the quality of the lubricant oil. The dewaxing
plant operates in a steady state and a transient state. The steady state operation involves periodic filter washing which results in the tank level going up
and down periodically. In this state the operator roughly estimates the tank
outflow rate by observing the tank level for several hours. The operator adjusts
the tank outflow rate by observing change in tank level and the time until the
next filter washing. During the steady state operation the tank level is main-
tained around 50% of the tank capacity. However, the problem of maintaining
a constant tank level gets aggravated during the transient state, i.e., switching
from one feed oil to another. The switching process from feed oil A to feed oil
B may result in lower or higher inflow rate and the tank outflow rate has to be

adjusted to prevent the tank from becoming empty or overflowing. Depending upon how the tank inflow rate will change and time of the switch, the operator has to adjust the tank level target before and during the switching process (which takes four hours).

In order to model this control problem the neuro-fuzzy hybrid control system design is based on the control processes employed by the human operator responsible for maintaining the tank level. The neuro-fuzzy hybrid consists of three components, namely, rough target component, correction component, and prediction component as shown in Figure 4.16. The neural networks NN_1, NN_i and NN_n in Figure 4.16 are used for different feed oils. The graphs underneath the neural networks in Figure 4.16 represent the predicted tank inflow change rate for particular feed-oil switching and the corresponding tank level target setting (to account for the predicted tank inflow change rate during feed-oil switching).

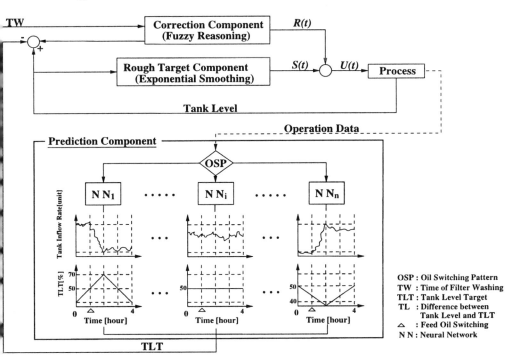

Figure 4.16.   Neuro-Fuzzy Control System for Dewaxing Plant © 1996 IEEE

The rough target component calculates the tank outflow rate as an average of the operation data. The objective of the correction component is to determine the correction to be employed to the tank outflow rate at any time t based on the tank level at time t, and the desired tank level target. The correction component consists of two groups of fuzzy control rules, one corresponding to when filter washing occurs, and the other corresponding to when there is no filter washing. These fuzzy control rules are used in the steady and transient state. Fuzzy control rules in filter washing involve two input variables, TL (the difference between the current tank level and the Tank Level Target (TLT)), and δTL (the rate of change of TL). The non-filter washing fuzzy control rules involves an additional input variable, TW (the time until the next filter washing). The fuzzy control rules consist of fuzzy inputs and crisp output representing the Correction (C)). The fuzzy control rules with two inputs and fuzzy sets are shown in Figure 4.17. PB (Positive Big), PS (Positive Small), ZE (ZEro), NB (Negative Small), and NB (Negative Big) represent the membership functions of the fuzzy inputs.

		$\triangle$ TL							TW		
		PB	PM	PS	ZE	NS	NM	NB	PB	PS	ZE
	PB	C^{+++}	C^{+++}	C^{++}	C^{+}	C^{0}	C^{-}	C^{--}	C^{++}	C^{+}	C^{0}
	PS	C^{+++}	C^{++}	C^{+}	—	C^{0}	C^{-}	C^{--}	—	—	—
TL	ZE	—	—	—	C^{0}	—	—	—	—	—	—
	NS	C^{++}	C^{+}	C^{0}	—	C^{-}	—	—	—	—	—
	NB	C^{++}	C^{+}	C^{0}	C^{-}	C^{---}	C^{---}	C^{---}	C^{0}	C^{-}	C^{-}

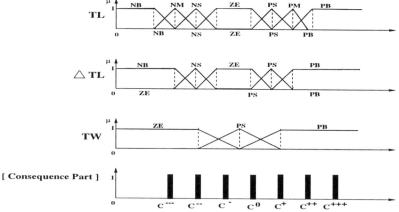

Figure 4.17.   Fuzzy Control Rules and Fuzzy Sets in Nonfilter Washing © 1996 IEEE

Finally, the prediction component involving a back propagation artificial neural network is used during feed oil switching. The main objective of the prediction component is to predict the tank inflow rate for different types of feed oils. A number of artificial neural networks are used, one for each feed oil. The input vector to a artificial neural network is made up of input features like charge rate of the feed oil, the primary solvent rate, the secondary solvent rate, the filter solvent rate, and the time. The artificial neural network is trained with 240 input patterns (one for each minute). Two separate pieces of information can be derived from the predicted tank inflow rate, i.e. whether the tank inflow rate is increasing, decreasing, or no change, and the time of the highest and the lowest level of the tank inflow rate. Based on these two pieces of information the tank level target is determined.

For example, consider the predicted tank inflow rate pattern shown underneath neural network NN_1 in Figure 4.16. The tank inflow rate pattern underneath the neural network NN_1 shows that the tank inflow rate is decreasing and it is at its lowest level one hour after the feed-oil switching. In order to account for this decreasing inflow rate pattern, the tank-level target value is changed (one hour before the feed-oil switching starts) for the fuzzy controller such that the tank level target is at the highest level (i.e. 70% of the tank capacity) one hour after the feed-oil switching. The tank-level target pattern illustrating this tank-level target setting behavior is shown underneath tank inflow rate pattern predicted by NN_1.

In another control application Morooka (1995) for shape control of rolling mills for producing thin strips of steel, stainless, aluminum, and other alloy metals, neural networks and fuzzy logic are used in a hybrid architecture as shown in Figure 4.18. The neural networks are used for recognizing the shape patterns from the sensor signals. Based on recognition of shape pattern of different strips, fuzzy control is applied on the work rolls in the rolling mill to produce uniformly rolled strips.

4.7.3   Neuro-Genetic Hybrid Control for Fruit Storage

In neuro-genetic hybrid architectures for control applications, neural networks are used for predicting the fitness parameters which are used by the genetic algorithms for optimizing the the control parameters or control action taken by the controller.

Morimoto et al. (1995) optimize a fruit storage process using a neuro-genetic hybrid architecture. The commercial price of a fruit (like apple) is primarily determined by its freshness. In order to retain its freshness after cultivation (especially in protected cultivation), the fruit has to be stored under proper environmental conditions. Relative humidity under storage is an important environmental condition which can affect the freshness of a fruit. Low humidity can induce severe wilting of the fruit surface and result in weight loss. On the other hand, high humidity can result in diseases related to mold. Thus the

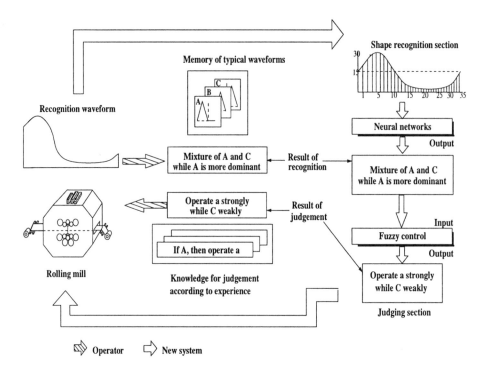

Figure 4.18.   Shape Control System of Rolling Mills © 1995 IEEE

optimization problem lies in the fact that the relative humidity has to controlled in a manner that it neither causes wilting nor causes disease in the fruit.

In order to solve it as an optimization problem, Morimoto et al. (1995) divide the storage process into 4 stages, wherein, 4-step set points of the relative humidity (h_1, h_2, h_3, h_4) are determined so as to minimize the fitness function $F(h)$. $F(h)$ is defined as:

$$F(h) = a.W_{loss}(h) + b.D_{dis}(h)$$

where, $W_{loss}(h)$ represents the weight loss of fruits caused by evapotranspiration, $D_{dis}(h)$ represents the degree of disease caused by excessive high humidity, and two coefficients, a and b, are weighting factors for weight loss and degree of disease respectively. The block diagram of the control system for optimization of the fruit storage system is shown in Figure 4.19. Genetic algorithms are used to determine the optimal value of 4-step relative humidity, h_1, h_2, h_3, and h_4. They are represented as 6-bit binary strings as shown in Figure

Skilled grower's techniques

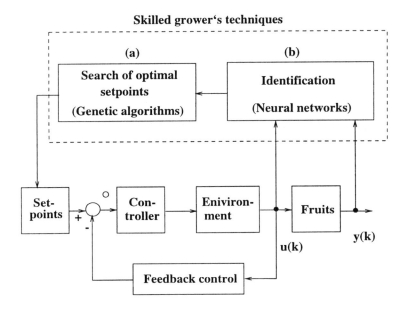

Figure 4.19.   Fruit Storage Control Process © 1995 IEEE

4.20. The back propagation neural network is trained on current time series of relative humidity $u(k)$, number of days after storage, and previous time series of weight loss $(y_{(k-1)})$ as inputs, and the current weight loss y_k as output. Thus, the neural network can predict on-line future weight loss given the current relative humidity, number of days after storage, and previous weight loss as inputs. The degree of disease based on relative humidity is determined from a non-linear model developed from many experiments. The predicted weight loss is then used for determining the fitness function for the genetic algorithm. The optimal control values are achieved through high crossover and mutation in the 10th generation. Morimoto et al. (1995) expect to use this optimization in a real system for fruit storage process control.

4.7.4 Symbolic-Neuro Hybrid Control for Robotic Motion

Applications in robotic motion are known to deal with declarative and reflexive skills. These skills are modeled using symbolic and soft computing paradigms. Handelman et al. (1990) describe an integrated approach for intelligent robotic control where rule based modules are used for declarative task knowledge and neural networks are used for reflective task knowledge. Motions involving declarative task knowledge are those which are characterized by inference,

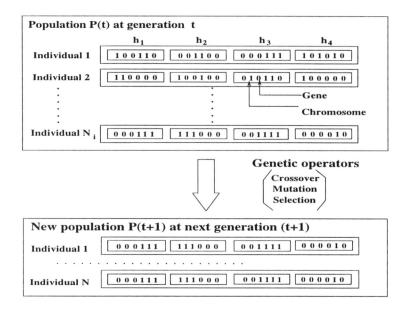

Figure 4.20.   Fruit Storage Optimization © 1995 IEEE

comparison, and evaluation, and provide insight into not only how something is done, but why. In contrast, motions involving reflective task knowledge relate specific responses to specific stimuli, are automatic, and require little or no thought. The system developed by them is used to teach a two-link robot manipulator a tennis like swing.

Fukuda and Shibata (1994) define two levels of hierarchical control for robotic motion, namely, learning level, and adaptation level as shown in Figure 4.21. The objective of the learning level is to develop the high level symbolic control strategy to be used by the Neural Servo Controller (NSC) at the adaptation level. The learning level involves recognition followed by planning as the two major activities. Recognition is performed initially through visual sensing and subsequently through active sensing in order to get detailed characteristics (e.g. object weight, stiffness, etc.). The output of the neural networks is symbolic identification of an object or its characteristics. This output serves as a flag to the symbolic knowledge base which is then used to determine the symbolic control strategy. The symbolic control strategy, along with the initial weights of the neural network of the NSC are fed to neural servo controller. The neural servo controller in the adaptation level then compensates for the vagueness of the symbol control strategy (e.g. apply strong force) to adapt the control

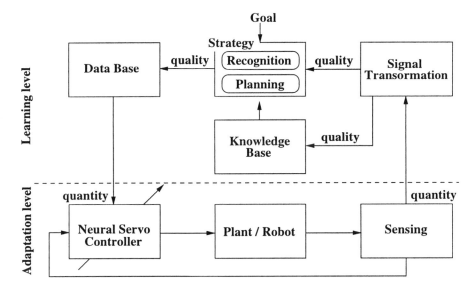

Figure 4.21.   Hierarchical Hybrid Control Architecture for Robotic Motion

action to the current status of the dynamic process. Thus the symbolic control strategy serves as a meta-knowledge for the NSC. Shiabata and Fukuda have demonstrated that this hybrid arrangement performs much better than the conventional controllers.

In this section and in section 4.7.2 and 4.7.3 respectively combination approaches involving fuzzy-neuro and neuro-genetic combination systems have been described. The next section looks at combination approaches involving fuzzy-neuro-genetic control.

4.7.5   Fuzzy-Neuro-Genetic Hybrid Control

A number of intelligent control applications have been developed by exploiting the strengths of fuzzy systems, neural networks, and genetic algorithms (Chiaberge et al. 1995; Nobre 1995). In these hybrid applications, fuzzy systems are used because they provide a human-like reasoning interface, and assist in capturing domain knowledge. Neural networks are used for tuning the performance of the fuzzy controller. Finally, genetic algorithms are used for optimizing the initially defined fuzzy control rules or optimizing a neuro-fuzzy controller with fuzzy neurons.

In the last chapter, optimization of a neuro-fuzzy network using genetic algorithms was described. Nobre (1995) has extended that concept by integrating

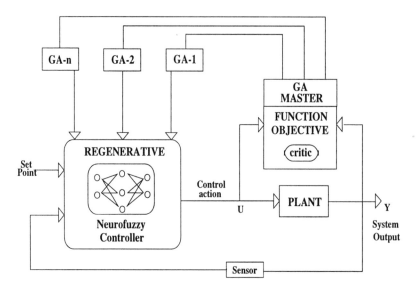

Figure 4.22.   Fuzzy-Neuro-Genetic Control © 1995 IEEE

GA with neuro-fuzzy controllers for developing adaptive control applications. The neuro-fuzzy-genetic controller architecture is shown in Figure 4.22. Instead of using a neural network critic (Lee 1991), an on-line GA critic is employed using reinforcement learning. The neuro-fuzzy controller in Figure 4.22 is developed using logical neurons. The fitness function of the on-line GA critic is a function of the control action (U), system output (Y), controller parameters, and performance measure. The fitness function is used by the on-line GA critic to adjust the connection weights as well as the structure of the neuro-fuzzy net. GA-1, GA-2,....,GA-n in Figure 4.22 imply parallel implementation of the GA critic and parallel search method.

Chiaberge et. al. (1995), propose a Hierarchical Hybrid Fuzzy Controllers (HHFC) architecture shown in Figure 4.23 for complex control problems. Their motivation for such an integrated architecture comes from the fact that for complex control problems, the number of state variables is high. This can result in an explosion of number of fuzzy control rules which can create difficulties in managing the design as well as the computational resources required for such complex control problems.

In the HHFC architecture the control problem is broken up into a hierarchical set of control tasks. The underlying intelligent paradigm for accomplishing each control task is fuzzy logic. That is, each fuzzy controller is a set of fuzzy control rules of the type:

$$IF\ (x_1\ is\ A_1),......(x_n\ is\ A_n)\ THEN\ F\ =\ f_1$$

$$IF\ not((x_1\ is\ a_1),......,(x_n\ is\ A_n\ ))\ THEN\ F\ =\ f_2$$

The consequent of the fuzzy control rules is a Weighted Radial Basis Function (WRBF) which can be modeled as a neural network or linear controller. Genetic algorithms are then used to optimize the HHFC representation. Each fuzzy controller as shown in Figure 4.23 can receive inputs from the external environment or from other fuzzy controller in the hierarchy. Thus in the HHFC architecture, the problem solver can select different control strategies. Chiaberge et al. (1995) have applied the HHFC architecture in number of complex control problems including bidirectional inverted pendulum in polar coordinates, and others.

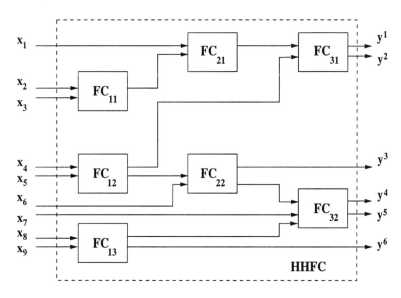

Figure 4.23.   Fuzz-Neuro-Genetic Control © 1995 IEEE

As shown in the HHFC architecture, the inputs to a fuzzy reasoning unit can be inputs from the environment or inputs from other fuzzy reasoning units. In order to determine the correct hierarchical configuration, the designer must understand the relationship between different input variables. A set of closely related input variables should form part of the same fuzzy reasoning unit. however, because of the continuous/numerical values of the input variables it becomes difficult to determine such relationships.

Shiabata et al. (1993) have proposed integration of fuzzy neural networks in the hierarchical control architecture described in the previous section. They have added another level, i.e. a skill level between the learning level and the adaptation level. The skill level is useful in the same task and different environment situations. The skill level employs a fuzzy neural network which is used to provide quantitative control references as output to the adaptation level in order to save time in the same task but different environments situation. The inputs to the fuzzy neural network at skill level are numerical values sensed by the sensors and some symbols which reflect the control strategy produced at the learning level. Genetic algorithms in the hierarchical control architecture are also used for optimization of the membership functions, neural network and fuzzy neural network structure, and learning connection weights. Fukuda et. al. (1995) use a combination of fuzzy reasoning, back propagation neural neural network, and genetic algorithms to determine the optimal hierarchical structure of fuzzy reasoning units for multi-input complex control problems. They start with an initial random hierarchical structure of fuzzy reasoning units. Depending upon valid input/s to a fuzzy reasoning unit in the random configuration, the genetic bit string is encoded with the numbers of the fuzzy reasoning units. Thus a bit string represents the hierarchical configuration of fuzzy reasoning units. Crossover and mutation are used to develop a new hierarchical configuration. In a given hierarchical configuration, back propagation technique is used to compute the overall mean and maximal square error of the top unit (in the hierarchical configuration) as well as error for each fuzzy reasoning unit. It is also used to tune the consequent part of the fuzzy rules in a fuzzy reasoning unit. The fitness of the hierarchical configuration is computed based on the mean and maximal squared error of the top unit, number of rules, and number of membership functions. The process of generating new hierarchical configuration, computing the squared errors, and tuning the fuzzy rules is continued till the best fitness value becomes less than the target one or the search reaches the limit set on number of generations.

This brings us to the end of combination based intelligent systems in this chapter. In the next chapter intelligent transformation systems are described in the context of knowledge discovery and data mining.

4.8 SUMMARY

Intelligent Combination systems are based on the premise that it is beyond the reach of a single intelligent methodology to cover the entire spectrum of human cognition and information processing. The limitations expressed by various intelligent methodologies when applied to complex real world problems support this premise. This chapter describes various applications which combine two or more intelligent methodologies to solve a complex problem. These intelligent approaches include neuro-symbolic combination systems, neuro-fuzzy combination systems, neuro-fuzzy-case combination systems, symbolic-genetic combination systems, and combination systems in intelligent hybrid control. While describing these approaches hybrid applications in areas like fault diagnosis, production scheduling, load forecasting, number & character recognition, condition based maintenance, and fashion shoe design have been covered. Applications in intelligent hybrid control include cart-pole balancing, dewaxing lubricant oil in a petroleum plant, shape control of steel rolling mills, and fruit storage control have also been described.

References

Becraft, R., Lee, P. L., and Newell, R. B. (1991), "Integration of Neural Networks and Expert Systems for Process Fault Diagnosis," *International Joint Conference on Artificial Intelligence*, pp. 832-37.

Cho, S-B. (1994), "A Neuro-Fuzzy Architecture for High Performance Classification," *Advances in Fuzzy Logic, Neural Networks and Genetic Algorithms*, pp. 67-84.

Chiaberage, M., Bene. G. D., Pascoli, S. D., Lazzerini, B., Maggiore, A., and Reyneri, L.M. (1995), "Mixing Fuzzy, Neural and Genetic Algorithms in an Integrated Design Environment for Intelligent Controllers," *1995 IEEE International Conference on Systems, Man, and Cybernetics*, vol. 4 pp. 2988-2993.

Fukuda, T., Hasegawa, Y., and Shimojima, K. (1995), "Structure organization of hierarchical fuzzy model using genetic algorithm," *1995 IEEE International Conference on Fuzzy Systems*, vol. 1 pp. 295-299.

Funabashi, M. et al. (1991), "A Fuzzy Model Based Control Scheme and its Application to a Road Tunnel Ventilation System," *Proceedings of IEEE Conference on Industrial Electronics, Control, and Instruments*, pp. 1596-1601.

Funabashi, M., and Maeda, A. (1995), "Fuzzy and Neural Hybrid Expert Systems: Synergetic AI," *IEEE Expert*, pp. 32-40.

Garga, A. (1996), "A Hybrid Implicit/Explicit Automated Reasoning Approach for Conditioned Based Maintenance," presented at *The ANSE Intelligent Ships Symposium II*, Philadelphia, November 25-26.

Gutknecht, M., and Pfeifer, R. 1990, "Experiments with a Hybrid Architecture: Integrating Expert Systems with Connectionist Networks," *General Conference on Second Generation Expert Systems*, Avignon, France, pp. 287-299.

Gutknecht, M., Pfeifer, R., and Stolze, M. (1991), "Cooperative Hybrid Systems," *proceedings of 12th International Joint Conference on Artificial Intelligence*, pp. 824-829.

Hamada, K., Baba, T., Sato, K., and Yufu, M. (1995), "Hybridizing a Genetic Algorithm with Rule-Based Reasoning for Production Planning," *IEEE Expert*, pp. 60-67.

Handelman, D.A., Lane, S.H., and Gelfand, J.J., 1990, "Integrating Neural Networks and Knowledge-Based Systems for Intelligent Robotic Control," IEEE Control Systems Magazine, pp. 77-86.

Hendler, J.A., 1989. Problem Solving and Reasoning: A Connectionist Perspective. In *Connectionism in Perspective*, pp. 229-244.

Hilario, M. (1995), "An Overview of Strategies for Neurosymbolic Integration," *Working Notes on Connectionist-Symbolic Integration: From Unified to Hybrid Approaches, IJCAI95*, Montreal, Canada, pp. 1-6. Hiri, T., 1990, "Applying Neural Computing to Expert System Design: Coping with

Hiri, T., 1990, "Applying Neural Computing to Expert System Design: Coping with Complex Sensory Data and Attribute Selection," In *Third International Conference on Foundations of Data Organization and Algorithms*, pp. 474-488.

Holmblad, L. P., and Ostergaad, J. J. (1982), " Control of a Cement Kiln by Fuzzy Logic," *Fuzzy Information and Decision Processes*, M. M. Gupta and E. Sanchez , Eds. Amsterdam: North Holland Publishing Company, pp. 389-399.

Karr. C., Sharma, S., Hatcher, W., and Harper, T. (1992), "Control of an exothermic chemical reaction using fuzzy logic and genetic algorithms," *Proc. of the International Fuzzy Systems and Intelligent Control Conference*, pp.246-254.

Klusch, M. (1993), "HNS: a hybrid neural system and its use for the classification of stars," *Proceedings of International Joint Conference on Neural Networks*, pp. 309-319.

Lee, C-C. (1991), "A Self-learning Rule-Based Controller Employing Approximate Reasoning and Neural Net Concepts," *International Journal of Intelligent Systems*, pp. 71-93.

Lim, J. H., Lui, H. C., Tan, A. H. and Teh, H. H., 1991, "INSIDE: A Connectionist Case-Based Diagnostic Expert System that Learns Incrementally," *International Joint Conference on Neural Networks*, Singapore, pp. 1693-1698.

Maeda, A., Ichimori, T., and Funabashi, M. (1993), "FLIP-net: A network Representation of Fuzzy Inference Procedure and its Application to Fuzzy Rule Structure Analysis," *Proceedings of Second IEEE International Conference on Fuzzy Systems*, pp. 391-395.

Maeda, A., Ashida, H., Taniguchi, Y., and Takahashi, Y. (1995), "Data Mining System using Fuzzy Rule Induction," *1995 IEEE International Conference on Fuzzy Systems*, vol. 5, pp.45-46.

Main, J., Dillon, T., and Khosla, R. (1996), "Use of Fuzzy Feature Vectors and Neural Networks for Case Retrieval in Case Based Systems," *1996 Biennial Conference of the North American Fuzzy Information Processing Society - NAFIPS*, Berkeley, California, USA, pp. 438-443.

Mamdani, E. (1974), Applications of Fuzzy Algorithms for Simple Dynamic Plant," *Proceedings of IEE*, vol. 121, no. 12, pp. 1585-1588.

Mamdani, E., and Assilian , S. (1975), "An Experiment with Linguistic synthesis with a Fuzzy Logic Controller," *International Journal of Man-machine Studies*, vol. 7.

Morimoto, T., Baerdemaeker, J.D., and Hashimoto, Y. (1995), "Optimization of Storage System of Fruits Using neural Networks and Genetic Algorithms," *1995 IEEE International Conference on Fuzzy Systems*, vol. 3, pp. 289-294.

Morooka, Y., (1995), "Shape Control of Rolling Mills by a Neural and Fuzzy Hybrid Architecture," *1995 IEEE International Conference on Fuzzy Systems*, vol. 5, pp.47-48.

Nobre, F. S. M., (1995), "Genetic-Neuro-Fuzzy Systems: A Promising Fusion," *1995 IEEE International Conference on Fuzzy Systems*, vol. 1, pp. 259-266.

Ostergaad, J. J. (1977), "Fuzzy Logic Control of a Heat Exchange Process," *Fuzzy Automata and Decision Processes*, M. M. Gupta, G. N. Saridis, and B. R. Gaines, Eds. Amsterdam: North Holland, pp. 285-320.

Senjen, R., Beler, M. D., Leckie, C., and Rowles, C. (1993), "Hybrid Expert Systems for Monitoring and Diagnosis," *1993 IEEE 9th Conference on Artificial Intelligence for Applications*, pp. 235-241.

Shibata, T., Fukuda, T., and Tanie, K. (1993), "Synthesis of Fuzzy, Artificial Intelligence, Neural networks, and Genetic Algorithm for hierarchical Intelligent Control - Top-down and Bottom-Up Hybrid Method," *Proceedings of International Joint Conference on Neural Networks*, vol. 3 pp. 2869-2872.

Shibata, T., and Fukuda, T. (1994), "Hierarchical Intelligent Control for Robotic Motion," *IEEE Transactions on Neural Networks*, vol. 5, no. 5, pp. 823-832.

Steels, L., 1989, "Artificial Intelligence and Complex Dynamics," *Concepts and Characteristics of Knowledge Based Systems*, Eds., M. Tokoro, et al., North Holland, pp. 369-404.

Srinivasan, D., Liew, A. C. and Chang, C. S. (1994), "Forecasting Daily Load Curves Using a Hybrid Fuzzy-Neural Approach," *IEE Proceedings on Generation, Transmission, and Distribution*, vol. 141, no. 6, pp. 561-567.

Sugeno, M, (1977), "Fuzzy measures and Fuzzy Integrals: A Survey," *Fuzzy Automata December Proceedings*, North Holland, pp. 89-102.

Tani, T., Murakoshi, S., and Umano, M. (1996), "Neuro-Fuzzy Hybrid Control System of Tank Level in Petroleum Plant," *IEEE Transactions on Fuzzy Systems*, vol. 4, no. 3, pp. 360-368.

Yasunobu, S., and Miyamoto, S. (1983), "Fuzzy Control for Automatic Train Operation Systems," *Proceedings of 4th International Conference on Transportation Systems*, pp. 39-45.

Zhongxing, Y., and Liting, G. (1993), "A hybrid cognition system: application to stock market analysis," *International Joint Conference on Neural Networks*, vol. 3, pp. 3000-3003.

5 KNOWLEDGE DISCOVERY, DATA MINING AND HYBRID SYSTEMS

5.1 INTRODUCTION

In the last two decades digitial revolution has invaded business enterprises. Computers have enabled organizations to store gigabytes of data related to stock markets, electricity consumption profiles, troubleshooting and diagnostic data, etc.. As outlined by Fayyad and Uthuruswamy (1996a), in scientific endeavors, data represents observations carefully collected about some phenomenon under study. In business, data captures information about critical markets, competitors, and customers. On the other hand, in manufacturing, data captures performance and optimization opportunities, as well as the keys to improving process and troubleshooting problems. The reason organizations store or collect all this data is to enable them to extract some useful knowledge (at a later date!!) which can make them more productive, efficient, and competitive. The terms, Knowledge Discovery in Databases (KDD), and Data Mining have emerged in the last five years from this need of extracting useful knowledge. KDD is a nontrivial process of identifying valid, novel, potentially useful, and ultimately understandable knowledge from data (Fayyad et al. 1996b). Data mining is a step in the KDD process that consists of applying data analysis and discovery (learning) algorithms that produce a particular enumeration of patterns (or models) over the data. In fact, the data mining process can fit into the framework of intelligent transformation systems described in chapter 3. It will be noticed in this chapter that the transformation process is largely a bottom up knowledge engineering strategy.

Useful knowledge can be summarized information, rules (or models) explaining competition or manufacturing process behavior, predictive knowledge for predicting electricity load profiles, classification, and clustering knowledge for determining classes, categories, and groups of various customers. Whereas, organizations have gone to great lengths in determining the most efficient means of storing data, not until recently has much work been done in the area of knowledge discovery and data mining. In this chapter, some applications which engage in KDD and data mining, and involve use of neural networks, fuzzy, and genetic algorithms are described.

5.2   KDD PROCESS

The KDD process (as shown in Figure 5.1) according to Fayyad et al. (1996b) and Fayyad (1996) involves nine steps:

- Understanding the domain.

- Data selection.

- Preprocessing of data.

- Data reduction[1].

- Selection of data mining method.

- Selection of data mining algorithm.

- Data mining.

- Interpretation of knowledge.

- Acting on the knowledge.

These nine steps lead to knowledge and are briefly described in the rest of this section.

5.2.1   Understanding the domain

Like in any software development process it is important to first gain an understanding of the application domain, and determine the goals of the KDD

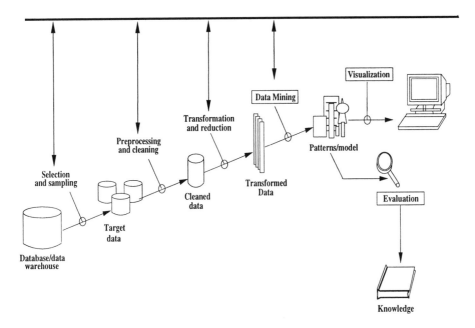

Figure 5.1.   The KDD Process © 1996 IEEE

process from the customer's viewpoint. The goals of the KDD process will later on determine the data mining method employed. For example, whether the goal is simply summarization of data or prediction, classification or clustering, or it is a combination of two or more of these data mining methods.

5.2.2   Data Selection

This step determines the target data set to be mined. The target data set may be a subset of the total data set in domain and may involve some or all the variables in the domain.

5.2.3   Preprocessing of Data

The goal here is to remove noise from the target data set and determine the strategies for dealing with missing data. For example, in a time-series prediction domain, the domain expert may consider profiles with abnormal patterns as noisy for the target data set. In other words, if the profile with abnormal pattern is included in the target data set it may unfavorably affect the mining process. The strategy for computing missing value may involve replacing it with a value from a similar database record or may involve a separate computation.

5.2.4 Data Reduction

Data reduction involves reduction in the dimensionality of the data. The reduction in dimensionality can be done through determination of the principal components or the most important variables, or applying a prior knowledge for eliminating the irrelevant variables. Also, depending upon the goals of the KDD process different views or perspectives of the data may be considered. For example, customer data in an organisation is looked at differently by different people in an organization. For the purpose of KDD , one view may be considered more effective than the other.

5.2.5 Selection of Data Mining Method

The selection process is based on matching of the goals of the KDD process with with a particular data mining method (e.g. summarization, prediction, classification, or clustering). As stated earlier, this matching process can result in combination of two or more data mining methods.

5.2.6 Selection of Data Mining Algorithm

The penultimate step to implementation of a data mining algorithm is the selection of the algorithm itself. For example, time-series prediction can be done using supervised back-propagation neural networks or unsupervised Kohonen self-organizing maps. So some exploratory analysis in terms of which algorithm will best model the domain as well serve the needs of the user needs to be done. This may involve development of a few prototypes and finally deciding on one of them (as approved by the user) to be implemented on the entire data set.

5.2.7 Data Mining

In this step the selected data mining algorithm is actually implemented. It deals with refinement of the learning or other parameters of the data mining algorithm in order to learn a optimal representation of the domain.

5.2.8 Interpretation of Knowledge

For user understanding, interpretation may involve visualization of the results (e.g. clusters), or even further extraction of symbolic rules from a neural network used for classification.

5.2.9   Acting on Knowledge

The desired prediction, classification or other forms of mined knowledge is either used directly, embedded in another system for further action, or simply documented. The mined knowledge before being acted upon may have to be validated with existing knowledge or previously extracted knowledge.

In the sections that follow, applications involving the KDD process are described.

5.3   KDD APPLICATION IN FORECASTING

In the recently deregulated electrical power industry (The work described in this application has been done by the first author (Dr Rajiv Khosla) for an electricity distribution company in Melbourne, Australia in 1995-96) in Melbourne, Australia there is a greater need to develop pricing, forecasting, and costing systems which are more responsive in terms of time and information needs of the industry. These systems should facilitate intelligent analysis of data as well as effective management of data keeping in view the existing as well as future needs of the industry. Here the development of one such system, namely, forecasting system is described in the format of the KDD process enumerated in the preceding section.

5.3.1   Understanding the Forecasting Domain

The forecasting system is used for short term one year, long term five year forecasts, and for forecasting customer energy usage profile. The one year and five year forecasts use time series and regression analysis respectively. The one year time series based forecasts are used for budget year estimates. The five year forecasts are used for energy, revenue, customer number (growth) forecasting. On the other hand, the customer energy usage profile forecast or the forecasting profile system is suited for pricing, purchasing, distribution planning, and determination of distribution charges.

Some of the problems identified with the existing forecasting system are:

- the existing structure of data in the forecasting system is incompatible with future needs of the company;

- it is difficult and time consuming to extract relevant information from the existing databases of the forecasting system;

- there is a need to develop multiple views of data in order to facilitate different comparisons of data and needs of different users;

- there is a need to tailor the forecasting system modeling to meet the needs of pricing, and

- the existing regression methods in the deregulated environment for forecasting have been found to be ineffective.

Based on the above problems, some of the goals of forecasting system are:

Goal 1. Given the complexity of the forecasting system data needs to be structured and transformed in a manner which meets the needs of its users.

Goal 2. Develop different perspectives/views of customer data from a user 's perspective for forecasting and pricing systems.

Goal 3. Model relationships between electricity load profile variables.

Goal 4. Use the model realized from Goal 3 to predict 24 one hour load profiles of various customers.

The subsections that follow describe the realizations of goal 1, 2, 3 and 4 respectively.

5.3.2   Forecasting System Data Selection

The data related to EOPET (ratio between Off Peak and Peak Energy consumption of a customer), LF (Load Factor) and 48 one hour historical load profile points is extracted from the forecasting database system. The data is used for computing EOPET and LF variables.

5.3.3   Preprocessing of Forecasting System Data

The preprocessing of data involves the following steps:

- Normalization of data.

- Eliminating noise based on a prior knowledge

- Strategy for computing missing data.

A large variation in the values of different variables in the target data set can adversely affect any learning algorithm. In this application, all the values have been normalized between 0 and 1 for all the variables. Noise in this domain is present in the form of abnormal profiles. For example, certain load profiles consist of deep valleys. On analysis, it was found that these customers were partially supported by their own captive power plants. On the other end of the spectrum there are certain straight line profiles which had very little variation over the 24 hour period. Thus, with the help of the domain expert some rules have been designed to eliminate such profiles from the target data set.

In case of missing data (e.g. EOPET), the missing variable value is replaced by its average value computed at the end of the preceding year.

5.3.4   Data Reduction using Data Warehouse & O-O Methodology

This step is necessitated by the varying needs of the users, and the complexity and quantum of data in the forecasting database. The data reduction process involves two steps:

- Integration of data warehouse and object-oriented methodologies for abstracting data through time and based on a prior knowledge.

- Reducing the number of profile points.

Data warehouse methodology partitions the data based on subject orientation, granularity, and time (Inmon and Kelley 1993). All these criteria help to build data repositories at different levels of abstraction thus restricting databases to manageable size. On the other hand, object-oriented methodology provides an intuitive means for capturing a prior knowledge of the users in conceptual data modeling terms. The object-oriented methodology compliments the data warehouse methodology in that it not only captures different aspects related to data structuring but also facilitates capturing of operations on data. Thus integration of object-oriented methodology with data warehouse methodology assists in exploiting the strengths of the data warehouse methodology and object-oriented methodology.

Figure 5.2 shows the transformation of raw data to operational and strategic data as a result of applying data warehouse and object-oriented methodologies. The strategic, operational, and raw data levels meet differing needs of users who interface with the forecasting system at different levels. These three levels abstract the data in conceptual terms as well as through the time dimension.

As can be seen in Figure 5.2 the data in the FPS model has been abstracted and partitioned based on *ASIC* (industry type related codes), *Contestability (contest) Level, Shift Type (A), Season, Month, Weekday, Working Day, Non-Working Day, Holiday* and *Weekend* objects. The *Holiday* and *Weekend* objects are partitioned (or disaggregated) into *Monday* to *Friday*, and *Saturday* and *Sunday* respectively. The data partition ing based on *Shift Type* (or working shift patterns) has been done to meet the needs of pricing. The season partitioning has been done to account for seasonal profiles and also seasonal customers. The contestability level relates to customers who can be contested by other distribution companies. The ASIC codes divide customers into various

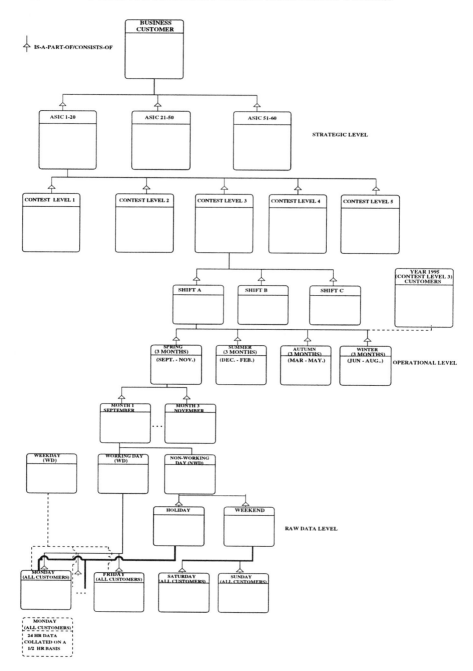

Figure 5.2.   Forecasting Profile System (FPS) - Conceptual Data Model

groups depending upon a number of factors, like type of industry, energy usage, etc.. The working day and non-working day data partitioning has been done to account for average working and non-working day profiles which are used consistently in the forecasting system. It may also be noted that the atomic level of profile data in the forecasting system is 1/2 hour data.

The shift type (working shifts), season, month, Working Day(WD), and Non-Working Day (NWD) provide multiple views of the profile data for detailed analysis as well as for customizing the pricing offers to various customers. Besides, these multiple views also reduce the possibilities of corruption of data (and thus enhance security). For example, one does not have to access the yearly database associated with *Year 1995 Customers* object for accessing seasonal, monthly, and working day, and non-working day customer data. The relationship link between objects at various levels are preceived as aggregational and has been represented accordingly. The target data set used in this application belongs to the *Contestability Level* objects at the strategic level.

Table 5.1. Sample Target Data

EOPET	LF	WD1	WD2	WD3	WD4	WD5	WD6	WD7	NWD1	NWD2	NWD3	NWD4	NWD5
0.507319	0.580859	0.68362	0.578654	0.459261	0.478681	0.635838	0.606778	0.578354	0.687512	0.508801	0.48391	0.494779	0.538172
0.367668	0.080496	0.121516	0.120773	0.124569	0.130276	0.096035	0.089014	0.125896	0.00522	0	0	0	0
0.428486	0.574136	0.544225	0.548998	0.592319	0.618392	0.749491	0.672489	0.607231	0.391259	0.390737	0.389782	0.361492	0.371486
0.493586	0.464634	0.380686	0.393458	0.374231	0.478242	0.472506	0.54462	0.47731	0.499199	0.467365	0.466207	0.447526	0.447158
0.475659	0.72865	0.680685	0.703033	0.710575	0.761019	0.830065	0.804732	0.749898	0.72514	0.606993	0.584335	0.579803	0.594225
0.362821	0.507027	0.617217	0.62293	0.631006	0.638223	0.710543	0.662171	0.632487	0.147289	0.139727	0.129639	0.071657	0.118744
0.408793	0.421704	0.288029	0.276683	0.272309	0.278281	0.536043	0.604077	0.433894	0.267687	0.237747	0.46123	0.391909	0.320909
0.241815	0.288761	0.055382	0.055738	0.067343	0.095954	0.567458	0.654727	0.202609	0.060063	0.069542	0.370131	0.079571	0.058906
0.245389	0.405906	0.058944	0.059725	0.126091	0.255355	0.780062	0.878223	0.301664	0.061575	0.113458	0.529791	0.147446	0.061672
0.3355	0.479925	0.224084	0.144296	0.303897	0.618815	0.794605	0.7481	0.443529	0.171041	0.362127	0.471625	0.261526	0.203331

The original target data set at the contestability level consisted of 50 data points. In order to reduce its dimensionality, a prior knowledge has been used to group profile data points for working and non working days. As a result, the final target data set vector is reduced from 50 data points to a more manageable 17 data points as shown in Table 5.1.

This step accomplishes goal 1 in terms of structuring the FPS data and goal 2 in terms of projecting and developing different perspectives and views of FPS data.

5.3.5   Selection of Data Mining Method

This application employs a combination of summarization, perspectives, modeling, and prediction methods for data mining. In order to achieve goals 3 and 4, modeling and prediction data mining methods have been identified.

5.3.6   Selection of Data Mining Algorithm & Data Mining

On exploratory analysis, it was found that the backpropagation algorithm is unsuitable for data mining on the target data set because of its long training times and lack of visualization features. Instead, Kohonen Self-Organization Maps (SOM) is chosen for its shorter training times, better visualization features, and good ability to deal with continuous values.

In the initial prototype, different grid sizes (8 X 8, 8 X 10, 15 X 10), and training cycles (ranging from 70 to 20000) have been used. The best results are achieved with grid size (shown in Figure 5.3) of 15 X 10, and 10,000 training cycles.

5.3.7   Interpretation of Knowledge & Acting on Knowledge

The clustered load profiles of various customers are shown in Figure 5.3. In Figure 5.3, symbols $a_1, a_2, \ldots .. a_{47}$ represent unique clusters. For purpose of inxxunsupervised learning, Kohonen clusters making prediction, the known variables of various customers are fed as inputs to the trained Kohonen map. The content addressable properties are used to retrieve the rest of the pattern or the values of rest of data points from the Kohonen map. The rest of the data points corresponded to the predicted load profile of the customer. Figure 5.4 shows a comparison between a predicted load profile and the actual unseen load profile. The prediction error has been calculated in terms of dollar cost as well as on a per customer per hour basis. The visualization features of Kohonen map (the grid in Figure 5.3) were used for validation purposes by the domain expert. The results from this process have been a major improvement over other statistical regression methods.

5.4   FINANCIAL TRADING APPLICATION

Financial trading is a volatile and noisy domain. Computer applications for decision support in this area need to continuously learn new knowledge in order to account for volatile character of the currencies. They also need to be robust to account for noisy and non-deterministic character of the domain (Goonatilake et al. 1994). For example, if a deterministic rule learnt or induced from the past data states that, *if the Australian dollar/ US dollar exchange rate*

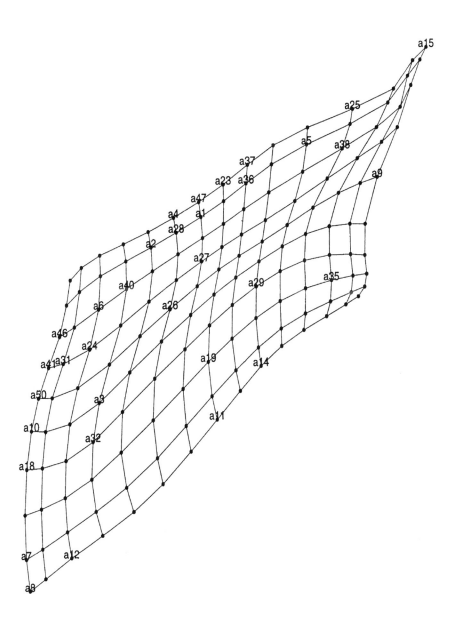

Figure 5.3.   Kohonen Grid with Clusters

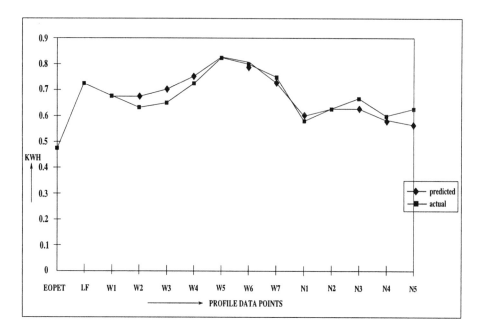

Figure 5.4.   Comparison of Predicted and Actual Load Profiles

is 0.8 and the volume of contracts traded is 15345 then the market will rise, it
is probable that these specific conditions will never be matched in the future.
That is, a suitable decision support system needs to account for the fuzzy nature
of the domain. Further, because of the monetary nature of the domain, the
financial trading managers need proper explanations before they can accept a
decision (e.g. buy or sell a currency) made by computer based decision-support
system.

Keeping in view, the pre-requisites outlined in the previous paragraph, Goonatilake et al. (1994) have combined fuzzy logic, clustering, and genetic algorithms
to learn fuzzy rule bases from historical financial trading data bases. They
have applied the induced fuzzy rule bases for predicting a buy/sell decision of
British Pound against US dollar.

In terms of the KDD process, the financial trading system can be divided
into five parts:

- Data selection.

- Preprocessing & data reduction.

- Selection of data mining algorithm & data mining.

- Interpretation/evaluation of knowledge.

- Acting on the knowledge.

5.4.1 Financial Trading Data Selection

The financial trading data from 1982 to 1987 has used as the target data set. The target data set consists of four variables, namely, closing price, volume, open interest, and price after 10 days. The closing price, volume, and open interest are independent variables, whereas, price after 10 days is a dependent variable (dependent on the three independent variables).

5.4.2 Financial Trading Data Preprocessing & Reduction

The three independent variables are preprocessed by computing the moving-average difference of the three independent variables. The Moving-Average Difference (MA_{diff}) for closing price, volume, and open interest is computed using the formula:

$$MA_{diff} = SMA - LMA/SMA$$

where, SMA is the Short Moving Average, and LMA is the Long Moving Average.

The short moving average is generally 1 to 10 days, and the long moving average is 1 to 200 days or 1 to 50 days (see Figure 5.5).
A strategy for buying and selling a currency can then be determined as follows:

IF the MA_{diff} is POSITIVE then the price is likely to RISE [action: BUY]

IF the MA_{diff} is NEGATIVE then the price is likely to FALL [action: SELL]

In the next step, the three independent variables, closing price, volume, and open interest are transformed by:

- clustering them into discrete bins, and

- fuzzyfying the discrete bins into fuzzy sets or fuzzy bins.

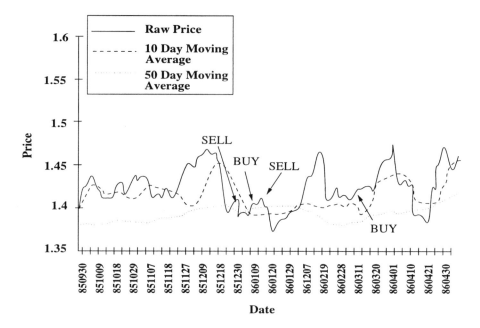

Figure 5.5. Trading Signals from Moving Average Model © 1994 Springer-Verlag

The clustering algorithm, namely, Single Linkage Clustering Method (SLINK), based on the nearest neighbor techniques forms clusters from the time series raw data for an independent variable. The clusters are then mapped into the linguistic categories (e.g. low, medium, high) given by the user for that variable using a heuristic cluster selection algorithm (Goonatilake et al. 1994). The numeric ranges of the clusters in a linguistic category then becomes the numeric range of the linguistic category. However, in this process the boundaries of the linguistic categories are crisp rather than fuzzy. Triangular fuzzy membership functions are used to convert the discrete categories into fuzzy sets. In order to do the conversion the Universe of Discourse (UoD) is discretized into the arbitrary number of bins. The minimum numeric value of UoD corresponds to the minimum value in the cluster with lowest values, and the maximum numeric value corresponds to the maximum value of the highest cluster in the highest linguistic category. For example, let us say the variable volume, v, has chosen clusters in the range of 100 to 1000 contracts covering *low*, *medium*, and *high* categories. Then the UoD for the volume variable is (100,....., 1000). If the UoD is discretized into say, 10 bins, then the 10 corresponding bin ranges will be:

$[100 - 190, 190 - 280, 280 - 370, 370 - 460, 460 - 550, 550 - 640, 640 - 730, 730 - 820, 820 - 910, 910 - 1000]$

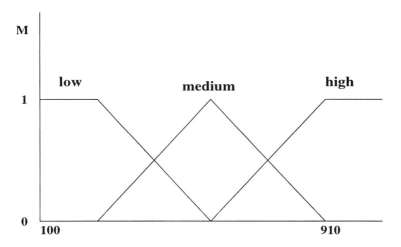

Figure 5.6.   Fuzzy Membership Function for Volume

The corresponding 10 fuzzy membership values for fuzzy sets *low*, *medium*, and *high* can then be computed from Figure 5.6. Thus a raw data value of 109 contracts will fall into the 100 - 110 bin, and the corresponding fuzzy values for that bin are: low - 1.0, medium - 1.0, and high - 0.0.

Likewise, the trading decisions (BUY or SELL) are binned in the range (-3, + 3) and fuzzified using fuzzy sets, *SELL, DO NOTHING*, and *BUY*. The negative values suggest a SELL decision, positive values suggest a BUY decision, whereas, 0 suggests a DO NOTHING decision.

5.4.3 Selection of Data Mining Algorithm & Data Mining

Genetic algorithms have been chosen to induce fuzzy rule bases from past data because of the need for explicit knowledge in financial trading. Packard's (1990) genetic algorithm which can be more appropriately referred to as genetic classifier system have been used. In Packard's system, the antecedent terms and their relationships (e.g., AND, OR , Don't Care) of a rule can be directly encoded as a chromosome. For example, a chromosome (*, *, (4,6), *, 2) representing an antecedent of a rule, implies that there are five independent variables in the domain, the value of the third variable can be 4 or 6, and first, second, and fourth variable have no conditional value (Don't Care). The fitness function is determined by the level of correlation between the states (values) of independent variables and the dependent variable Also, conditional sets or chromosomes with large number of Don't Cares are penalized while evaluating in order to encourage more variable participation and to guard against statistical flukes.

In the financial trading application, in order to account for the non-deterministic and fuzzy nature of the domain each member of the population is made of fuzzy rule bases (consisting of four rules each) rather than simply an individual rule. As stated by Goonatilake et. al. (1994), it is the aggregation effect of all the fuzzy rules (all of which may match the data) that results in a fuzzy system's ability to deal with brittleness. The evaluation of each member or fuzzy rule base is done by applying Mamdani's and Assilian's (1975) compositional rule of inference and then using the center of area (Barenji 1992) for defuzzification of the result. Thus, the result will have a value in the range (-3, +3). The result is then passed through a threshold (i.e., value less than - 2.2 means SELL decision, value greater than 2.2 means BUY decision). The BUY or SELL decision, the fitness of the rule base is determined by comparing the BUY or SELL decision with known 'best' decisions using the past data.

5.4.4   Interpretation of Trading Knowledge & Acting on the Knowledge

Two fittest fuzzy rule bases consisting of four rules each are mined from the past data. A fuzzy rule of the first fuzzy rule base looks like:

[G2[ma-diff-1-50-fuzzy-values []] AND [ma-diff-1-100-fuzzy-values []] AND [ma-diff-1-200-fuzzy-values [negative]] AND [vol-ma-diff-1-10-fuzzy-values []] AND [vol-ma-diff-1-20-fuzzy-values []] AND [oi-ma-diff-1-10-fuzzy-values []] AND [oi-ma-diff-1-20-fuzzy-values [neutral]] AND [price-fuzzy-vola-20 []] AND [price-fuzzy-vola-50 []] AND [price-fuzzy-vola-100 []] THEN [action [SELL]]]

In the above fuzzy rule the terms *ma-diff, vol-ma-diff, oi-ma-diff,* and *price-fuzzy-vola* stand for moving average difference, volume moving average difference, open interest moving average difference, and price fuzzy volatile respectively. The two fuzzy rule bases produced 61% correct trades as against 64.2% correct trades produced by the human trader. Since various parameters like political events are not included in the fuzzy model, the fuzzy rule bases can at best be used as a decision support tool whose decisions can be revised or adapted by the human expert. Goonatilake, et. al. (1994) also recommend use of profit in each rule base as an additional fitness measure to those used in this application.

5.5 LEARNING RULES & KNOWLEDGE HIERARCHIES USING KOHONEN NETWORK

In the previous section supervised neural networks have been used for rule extraction. However, it is possible that the output classes are not known in a problem domain. In these problem domains data clusters have to be learnt in order to extract meaningful knowledge. In this section a knowledge extraction method developed by (Suing 1992; Witten 1994; Dillon, Sestito, Witten, and Suing 1994) using unsupervised Kohonen networks is described for learning rules and building knowledge or concept hierarchies.

5.5.1 The Knowledge Extraction Method

The Kohonen network has been chosen because of its it is simple in structure, with no hidden layers or complex feedback to complicate the nature of the network's configuration (Suing 1992). Its learning algorithm is straightforward and no tedious computations are needed. It also offers the ability to find clusters in the data and performs an ordered or topology preserving mapping that reveals existing similarities in the input vectors. The unsupervised learning algorithm used in the Kohonen net is based on competitive learning and can be summarized as:

a. Initialize weights using a selection of normalized input vectors or patterns and set the initial neighborhood to be large.

b. Stimulate the net with a given input vector.

c. Calculate the Euclidean distance between the input vector and the weight vector associated with each output node and select the output with the minimum distance.

d. Update weights for the selected node and the nodes within its neighborhood.

e. Repeat from b.

The symbolic knowledge to be extracted from the Kohonen net is in the form of production rules and concepts and a related concept hierarchy. The steps involved are:

1. Preprocess data to transform it into the correct format.

2. Assign a given dimension to the output layer.

3. Determine the weights of the Kohonen network.

4. Delete the irrelevant attributes that are identified as having zero weight vector components to all the output nodes.

5. Use either the Threshold technique or the Breakpoint technique to determine the contributory and inhibitory inputs for the initial rules (section 5.5.2).

6. Using the antecedents for the rules, define clusters of output nodes, each cluster being associated with a given rule.

7. Assign real-world semantics to the clusters if possible, or alternatively, assign virtual labels. These define the initial set of concepts.

8. If a concept does not have any *don't care* inputs, obtain the final form of the production rule by affirming its contributory inputs and negating its inhibitory inputs in the antecedent.

9. For concepts with *don't care* inputs, form an initial rule containing the affirmation of the contributory inputs only. Explore these concepts to determine if they are higher lying concepts in the concept hierarchy.

   – If a concept is a higher lying concept in the hierarchy, use a production rule containing only affirmation of the contributory inputs to define the concept.

   – If a concept is not a higher lying concept, explore if it is a concept at a leaf in the hierarchy. If it is a leaf, form a production rule containing both the affirmation of the contributory inputs and the negation of the inhibitory inputs. If the definition given in the production rule does not match a known concept, mark the corresponding nodes as unidentified and do not process them further (section 5.5.3).

   At this stage, form a preliminary concept hierarchy.

10. Determine the number of unidentified nodes still requiring exploration. See if any of these can be assigned previously unrecognized concepts that are either intermediate or lowest level concepts. If so, using the method of Steps 8 and 9, determine their position in the hierarchy.

11. If the number of unidentified nodes is acceptable, stop. Otherwise, the number of unidentified nodes containing mixtures of instances of neighboring clusters is too high and not all known concepts have been identified. Change the dimension of the output layer and repeat from Step 3.

12. Carry out post-processing to delete negative attributes that are redundant to obtain the final production rules (section 5.5.4).

5.5.2 Extraction of Production Rules

Two different approaches were used to extract conjunctive rules in Step 5 from the Kohonen network: the Threshold technique and the Breakpoint technique.

5.5.2.1 Threshold Technique.
In a Kohonen network, the input with the largest weight link to an output node makes the largest contribution to the node. Therefore, the one or more largest components of the weight vector associated with a particular output can be taken as contributory to that output. However, with problems such as noise inherent in real-world domains, the component weights within a certain limit of the maximum value should be considered. All inputs i are included as contributory inputs in the antecedent of the rule for output node j if:

$$|W_{maxj} - W_{ij}| < T$$

where the threshold T is a positive number between 0 and 1. Note that if W_{maxj} is not sufficiently large, no input is considered contributory. In addition to determining those inputs with an excitatory effect on an output, it is useful to characterize the inputs with a strong inhibitory effect. A small weight vector threshold near zero delineates the group of inhibitory inputs. Each inhibitory input is treated by Anding its negation in the associated antecedent. The inputs that are not selected as either contributory or negated inputs are treated as *don't care* in the related rule.

5.5.2.2 Breakpoint Technique.
The existence of a clear breakpoint in the sorted weight vectors such that one component is at least β times larger than the next component is investigated. Those inputs with weight components greater than or equal to the breakpoint are the contributory inputs in the antecedent. For output node j:

- Sort $W_j = (W_1j, W_2j, ..., W_Nj)$ into a list
  $W_{sortj} = [W_j^1, W_j^2, ..., W_j^k, W_j^{k+1}, ..., W_j^N]$
  such that $W_j^{k+1} \leq W_j^k$ for $k = 1, ..., N$

- Determine a breakpoint k' such that $W_j^{k'}/W_j^{k'+1} \geq \beta$

- The contributory inputs in the antecedent of the rule are the inputs i corresponding to all $k \leq k'$.

- The inhibitory inputs that are negated in the antecedent of the rule and the *don't care* inputs are determined as for the Threshold technique.

If there is no clear breakpoint in the sorted list, one of three distinct cases applies. If all the weights are considerably larger than zero and are of a similar size, they can all be considered as contributory inputs. Otherwise, if all the weights are small and of a similar size, no input is considered to be contributory. In this case if all the weights are less than $Tinh$, then all the inputs are inhibitory and are negated in the antecedent. Finally, there could be a spread of weight component values in the range of 0 to 1 with no sharply defined break. This situation requires the Threshold technique.

5.5.3   Unidentified Nodes and Misclassification

When the clusters are defined, nodes not belonging to a given cluster could represent:

- intermediate concepts or higher lying concepts in the concept hierarchy that have not been identified, provided they have one or more *don't care* inputs, or

- a mixture of instances of topographically adjacent clusters and thus the nodes do not represent a separate conceptual existence. In case 1, additional clusters corresponding to these concepts are designated. In case 2, if the number of unidentified nodes is unacceptably large, a different dimensional output layer is specified and the process is repeated until the number of unidentified nodes has been reduced to an acceptable level. If some known concepts in the domain are not being picked up, then the net has insufficient discriminatory power. This may be solved by increasing the dimension of the output layer. However, for noisy data, a larger output layer could lead to more unidentified nodes.

5.5.4   Removal of Redundant Attributes

The final set of production rules might contain some negated attributes that are redundant. The final concept hierarchy is used to delete such attributes. When a particular concept is selected, all contributory attributes for the rules in the alternate branches to their leaf nodes are determined. These inputs or attributes are *contributory attributes for the alternative subhierarchies*. All the output nodes that form the leaf concept clusters for the current subhierarchy are also identified. The weight components from each of the *contributory attributes for the alternative subhierarchies to all the leaf output nodes for the current subhierarchy* are examined. If they are zero for a particular attribute, that attribute is irrelevant for discriminating between concepts in the current subhierarchy. Its negated form can be deleted from the production rules for the leaf concepts in the current subhierarchy.

5.5.5 Led Digit Domain Example

This domain contains data defining the 10 decimal digits on LED displays. Each input pattern is composed of 7 binary attributes only. No target output vector was available to assist the network during training. Neither the number nor nature of the output classes was indicated. Two separate training data sets were used, one comprising 1723 examples of clean data, and the other comprising 1895 examples with a noise level of 10%. As the Kohonen net requires a large data set for learning to occur, the relevant data set was repeatedly fed to the network to simulate approximately 50,000 examples.

5.5.5.1 Clean LED Data. A portion of the weight matrix produced after learning by the 7X7 dimension output layer Kohonen network is given in Table 5.2.

Table 5.2. Portion of the Weight Matrix (7 X 7 Dimension Output Layer) for Clean LED Data

Output node (row, col)	Input nodes						
	0	1	2	3	4	5	6
(6,0)	0.4472	0.0000	0.4472	0.4472	0.4472	0.0000	0.4472
(6,1)	0.4259	0.2232	0.2027	0.4259	0.4259	0.2232	0.4259
(6,2)	0.4082	0.4082	0.0000	0.4082	0.4082	0.4082	0.4082
(6,3)	0.4082	0.4082	0.0000	0.4082	0.4082	0.4082	0.4082
(6,4)	0.4082	0.4082	0.0000	0.4082	0.4082	0.4082	0.4082
(6,5)	0.4472	0.4472	0.0000	0.4472	0.0000	0.4472	0.4472
(6,6)	0.4472	0.4472	0.0000	0.4472	0.0000	0.4472	0.4472

Input nodes:			
0 - Up-center	1 - Up-left	2 - Up-right	3 - Mid-center
4 - Down-left	5 - Down-right	6 - Down-center	

Breakpoint Technique: Consider the Breakpoint technique applied to node (6,5) from Table 5.2. The input weight components sorted from maximum to minimum are:

$W_{sortj} = [W_0, W_1, W_3, W_5, W_6, W_2, W_4]$
That is,
$W_{sortj} = [0.4472, 0.4472, 0.4472, 0.4472, 0.4472, 0.0000, 0.0000]$

A breakpoint is determined from this by finding the input weight component that is more than twice the next one, giving the inputs 0, 1, 3, 5, and 6 as contributory inputs. For $Tinh = 0.05$ (reflecting clean data), inputs 2 and 4 are inhibitory. Therefore, the antecedent part of the rule is constructed as:

Up-center AND Up-left AND Mid-center
AND Down-right AND Down-center
AND NOT Up-right AND NOT Down-left

From an understanding of the LED digit domain, this output node can be associated with the lowest level concept labeled FIVE, giving the rule:

Up-center AND Up-left AND Mid-center
AND Down-right AND Down-center
AND NOT Up-right AND NOT Down-left
THEN FIVE

The contributory inputs in the antecedents with no *don't care* inputs produced by this technique for a 7X7 output layer of the network are listed in Table 5.3. In this table, all the unlisted inputs are negated. They are not included for ease of reading. The associated concepts or consequents of the corresponding rules are then identified and indicated in the curly brackets. The designation of output nodes into clusters is shown in Table 5.4.

Table 5.3. Contributory Inputs in the Antecedents using the Breakpoint Technique

Up-center AND Up-left AND Up-right AND Down-left AND Down-right AND Down-center	{ZERO}
Up-right AND Down-right	{ONE}
Up-center AND Up-right AND Mid-center AND Down-left AND Down-center	{TWO}
Up-center AND Up-right AND Mid-center AND Down-right AND Down-center	{THREE}
Up-left AND Up-right AND Mid-center AND Down-right	{FOUR}
Up-center AND Up-left AND Mid-center AND Down-right AND Down-center	{FIVE}
Up-center AND Up-left AND Mid-center AND Down-left AND Down-right AND Down-center	{SIX}
Up-center AND Up-right AND Down-right	{SEVEN}
Up-center AND Up-left AND Up-right AND Mid-center AND Down-left AND Down-right AND Down-center	{EIGHT}
Up-center AND Up-left AND Up-right AND Mid-center AND Down-right AND Down-center	{NINE}

Using the breakpoint technique with various output layer dimensions, the clusters representing the digits 0 - 9 were identified and generated as rules. The remaining nodes fell into two categories:

- Nodes with *don't care* attributes that can be interpreted as higher level concepts without a real-world significance for this domain.

- Lowest level concepts that are not identifiable as they represent a combination of the attributes of neighboring clusters. The same situation occurred

Table 5.4.   Mapping Produced in a 7X7 Output Layer using the Breakpoint Technique

ONE	ONE	ONE	ONE	FOUR	FOUR	FOUR
HLC1/37	HLC1/37	THREE	THREE	NINE	NINE	NINE
SEVEN	SEVEN	THREE	THREE	NINE	NINE	NINE
HLC3/7/8	HLC3/7/8	THREE	EIGHT	HLC8/9	HLC8/9	NINE
EIGHT	HLC6/8	SIX	EIGHT	ZERO	EIGHT	FIVE
EIGHT	EIGHT	SIX	EIGHT	ZERO	EIGHT	FIVE
TWO	EIGHT	SIX	SIX	SIX	FIVE	FIVE
Nodes marked as HLC1/3/7, HLC3/7/8 and HLC8/9 have *don't care* inputs and are abstract higher level concepts						

in the BIRD domain where some nodes became responsive to more than one cluster. From our knowledge of real-world domains, such nodes reflect mixtures of instances.

Threshold Technique: The mapping produced by the threshold technique for a 7X7 output layer of the network learning from the same data set (with no *don't care* inputs) is given in Table 5.5. The clusters representing the digits 0 - 9 were identified and generated as rules. The remaining nodes all have *don't care* attributes and the majority can be interpreted as abstract higher level concepts that do not have a real-world significance for this domain. Regardless of the technique used, the mappings produced in Tables 5.4 and 5.5 self-organize in the sense that similar clusters (such as ONE, FOUR, and SEVEN) are usually topologically adjacent.

Effect of the Output Layer Dimension: Summary results for different sizes of the output layer are given in Table 5.6. As the dimension increases, the percentage of lowest level identified nodes generally increases. A maximum percentage of lowest level classified nodes was reached for the 9X9 dimension. Larger dimensions also allowed for the identification of all the lowest level clusters in the domain. Determination of the most effective dimension of the output layer is carried out by progressively testing different sizes, as explained in the algorithm. Small dimensions can lead to some sort of contraction of representation since too much knowledge or too many clusters are being mapped to a small area. Also, only a few of the known clusters may be identified with a small output layer. Larger dimensions of the output layer help to illustrate the statistical distribution of the data set clearly. However, a very large dimension can be a waste of space and time during the learning process.

The LED digit example illustrates how the knowledge extraction method can be applied to extract meaningful rules. Although it may not be mean-

Table 5.5.   Mapping Produced in a 7X7 Output Layer for the Lowest Level Concepts Using the Threshold Technique

ONE	ONE	ONE	ONE	*	FOUR	FOUR
*	*	THREE	THREE	*	*	*
SEVEN	SEVEN	THREE	THREE	*	NINE	NINE
*	*	THREE	*	*	*	*
EIGHT	*	SIX	*	ZERO	*	FIVE
*	*	SIX	*	ZERO	*	FIVE
TWO	*	SIX	SIX	SIX	FIVE	FIVE

Nodes marked as "*" have *don't care* inputs and their contributory inputs are not necessarily the same.

Table 5.6.   Identified Lowest Level Output Nodes for Different Sizes of the Output Layer (T = Threshold Technique, B = Breakpoint Technique)

Rules	Number of nodes identified													
	4X4		5X5		6X6		7X7		8X8		9X9		10X10	
	T	B	T	B	T	B	T	B	T	B	T	B	T	B
ZERO							2	2	10	10	6	6	6	6
ONE			1	1	2	2	4	4	2	2	11	11	4	4
TWO			1	1	2	2	1	1	4	4	2	2	4	4
THREE			1	1	2	2	5	5	1	1	9	9	13	14
FOUR			1	1	3	3	2	3	2	2	9	9	2	2
FIVE					1	1	4	4	1	1	6	6	10	10
SIX		1			2	2	5	5	7	7	4	4	10	13
SEVEN			1	1	2	2	2	2	2	2	10	10	3	3
EIGHT		7		5		6	1	9	6	13	7	13	21	23
NINE		1	1	5	2	6	2	7	4	10	6	7	11	13
Unidentified nodes	16	7	19	10	20	10	21	7	25	12	11	4	16	8

ingful to extract a concept hierarchy in the LED digit example, Figure 5.7 shows a meaningful concept hierarchy extracted for the bird domain using the knowledge extraction method described in this section.

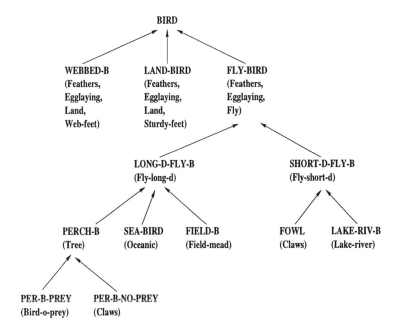

Figure 5.7.   Concept Hirarchy of the Bird Domain

5.6   RULE EXTRACTION IN COMPUTER NETWORK DIAGNOSIS APPLICATION

Computer networks have expanded in size and have become more complex. Along with these changes has come the recognition of network management as an important aspect of computer networking. Despite the fact that there are a number of network management systems, most of them deal only with problems at the lower layers of the network hierarchy (Nuansri, Singh, and Dillon 1995). Thus problems occurring at these layers can be easily solved while those at the upper layer, in particular the application layer, are relatively difficult to solve. The nature of problems at the application level significantly differs from those that occur at the lower levels. Lower layer problems are well-understood, while problems at the application layer are complex, application dependent, and distinct from one another. It is hard to model a significant part of the set of problems that may occur at the application level. Some problems may never have occurred before. In some cases, not all of the problems have been solved. This may be because of the difficulty of modeling the reasoning relating to a collection of knowledge, or because of the structure of the problems to be solved. Thus it is difficult to apply only expert system techniques to these complex domains. In addition, the behavior of applications is sometimes unpredictable and might depend on other events, hidden, or unknown at the time. Some applications can be considered as fundamental applications which are used by other applications. Thus problems that occur in these applications might induce other problems in the applications using them. These types of dependencies have to be taken into account when considering the problem solving mechanisms. To solve most of the upper layer problems, the original problems have to be traced. This means that each application which causes a problem, or which has a tendency to cause a problem, has to be investigated so that a problem solving method can be determined.

In order to reach an accurate diagnosis of the network, Nuansri, Singh and Dillon (1995) selected a testbed application, namely, the Domain Name System (DNS). DNS is, currently, probably the only tool that is generally required by almost all network applications, which rely on DNS services to translate host names into IP addresses and vice versa.

5.6.1 Domain Name System and its Problems

The DNS is a distributed database which provides a mechanism for naming resources in such a way that the names are usable in different hosts, networks, and protocol families. The DNS consists of three major components (Mock-apetris 1987): a domain name space and resource records; a name server; and a resolver. A domain name space is the specification for the DNS internal name space which is represented by a variable-depth tree structure. Each node of the tree represents part of the domain name system, called a *domain*. Resource records are data associated with the names in the domain name space. This data is distributed over the network and is used by name servers to provide services to network applications which require names and IP addresses to be resolved.

A name server is the repository of information that makes up the domain database. Each name server has complete information about some part of the domain name space for which it is responsible. This part of the information is called a *zone* and the name server has authority for that zone. Each zone is controlled by a specific organization which is responsible for distributing current copies of the zones to multiple name servers. This makes the zones available to clients throughout the Internet. The implementation of the DNS name server software used on most UNIX systems is the Berkeley Internet Name Domain (BIND) software. Under BIND, the name server is a process called *named*. Hereafter, it will be interchangeably used with *name server process*.

A resolver is a program, typically a system routine, that interfaces between name servers and user programs, or clients. It extracts information from name servers in response to a client request. More information about the DNS can be found in (Mockapetris 1987)

5.6.1.1 DNS Problems. There are a number of errors that occur in the DNS. These errors not only affect the DNS itself but also applications using its services. The DNS is, normally, not directly used by general users. Rather, it is used by other network applications. Thus errors occurring in the DNS directly affect the applications. To avoid such correlated problems, the DNS must be reliable at all times. However, this is almost impossible, and what we can do is minimize the number of errors that occur in the DNS so that other applications will not suffer because of them.

The *named* reports detected errors in a text format which can be logged in a log file. The error messages are usually used by domain name owners to monitor the domain name system. Although these error messages are important, in practice they are usually ignored. This might occur because of the DNS protocol and its architecture that tries its best to serve its client. The size of the error log file is another issue that discourages human attention. Along with real error messages reported, there are several non-error messages. Thus the log file can be very big. The size of the file and its contents are generally beyond the ability of human zone managers to scan to diagnose the DNS problems. Moreover, the use of these error messages is not straight forward as some problems produce more than one error message type while other problems are reported by only one error message. This leads to difficulties analyzing and diagnosing the problems. In order to make use of these error messages, some methodologies to analyze and find the cause of each error message are required.

5.6.2 Hybrid Diagnostic System for DNS Problems

Based on the result of DNS problem analysis, a hybrid system which is a combination of a neural network system, which can learn from data in the past, and a rule-based expert system which makes use of an output from the learning process has been developed. A tool called *BRAINNE*, an automated knowledge acquisition tool based on back propagation neural networks (Sestito and Dillon 1994), and an object-oriented expert system tool, *NEXPERT* (Nexpert 1989) are used in the application.

Thus, in the Automated Knowledge Based Diagnostic System for DNS Problems (AKBDS), BRAINNE is used as an automated knowledge acquisition tool that allows to extract essential knowledge using DNS faults (errors) and their causes while the NEXPERT is later used to infer with the derived rule-based expert system for the analysis and diagnosis part. The AKBDS can then be used to monitor the domain name system in a real situation by scanning error messages from the *named* log files. While doing this, the AKBDS will check the log files from time to time to obtain new rules. These new rules will be added to the rule-based system so that it is always up to date. Thus the rule-based system will cover as many potential errors as possible.

5.6.2.1   Knowledge Acquiring Process. The learning module of the system is required to learn from the errors and the desired result from the learning process is a relationship between an error or a group of errors and their correspondent causes. In this implementation, the supervised learning module is used. Thus the input pattern for the learning process consists of two components: the problems (error messages), and the known causes (faults). The first component can be easily obtained from a *named* log file, while the second component is not easy to obtain. This is because of the nature of the name server software which does not explicitly state the cause of any problem reported. Consequently, we have to create some mechanisms by which we can obtain these faults and match them with a given error message or a group of messages. To do this, two methods are proposed: extracting knowledge from experience and human experts, and forcing faults on a name server. The result from these knowledge acquiring processes is pairs of problems and correspondent causes which are later used as an input pattern for BRAINNE.

5.6.2.2   Extracting Knowledge from Experience and Experts. In this method errors that occurred in the past were collected from several log files to create pairs of an error or a group of errors along with the fault causing them. To do this, knowledge of DNS and its software is required. In addition, several human experts in this field were also consulted.

5.6.2.3   Fault Forcing Method. Although the previous method is simple and can provide the desired information, there is a flaw in it: the information obtained by this method is not guaranteed to be correct or complete. The alternative method is to attempt to create faults in the domain name system and to observe the corresponding errors so that an individual fault with an error message or a group of messages reported for each fault type can be matched. By using this methodology the fault forcing process can be repeated, if desired, so that the obtained information is confirmed. Once the result is confirmed, it is then used as an input pattern for the neural network training process.

However, there are also some types of faults that are difficult to obtain by this method also. Some problems are caused or depend on other events that are beyond our control. For instance, there are some error messages that occur because of the corruption of UDP packets used by the DNS. This is not a real DNS problem. Some errors are believed to occur because of some flaws (bugs) in the software being used. It is not easy to create these kinds of errors using this method. There are also some errors, which while not impossible to create, take quite a large amount of time to be reported, partly because they also depend on some other parameters of the environment e.g. a system configuration, or the underlying network protocols. These type of errors are ignored by this method at this stage.

5.6.2.4   Problem Analysis.   In the monitoring and diagnosing process, the system has to diagnose problems corresponding to the errors that were logged. The diagnostic method is based on the knowledge of DNS faults and errors that were obtained from the knowledge acquiring process.

From the study, DNS problems can be grouped into two main categories: problems caused by errors in configuration of DNS database files, and those that occurred as a result of general network problems. These groups of problems affect DNS to different degrees, and the methods used to deal with them, in order to solve the problems, are also different. The errors caused by network communications problems are not too difficult to deal with as the problem itself is not too complicated, whereas those caused by configuration errors are more complex. Most of the latter problems propagate around the network which sometimes make it difficult to trace back to the site where the problem originated.

5.6.2.5   Configuration Problems.   As a result of the study, it was found that the majority of the errors reported by *named* are caused by an incorrect DNS data due to wrong configuration and delegation. They are also caused by the problem of information inconsistency which is regarded as the major problem of the DNS. According to the distributed architecture and implementation of the DNS, information for a particular zone always exists at more than one place. When the information is not consistent amongst the name servers, diagnosis is difficult as the incorrect information is also propagated around the network along with the correct one. Moreover, some detected problems can be transient. These can occur when zone information has just been changed at one name server while some other participating name servers have not yet updated their data. In this case, this error will vanish when the information has eventually reached all relevant name servers.

5.6.2.6   Network Related Problems.   These problems occur when a name server cannot update zone data of a particular zone. They are normally caused by the inability of a secondary name server to contact the primary name server of that zone for a period of time. Unless the information is updated before its specified expiry time is reached, it will expire and the name server will stop giving answers to any query related to that zone. Like the first group of problems, these problems can also be transient. Examples of error messages in each category are given in (Nuansri, Singh, and dillon 1995).

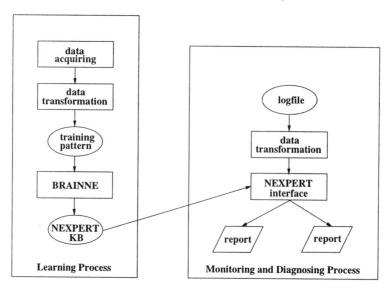

Figure 5.8.   Hybrid Diagnostic System for DNS Problems

5.6.2.7   System Implementation.   The system consists of several main components which are depicted in Figure 5.8 are data acquiring process, data transformation functions, BRAINNE, and an NEXPERT (Nexpert 1989) interface function. The data acquiring process is a process that is used to acquire an input pattern for a learning module, based on the methods described in the preceding sections. The output from the learning process is a set of rules, of which a part is shown below, representing the output from BRAINNE.

Rule   1 (covers 123 exs OK, 0 exs NOT_OK)

((1) forwarding loop == yes)
==> (0) configuration error

Rule   13 (covers 221 exs OK, 0 exs NOT_OK)

((7) Err/TO getting serial # == yes)
((11) connection refused == yes)
((8) masters for secondary zone unreachable == yes)
==> (3) no name server process running

The NEXPERT rule base derived from BRAINEE is shown below. The output from BRAINNEE is used build the Left Hand Side (LHS) or antecedent part, and HYPOthesis (HYPO) of the rule. The inference or reasoning is then done by NEXPERT by using the Right Hand Side (RHS) or consequent part of the rule.

(RULE= R1
(LHS= (Yes (forwarding_loop)))
(HYPO= configuration_error)
(RHS= (Execute ("forward_loop")))
)
(RULE= R13
(LHS= (Yes (Err/TO_getting_serial_#))
(Yes (connection_refused))
(Yes (masters_for_secondary_zone_unreachable)))
(HYPO= name_server_problem)
(RHS= (Execute ("check_ns")))
)

More details of the implementation can be found in (Nuansari, Singh, Dillon 1995).

5.6.2.8   System Testing.   After the rule base was created from rule sets of the learning process, the diagnostic system was then tested with real data from *named* log files. During the testing process, several new error messages were found and corresponding new rules were added into the rule base. This happened because some errors rarely occur and they were not found during the learning process. After the rule base was satisfied, the system was used to diagnose a daily *named* log file. It was found that the rule base covered most of the error types. As a result, the diagnostic system was able to diagnose almost all error messages. Only about 0.5% of unrecognized errors were found (unknown and rejected by the diagnostic system). However, some of them, such as "not enough memory," "network unreachable," and "file table overflow" are not DNS problems but they were reported due to the problem of the machine running a name server. When a new error is found, a corresponding rule can easily be added into the rule base. This is the advantage of using a rule-based system as part of the diagnostic process. The system does not require a retraining process in order to add new rules. The retraining process might be necessary when the system is applied to some applications that are not stable, or which are in the developmental and testing phase. Tables 5.7 and 5.8 list the the different types of errors derived from a log file of a name server called *munnari.oz.au* (*munnari.oz.au* is a major name server being either the primary or a secondary server for hundreds of zones) between the period September 19 - October 5 1995. These tables represent some statistical values obtained from running the diagnostic system. In Table 5.7, error messages are grouped according to the cause of the error. There are three subgroups: configuration error, network related error, and others. Table 5.7 lists the type of error, the number and the percentage of occurrences. In this figure, same error messages are counted regardless of from which name server or zone they were produced. Table 5.8 shows the number of zones that were involved in error in each category. Note that the total number of messages that were logged is not the same as the number of error messages. This is because the log file also contains other types of messages reported by *named* such as warnings or operational report messages.

An output from each diagnostic function is a report of the cause or the problem, and the original name server or zone that created that problem. It provides some other information that might be useful to correct the problem, including the error message or messages that were reported. Suggestions for the correction of some type of error are also given.

5.7 SUMMARY

Knowledge discovery in databases, data mining, and transformation systems have become significant in the 90's because of the need to extract meaningful knowledge from gigabytes of data related to stock markets, electricity consumption profiles, and trouble shooting and diagnosis stored in organizational databases. The KDD process developed by Fayyad et al. (1996b) for extracting and interpreting knowledge involves nine steps, namely, understanding the domain, data selection, preprocessing of data, data reduction, selection of data mining method, selection of data mining algorithm, data mining, interpretation of knowledge, and using the knowledge. Meaningful Useful knowledge can be summarized information, rules (or models) explaining competition or manufacturing process behavior, predictive knowledge for predicting electricity load profiles, classification, and clustering knowledge for determining classes, categories, and groups of various customers

Data mining is a step in the KDD process that consists of applying data analysis and discovery (learning) algorithms that produce a particular enumeration of patterns (or models) over the data. In fact, the data mining process can fit into the framework of intelligent transformation systems described in chapter 3.

A number of applications based on the KDD process, data mining, and transformation systems have been described in this chapter. In the forecasting

Table 5.7.   Log File Listing Type of Error and their Occurrences

Error Message Type	Number	Percentage of Occurrences
Configuration Errors		
lame delegation	42709	69.569
attmpted update to auth zone	1751	2.852
response from unexpected source	1705	2.777
forwarding loop	1418	2.309
contains our address	563	0.917
wrong serial number	528	0.860
cname error	51	0.013
database format error	8	0.003
unknown type	2	0.003
outside zone	2	79.38
Total	48737	
Network Related Errors	5546	9.034
Err/To	3569	5.813
timed out	1108	1.804
bad SOA	844	1.374
masters zone unreachable	38	0.061
connection refused	432	0.703
zone expired	11537	18.792
Total		
Others	1026	1.671
malformed response	80	0.130
not enough memory	4	0.006
address reuse	1110	1.808
Total		
Number of error messages	61384	
Number of messages logged	61390	

system, application meaningful knowledge has been extracted in terms summarization of data, clustering and prediction. Data warehouse, object-oriented methodology, and Kohonen's self organizing maps have been used for the purpose. Genetic-Fuzzy system has been described for learning predictive fuzzy knowledge in financial trading. Unsupervised and supervised transformation systems have been described for learning symbolic rules and concept hierarchies.

Notes

1. Fayyad et al. (1996c) actually call this step "Transformation and reduction of data".

Table 5.8.   Log File Listing Error Zones

Error Type	Number of Zones
lame delegation	4071
Err/To + masters for secondary zone unreachable	382
response from unexpected source	137
forwarding loop	106
contains our address	79
attmpted update to auth zone	62
cname error	11
Err/To + connection refused	8
Err/To + bad SOA	3
wrong serial number	3

References

Barenji, H. (1992), "Fuzzy Logic Controllers," *An Introduction to Fuzzy Logic Applications in Intelligent Systems*, Eds., R. Yager and L. Zadeh, Kluwer Academic Publishers.

Dillon, T.S., Sestito S., Witten, M., and Suing, M. (1994), "Symbolic Knowledge from Unsupervised Learning," *AAAI International Symposium on Integrating Knowledge and Neural Heuristics*, Pensacola, Florida, USA, pp. 47-57.

Fayyad, U., and Uthurusamy, R. (1996a), "Data Mining and Knowledge Discovery in Databases," *Communications of the ACM*, vol. 39, no. 11, pp. 24-26.

Fayyad, U., Piatetsky-Shaprio, G., and Smyth, P. (1996b), "From Data Mining to Knowledge Discovery in Databases," *AI Magazine*, AAAI Press, vol. 17, no. 3, pp. 37-54.

Fayyad, U. (1996c), "Data Mining and Knowledge Discovery: Making Sense out of data," *IEEE Expert*, vol. 11, no. 5, pp. 20-25.

Goonatilake, S., Campbell, J.A., and Ahmad, N. (1994), "Genetic-Fuzzy Systems for Financial Decision Making," *Advances in Fuzzy Logic, neural Networks and Genetic Algorithms*, Springer, pp. 202-223.

Inmon, W.H., and Kelley, C. (1993), *Rdb/VMS, Developing the Data Warehouse* , QED, Publication Group, Boston, USA.

Kohonen, T. (1990), "The Self-Organizing Map," *Proceedings of the IEEE*, vol. 78, no. 9, pp. 1464-80.

Mamdani, E., and Assilian , S. (1975), "An Experiment with Linguistic synthesis with a Fuzzy Logic Controller," *International Journal of Man-machine Studies*, vol. 7.

Mockapetris, P. (1987), "Domain Names - Concepts and Facilities," *RFC 1034, and RFC 1035, November 1987.*

Nexpert Object Manuals (1989), Neuron Data Inc. CA, USA.

Nuansri, N., Singh, S., and Dillon, T. (1995), "Implementation of a Hybrid System for DNS Diagnosis," *Technical Report, Department of Computer Science and Computer Engineering*, La Trobe University, Melbourne, Australia.

Packard, N. H. (1990), "A genetic Algorithm for the Analysis of Complex Data," *Complex Systems*, vol. 4, pp. 543-572.

Sestito, S., and Dillon, T.S. (1994), *Automated Knowledge Acquisition*, Prentice Hall, Sydney, Australia.

Suing, M. (1992), "Machine Learning/Automated Knowledge Acquisition using Unsupervised Neural Networks," *Honors Thesis*, Department of Computer Science and Computer Engineering, La Trobe University, Melbourne, Australia.

Witten, M. (1994), "Automated Knowledge Acquisition in Unsupervised Supervised Learning Modes," *Honors Thesis*, Department of Computer Science and Computer Engineering, La Trobe University, Melbourne, Australia.

6 ASSOCIATION SYSTEMS - TASK STRUCTURE LEVEL ASSOCIATIVE HYBRID ARCHITECTURE

6.1 INTRODUCTION

The hybrid architectures described in chapters 3, 4, and 5 are based on fusion, transformation, and combination of intelligent methodologies like expert systems, fuzzy logic, neural networks, and genetic algorithms. The concepts of fusion, transformation, and combination have been used in different situations or tasks, and by applying top-down and/or bottom-up knowledge engineering strategy. All these hybrid architectures have a number of advantages in that the hybrid arrangement is able to successfully accomplish tasks in various situations. However, these hybrid architectures also suffer from some drawbacks. These drawbacks can be explained in terms of the quality of solution and range of tasks covered (see Figure 6.1). Fusion, and transformation architectures on their own do not capture all aspects of human cognition related to problem solving. For example, fusion architectures result in conversion of explicit knowledge into implicit knowledge, and thus loose on the declarative aspects of problem solving. Thus, they are restricted in terms of the range of tasks covered by them. The transformation architectures with bottom up strategy get into problems with increasing task complexity. Therefore the quality of solution suffers when there is heavy overlap between variables, where the rules are very complicated, the quality of data is poor, or data is noisy. Also, because they lack explicit reasoning, the range of tasks covered by them becomes restricted. The combination architectures cover a range of tasks because of their inherent flexibility in terms of selection of two or more intelligent methodologies. However, because of lack of (or minimal) knowledge transfer among different modules the quality of solution suffers for the very reasons the fusion and transformation architectures are used.

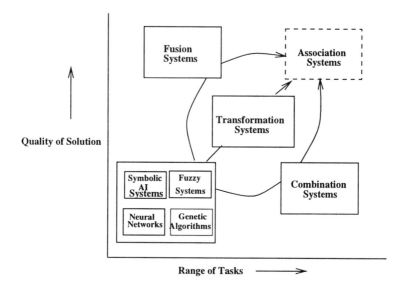

Figure 6.1.   Quality of Solution and Range of Tasks

The fact of the matter is that the fusion, transformation, and combination architectures have been motivated by and developed for different problem solving tasks/situations. Thus it is useful to associate these architectures in a manner so as to maximize the quality as well as range of tasks that can be covered. In this chapter, an associative hybrid architecture is developed in the context of generic tasks involved in problem solving.

The associative hybrid architecture incorporates fusion, transformation, and combination concepts described in chapters 3, 4 and 5 respectively. At the same time, it integrates the artificial and computational intelligence levels of problem solving involving symbolic, fuzzy, neural network, and genetic/evolutionary programming methods. The associative hybrid architecture is applicable to large scale complex data intensive and time critical problems. It is built around integration of various perspectives which characterize hybridization, and infor-

TASK STRUCTURE LEVEL

**OUTLINES THE GENERIC STRUCTURE IN TERMS OF TASKS AND
METHODS USED TO ACCOMPLISH THE TASKS IN VARIOUS DOMAINS
E.G. , DIAGNOSIS, PLANNING, MONITORING, CONTROL.**

COMPUTATIONAL LEVEL

**OUTLINES THE KNOWLEDGE CONTENT REQUIRED TO REALIZE
THE TASK STRUCTURE LEVEL ARCHITECTURE.**

**IT INCLUDES IDENTIFICATION OF KNOWLEDGE REPRESENTATION,
COMMUNICATION AND OTHER ONTOLOGICAL CONSTRUCTS.**

PROGRAM LEVEL

**OUTLINES THE DETAILS OF THE PROGRAMS, SUB-ROUTINES, DATA
STRUCTURES TO BE USED FOR IMPLEMENTING THE COMPUTATIONAL
ARCHITECTURE.**

Figure 6.2.   Three Levels of Associative Hybrid Architecture

mation processing, five information processing phases, generic tasks to be accomplished in each phase, methods employed to accomplish a generic task, and the knowledge engineering strategy employed.

The associative hybrid architecture based on the above contexts is realized at three levels as shown in Figure 6.2, namely, the task structure level, the computational level, and the program level. In this book task structure and computational levels of the associative computational level hybrid architecture are described. The associative hybrid architecture in this chapter is described at the task structure level. In the next chapter, the computational level is described.

6.2 VARIOUS PERSPECTIVES CHARACTERIZING PROBLEM SOLVING

The development of any intelligent problem solving architecture is characterized by various perspectives which form part of its structure and behavior. The development of intelligent associative hybrid architecture is characterized by the following perspectives:

- Philosophical perspective

- Neurobiological control perspective

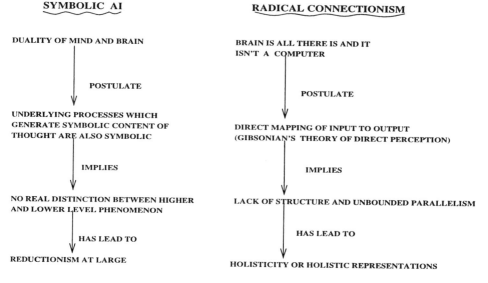

Figure 6.3.   Philosophical Considerations

- Cognitive science perspective

- Computational and artificial intelligence perspective

- Physical systems perspective

- Fuzzy systems perspective

- Forms of knowledge perspective

- Learning perspective

- User perspective

6.2.1   Philosophical Perspective

Some of the philosophical underpinnings of symbolic and connectionist systems are shown in Figure 6.3. The parallel and distributed properties of the connectionist models and their apparent proximity to our neural structure have encouraged some of the proponents of connectionism (Churchland 1990; McClelland and Rumelhart 1986a; McClelland and Rumelhart 1986b) to claim that human memory has a holistic character. The simultaneous consideration of multiple constraints in Parallel Distributed Processing (PDP) (McClelland

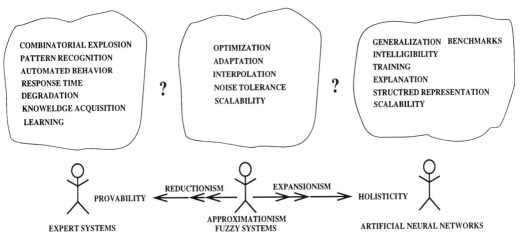

Figure 6.4.   Need for an Hybrid Architecture

and Rumelhart 1986a) models imply a holistic proposition for solving a problem at hand. According to its proponents connectionism is said to perform direct recognition while symbolic AI performs recognition by sequentially computing intermediate representations.

On the other hand, determination of goals, sub-goals, sub-sub-goals and creation of structured problem sub-spaces as proposed by (Newell 1972) is indicative of the reductionist philosophy at work. The proponents of the discrete symbol systems propose logical chunking as a method of problem solving (Laird et al. 1987; Newell 1980; Newell, 1981a; Newell 1981b).

The above philosophical comparison represents an inflexible and fundamentally opposing nature of the two problem solving philosophies. On the one hand, the symbolic systems, by pursuing a reductionist philosophy of logic and provability have become rigid and inflexible. Also, today using symbolic systems we can cope with highly specialized problems solved by human experts, but these very systems have problems coping with the easiest of problems which children can solve like walking, recognizing objects, etc. (Minsky 1991). Symbolic AI systems based on the production rule philosophy are constrained to deal with pattern recognition systems which are intuitively parallel or systems which involve approximate (partial matching) or fuzzy reasoning. In fact, the recent resurgence of fuzzy systems has exposed this rigidity in no uncertain manner and has become a intelligent methodology which is going to be used extensively in the 90s. However, for complex tasks, fuzzy systems need to be enhanced through optimization, noise tolerance, and adaptation.

The proponents of the reductionist view have implicitly assumed that the underlying processes at the micro level which generate the symbolic content of thought are also necessarily symbolic. This implicit assumption has been established as incorrect with the development of connectionist architectures (Chandrasekaran 1990; McClelland and Rumelhart 1986a; Smolensky 1990). On the other hand, the all embracing holistic philosophy has not delivered the goods as was initially expected when scaled up to complex tasks. Furthermore, the connectionist models are more concerned with modeling the dynamic rather than the structural characteristics of a system which can contribute to intelligent behavior.

The reductionist and holistic views thus have in a sense restricted the search for more generic information processing architecture which is composed of abstraction, structure and logic embodied in symbolism (Fodor and Plyshyn 1988), and the underlying parallel and distributed processes.

6.2.2   Cognitive Science Perspective

The cognitive science perspective has been discussed in chapter 1. It is briefly revisited here. One of the benefits of the symbolic-connectionist debate from a cognitive science viewpoint (Chandrasekaran 1990; Churchland 1990; Fieldman 1988; Fodor 1983; Fodor et al. 1988; Newell 1990; Shepard 1988; Smolensky 1990) has been that it has addressed a number of issues related to information processing, learning and knowledge representation. The aspects which emerge from an information processing viewpoint are the granularity of the processes (Chandrasekaran 1990; Fodor 1983; Lallement, Hilario and Alexandre, 1995; Lallement and Alexandre 1995b), notion of time (Newell 1990), and the levels of information processing (Fodor 1983; Fodor et al. 1988; Newell 1990). The macro and micro level granularity of the processes associated with human cognition have lead researchers to think that the symbolic and connectionist theories on their own are likely to be inadequate to cover the range of phenomena in cognition. Furthermore, the notion of time associated with these processes (which varies from 100ms to a few seconds/minutes (Chandrasekaran 1990; Newell 1990) further establishes that one has to account for sequential as well as automated form of reasoning mechanisms in an information processing architecture. Besides, according to Newell (1990), *the fact that human beings are grounded in the world implies additional constraints that must be taken into account in constructing our theories.* That is, it is essential that the information processes theories which emerge from such studies need to be grounded in the real world in which human beings interact with the physical systems surrounding them.

6.2.3   Neurobiological Perspective

The investigations into the visual nervous system of mammals reveals that the visual system works at different levels of abstraction (Rao 1995). The model designed by (Rao 1995) posits a role for the hierarchical structure of the visual cortex and its reciprocal connections between adjoining visual areas in determining the response properties of visual cortical neurons. Besides, the past studies (Beale and Jackson 1990) have shown that the complex cells in the visual cortex perform higher level functions whereas the simple cells and the ganglion cells perform lower level functions. The hierarchical arrangement of the complex cells, simple cells and the ganglion cells allows the visual system to extract more and more abstract information from the initial electric signals. Secondly, research into the functional principles of the brain shows that it exercises central and hierarchical control at different levels (sensory, motor, etc.) to provide both stability and adaptability (Beale and Jackson 1990). Finally, the highly connected and parallel design of the brain allows it to work on many different things at once (e.g. vision and speech). The brain, with its parallel design, is able to represent a host of different external stimuli/items of information in a distributed form and is able to process these items of information in a parallel manner.

6.2.4   Physical Systems Perspective

The underlying principles of abstraction and hierarchical control described in the previous section undoubtedly form an intrinsic part of man made systems like power systems, organizational systems, telecommunication systems, process control systems, etc. For example, in power systems, the power network is hierarchically decomposed into transmission, sub-transmission and distribution levels. The structure of these systems influences their behavior. Humans, while reasoning with these systems likewise engage in structural and/or behavioral decomposition of the problem and perceive the solution at different levels of abstraction.

Thus it can be said from the neurobiological and physical systems perspectives that the principles of abstraction and hierarchical control not only form a part of our neural hardware but also form a structural and behavioral part of our information processing strategy.

6.2.5   Fuzzy Systems Perspective

Fuzzy systems as have been described in the preceding chapters help to model imprecision in data. It enables humans to model or group real-valued attributes into fuzzy or non-crisp sets. These fuzzy sets are used to develop fuzzy rules and fuzzy reasoning. Fuzzy reasoning is a form of human-like approximate reasoning where we consider the degree to which a rule agrees to our current understanding of reality rather than the probability that it is a true description of that reality (Aminzadeh and Jamshidi 1994). From a information processing viewpoint, fuzzy systems have been fused with neural networks through fuzzification of input data, and incorporating fuzzy logic operations within neural networks. A number of applications in the process control, pattern recognition, and fault diagnosis area have been built with fuzzy-neural fusion and fuzzy expert systems. These applications have established that fuzzy systems form an important part of human information processing for modeling uncertainty and imprecision.

6.2.6   Artificial and Computational Intelligence Perspective

In Bezdek (1994) a model outlining three levels of intelligent activity, namely, computational, artificial, and biological is described. The computational level (or the lowest level), according to the model deals with numeric data and involves pattern recognition, whereas the artificial intelligence level augments the computational level by adding small pieces of knowledge to the computational processes. The biological intelligence level (the highest level) processes sensory inputs and, through associate memory, links many sub-domains of biological neural networks to recall knowledge (Bezdek 1994; Medskar 1995). The computational and the artificial levels according to (Bezdek 1994) lead to the biological level. An integration of the computational and artificial intelligence levels can thus be considered as a way of modeling biological intelligence.

6.2.7   Learning Perspective

Learning forms an important ingredient of any intelligent system. Based on the cognitive science studies various types of learning techniques have been identified including learning of procedures, concept learning, reinforcement learning and others. One of the motivations for surge in interest in the connectionist models has been their ability to learn and adapt. However, most of the connectionist models are known to adopt the "Learning from scratch," philosophy. This is probably not adequate for complex domains, a point already made by some prominent researchers in this area. To quote from Arbib (1987, page 70):

Humans are intelligent because evolution has equipped them with a richly structured brain. This structure, while serving a variety of functions, in particular enables them to learn. A certain critical degree of structural complexity is required of a network before it can become self-modifying – no matter how sophisticated its reinforcement rules – in a way that we could consider intelligent.

Furthermore, to quote from Malsburg (1988, page 26):

However, some fundamental problems remain to be solved before flexible robots can be realized. One of these problems is scalability to realistic size. Learning strategy based on exhaustive search of full combinatorial phase spaces blow up too quickly. The solution to this problem will very likely have to be based on the introduction of a clever a priori structure to restrict the relevant phase spaces.

6.2.8 Forms of Knowledge Perspective

Symbolic, connectionist, and fuzzy research communities have developed a rich set of representational mechanisms which have enabled cognitive scientists to characterize human cognition. Broadly, the knowledge representation mechanisms can be categorized into three forms (see Figure 6.5):

- sub-symbolic knowledge

- symbolic and non-formal or fuzzy, and

- symbolic and formal

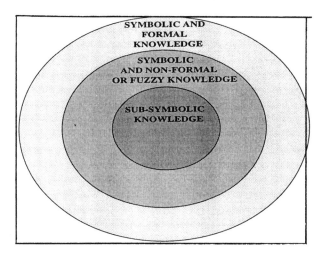

Figure 6.5. Forms of Knowledge

Sub-symbolic knowledge can be seen as sub-conceptual, intuitive or tacit knowledge in a particular domain which necessarily cannot be articulated in terms of rules. For example, a five-year old child learning to play baseball or cricket finds it difficult to express in natural language the patterns of response learnt by him in hitting a ball thrown at him in different patterns. These types of problems are best expressed with sub-symbolic knowledge and may involve parallel processing for inference. Symbolic and non-formal knowledge can be seen as conceptual knowledge represented by frames and objects. It can also be represented by macro level heuristics/rules or fuzzy rules which are an outcome of experience of the human expert and generally cannot be formalized in a rigorous fashion (e.g. heuristics taught by a coach to a baseball or cricket player). Symbolic and formal knowledge can be seen as knowledge represented by a more rigorous formalism (e.g. physics, applied mathematics, etc.) such as a mathematical/structural/behavioral model or formal logic representation.

The different forms of knowledge also represent knowledge of varying granularity. The sub-symbolic knowledge layer has fine granularity. The fine granularity is based on the highly distributed nature of knowledge at the sub-symbolic or micro level. The granularity becomes coarser as one moves from the sub-symbolic knowledge level to symbolic and formal knowledge level.

6.2.9   User Perspective

Humans form an important part of solution to most real world problems. Thus it is imperative that any architecture which is derived out of using the two methodologies should result in reducing the cognitive barriers between the human experts and the computer. This is vital for the following reasons:

- **Acceptability** - without cognitive compatibility, the system's behavior can appear surprising and unnatural to the user. In other words, architectures with low cognitive compatibility will lead to low acceptability

- **Effectiveness** - without cognitive compatibility, effective interpretation of user's or expert's problem solving behavior is at risk which may result in unsatisfactory performance.

Thus user intelligibility should form an important aspect of system architecture as it will not only facilitate higher acceptability and better performance but also facilitate high level of user interaction.

Further, issues like scalability and cost effectiveness (reduced development time, reduced memory requirements, fast execution, etc.) are equally important to enable use of hybrid architectures on large scale problems. Scalability requires that an architecture should permit an open-ended approach for system development. That is, any incremental changes in the systems can easily flow through the architecture with minimal damage to reliability, speed of operation, etc. of the whole system. The existing symbolic systems for complex large scale problems invariably have deep solution hierarchies, whereas neural network systems for such problems have to cope with issues related to precision and reliability.

The various perspectives and their characteristics are shown in Figure 6.6.

6.3   TASK STRUCTURE LEVEL ARCHITECTURE

The Task Structure Level (TSL) architecture developed by Khosla and Dillon (1992a, 92b, 93, 94a, 94b, 94c, 94d, 95, 97) is shown in Figures 6.7 to 6.11 respectively. It has been derived from the task structure level perspectives and their characteristics/constraints described in the preceding sections. It is assumed that the problem domain under study is data intensive, and notion of time is important to the solution of the problem.

The TSL architecture is defined in terms of the:

- information processing phases

- tasks to be accomplished in each phase

- Top Down (TD) or Bottom Up (BU) knowledge engineering strategy adopted to accomplish a task

- intelligent methods used to accomplish the tasks in different phases

- hybrid arrangement (Combination, Transformation, or Fusion) of the intelligent methods in each phase, and

- constraints which need to be satisfied for accomplishing a task

Five information phases have been identified. These are Global Preprocessing Phase, Decomposition Phase, Control Phase, Decision Phase, and Postprocessing Phase. The information processing phases have been derived from the neurobiological, cognitive science, and other perspectives which impose constraints like structure, control, granularity of processes, notion of time, conscious and automated behavior, and others.

6.3.1   Global Preprocessing Phase

The global preprocessing phase involves two tasks:

- input conditioning

- noise filtering

Input conditioning and global noise filtering is required for making the inputs from external environment fit for processing. In this phase heuristical domain knowledge (content of which is symbolic or fuzzy) is used by employing symbolic or fuzzy methods for filtering out noise which is peculiar to the domain

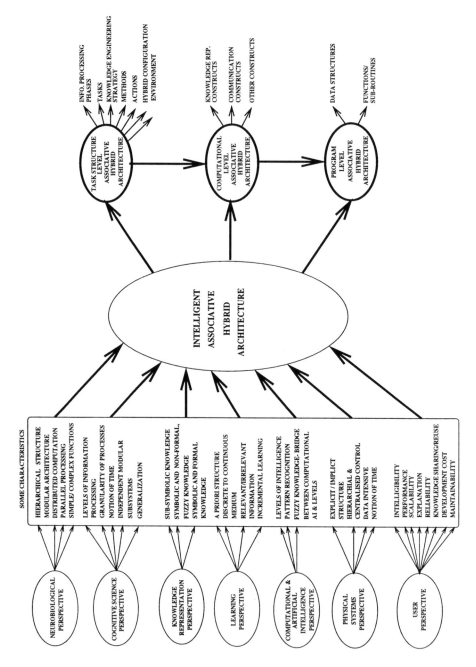

Figure 6.6.   Perspectives for Hybridization, and Information Processing

as a whole or common to many domain concepts/classes. Otherwise, if the domain knowledge is incomplete/insufficient or because of the complexity of the domain itself neural networks and/or other mathematical methods like Fast Fourier Transforms (FFT) can be used in conjunction with the symbolic or fuzzy methods as shown. The neural networks and/or other mathematical methods are used to transform the numerical data into qualitative or symbolic classes) which can then be reasoned with using symbolic or fuzzy reasoning.

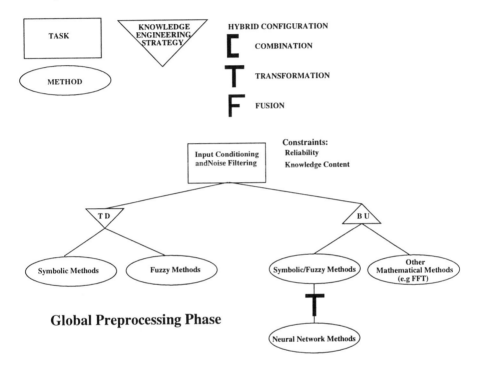

Global Preprocessing Phase

Figure 6.7.   TSL Architecture - Preprocessing Phase

6.3.2   Decomposition Phase

The different levels of abstraction used in information processing represent aggregations or generalizations of varying granularity. The goal of the decomposition phase is to decompose the problem under study into set of abstract concepts/classes. In order to deal effectively with the complexity, the idea is to intially define the solution in terms of the primary subsystems of the domain. These subsystems are abstract in the context of the problem under study as

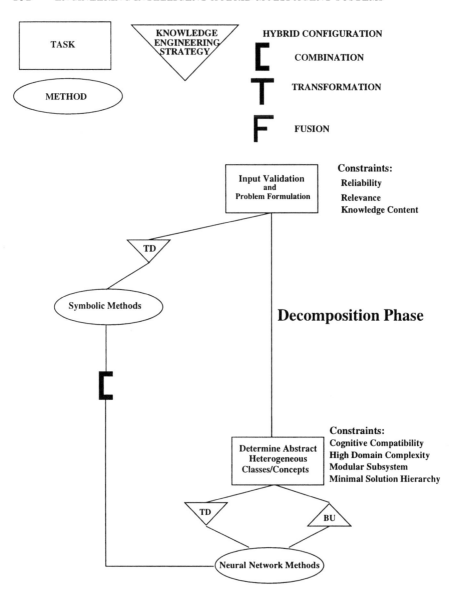

Figure 6.8.   TSL Architecture - Decomposition Phase

they do not provide a direct or immediate solution to the problem. Two distinct tasks are performed in the decomposition phase:

- input validation & problem formulation

- determination of abstract classes

The input validation in this phase can involve context validation of the input data i.e. whether the input data is consistent with the domain space being modeled and is consistent with the input variables in the domain space. The 'Relevance constraint shown in Figure 6.8 implies that it is essential to determine the contextual relevance of data.

Problem formulation can involve conversion of input data into a state which facilitates its use by a particular method, imposing constraints of time on the solution and nature of communication with its environment given a particular context, and determining the execution (or plan). Input validation and problem formulation tasks need to be accomplished reliably as the quality of the solution impinges on it. The knowledge content of this task is invariably symbolic and thus symbolic methods are used to accomplish this task. The second distinct task in the decomposition phase is determination of abstract concepts/classes which are not directly related to the problem to be solved and are not explicitly used or do not explicitly exist in the domain/problem space being addressed. They are abstractions of the existing domain classes and are used to reduce the complexity of problem on hand. These abstract concepts elementarily classify a domain and at the same time reduce the complexity of the problem in hand. From a cognitive view point, humans are very quick and accurate in elementary classification in various domains. In other words, elementary classification or decomposition generally forms a part of our unconscious or automated behavior which is best reflected by neural networks. More so, constraints like concept autonomy, independent subsystems (Aminzadeh and Jamshidi 1994; Fodor et al. 1983), and minimal solution hierarchy require consideration of multiple number of features simultaneously. Thus, the TU knowledge engineering strategy is to use neural networks for accomplishing this task. At the same time neural networks can also be used to learn these abstract classes using neural networks in a BU manner. It may be noted here that neural networks used in different phases of the TSL architecture represents different abstractions of a problem domain and thus may use different input features or types of data in order to learn the different abstractions. This point will become more clear when the knowledge content of these methods is discussed at the computational level in the next chapter.

The intention of this phase is to create abstract classes in a domain which are independent of each other. However, it is possible that in some situations or problems this may not be the case. That is, in some situations a certain degree of interaction may be required between the abstract classes to solve a given problem. In such situations, the decomposition phase also assumes the role of global control and resolution of conflicts arising out of the interaction between these abstract classes.

6.3.3 Control Phase

Control Phase operates within each abstract class. In the control phase the following tasks are accomplished:

- local noise filtering

- input validation and problem formulation

- viability

- determination of decision level classes

- resolve conflicting outcomes of decision level classes

6.3.3.1   Local Noise Filtering, Input Validation & Problem Formulation. The local noise filtering, input validation, and problem formulation tasks are similar to those done in the global preprocessing phase and decomposition phase except that instead of doing these tasks at the global level, they are now done at the local level within each abstract class.

However, in a number of incremental learning and adaptive systems (e.g. robotic control) this high level knowledge needs to be extracted for new situations for input validation, problem formulation (and planning). Thus BU knowledge engineering strategy will have to employed as shown in Figure 6.9 where new skills are being learnt all the time. Neural networks can be used for rule extraction in and genetic algorithms can be used for optimizing the neural network structure, and learning the connection weights.

6.3.3.2   Determine Decision Level Classes. The third task is to determine the decision level classes within each abstract class. Decision level classes are those classes inference on which is of importance to a problem solver. These decision-level classes generally explicitly exist in the problem being addressed. These decision level classes could represent a set/group of network components or individual components in a telecommunication, electric power network problem, or a possible set of control actions in a control application. In fact, tasks like context validation, and problem formulation assume more significance in this phase as the modeling of the underlying domain control structure is context dependent. It will be shown during the application of the TSL architecture in chapter 11 that the domain structure can be modeled in different ways based on different contexts. The knowledge content in this phase for local noise filtering,

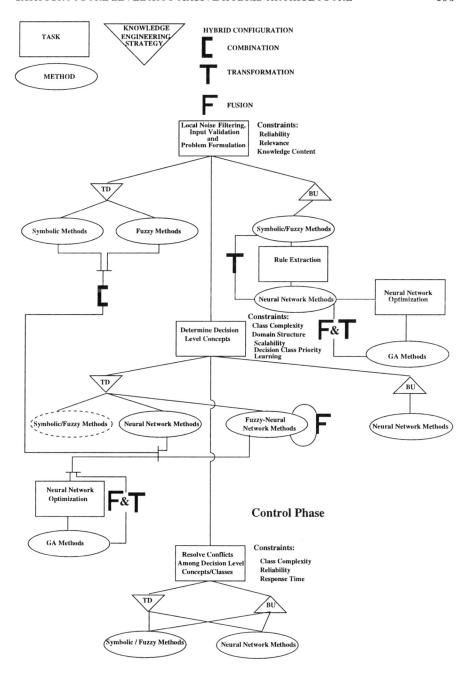

Figure 6.9.   TSL Architecture - Control Phase

input validation, and problem formulation tasks can be symbolic and/or fuzzy because we are now determining differences and variations within a particular subsystem or class.

The granularity of a decision level class can vary between coarse and fine. The coarsity and the fineness of a decision level class depends on the context in which problem is being solved and the decision level class priority in a given context. In one context, a decision level class may be less important to a problem solver, and thus a coarse solution may be acceptable, whereas, in another context the same decision level class may assume higher importance and thus a finer solution may be required. That is, if the decision level class priority is low, then its granularity is coarse, and the problem solver is satisfied with a coarse decision on that class. Otherwise, if the decision level class priority is high then the decision-level class has fine granularity and problem solver wants a fine set of decisions to be made on the decision-level class which would involve a number of microfeatures in the domain space. Thus in a given context if the decision-level class has coarse granularity the symbolic and/or fuzzy methods can do the job [1].

However, if the the decision level class has fine granularity then neural networks, fuzzy-neural networks can be used. From a cognitive viewpoint, neural network methods can be seen as learning the decision-level classes from the micro features and determine which decision-level classes in the next phase (i.e. decision phase) need to be activated. Genetic Algorithms can be used for optimization of neural networks in terms of optimizing the training data set, learning the connection weights, and topological refinement. The use of fuzzy-neural networks implies fusion of fuzzy logic with neural networks in terms of fuzzy membership functions and/or inferencing procedures. The reason for symbolic and fuzzy nature of knowledge will become more clear in the next chapter when learning knowledge and types of input features in this phase are discussed. The BU knowledge engineering strategy can be employed for learning the fuzzy membership functions or extracting new decision class/es using neural networks.

Further, because of the difference in the nature of tasks involved in the control phase and decomposition phase, the nature of communication between decomposition phase and control phase is different from the nature of communication between control phase and decision phase.

6.3.3.3   Resolve Conflicting Outcomes of Decision Level Classes. It is possible that the decisions made by a decision level class may conflict with the decisions by another decision level class. For example, let two decision level classes represent decisions on two telecommunication network components.

Then if each decision level class predicts that the other network component is faulty, then their is a conflict.

The conflict may be resolved by looking at their structural indisposition in the problem domain space or through other means. The conflicts can also occur with respect to previous knowledge or in situations that involve temporal reasoning. In case of temporal reasoning the previous result may become invalid or conflict with the result based on new data. This high level task is generally accomplished accomplished through symbolic methods unless the use of neural networks is necessitated because of the large number of variables involved.

6.3.3.4 Viability. In some real time systems it may become necessary to compute the computational resources and the time required by different decision level classes to determine the solution. Thus, certain decision level classes may not be considered viable under these constraints and thus may not be activated. For accomplishing this task symbolic or fuzzy methods are used.

6.3.4 Decision Phase

In the decision phase the following tasks are undertaken:

- noise elimination and input validation

- determine or predict specific classifications with each decision-level class

Noise elimination, and input validation tasks are similar to those in the previous phases except that they are now accomplished at the decision class level.

6.3.4.1 Determine or Predict Specific Classifications. The task is to determine or predict the specific classifications within each decision-level class or instances of a decision-level class. These predictions may be a specific fault in a component or a group of components or predicting the outcome of a control action by a controller.

Cognitively, it is in the decision phase that humans engage in generalization. For example, in a power system control center, the operator is accurate and quick in detecting whether a set of alarms are circuit breaker(CB) alarms or communication alarms, or whether they are 220kv CB alarms or 66kv CB alarms. However, the operator generalizes on whether these set of alarms depict a bus fault or multiple line fault. That is, if a bus fault occurs as a result of eight CB alarms in a bus section, the operator is likely to generalize the same fault for six CB alarms in real time. This is largely due to fine granularity and the distributed nature of micro features in this phase. Besides constraints like response time and others as shown in Figure 5.10 also need to be considered in this phase. Thus neural networks or fuzzy-neural networks (Tang, Dillon

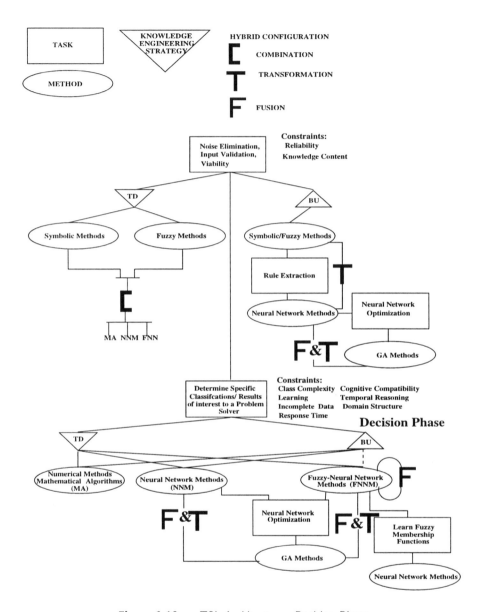

Figure 6.10.   TSL Architecture - Decision Phase

and Khosla 1995a, 95b) which have prediction, generalization, and noise tolerance properties can be used in this phase. can be used in this phase in order to determine specific classifications. These generalizations, however, need to validated. This is done in the post-processing phase.

Mathematical algorithms can also be used in this phase if the prediction or classifications can be mathematically modeled.

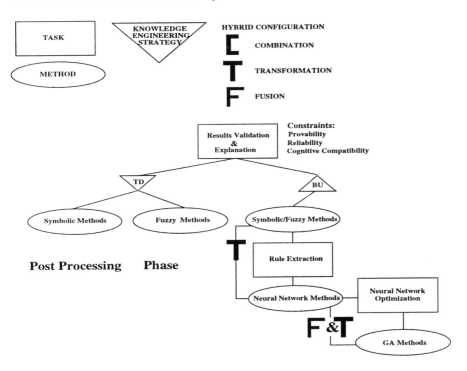

Figure 6.11.   TSL Architecture - Post Processing Phase

6.3.5   Postprocessing Phase

Postprocessing phase (shown in Figure 6.11) represents logic and provability which are the hallmarks of our conscious reactions to the external environment. Thus, in the postprocessing phase symbolic methods are used to validate or evaluate decisions made in decision phase. This is done by using symbolic or fuzzy cross checking modules for various decision level classes. These modules can be fault propagation model/s of or other application dependent cross-checking modules. In case these models are not available they can be built for each decision level class by employing BU knowledge engineering strategy and using neural networks.

In situations where these models are difficult to build the postprocessing phase can be seen as a feedback loop (e.g. in control applications). That is, decision validation can be provided by a human agent or the environment in which the system is operating. For example, a control action taken by the inverted pendulum control system may result in the pole balancing or falling over. This result is the feedback from the environment validating or invalidating the control action by the inverted pendulum control system.

6.4   SOME OBSERVATIONS

The TSL architecture described in the preceding section has outlined the information processing phases at different levels of information processing, tasks in each phase, knowledge engineering strategy, constraints on accomplishment of the tasks, symbolic, fuzzy, neural network, and genetic programming methods employed to accomplish different tasks, and the hybrid arrangement between various intelligent methods. The information processing in the TSL architecture can be seen to represent a deliberative reasoning structure at the global level, whereas reasoning within each phase is a mixture (with varying granularity) of analytical and automated methods. Thus it does not entirely rely on the explicit deliberation of classical architectures like SOAR (Laird et al. 1987), Robo-SOAR (Laird et al. 1991), THEO (Mitchell 1990), and their descendants which have been found wanting in terms of their real-time behavior. It may be noted that one of the roles of the BU knowledge engineering strategy is to enhance the adaptive properties of the TSL architecture. Further, the TSL architecture is independent of the type of neural network/symbolic/fuzzy/genetic methods to be used or the type of learning techniques/algorithms to be employed.

The TSL architecture can be seen as representing an open loop system as well as a closed loop system. In problems where logical and provable models of the various problem components exist the TSL architecture can be seen as an open loop system. In these systems, it does not need any feedback from its environment to solve the problem. However, in problems where logical and provable models of the various problem components do not exist or are too complex to be built the TSL architecture can be seen as a closed loop system. In these systems feedback is provided by the human agent (user) or other external agent especially to validate the decisions made in the decision phase.

The preceding paragraphs in this section have made observations on what can be derived from the TSL architecture. There are some important problem solving aspects which have not been outlined in the TSL architecture. It has not been determined in the TSL architecture, what are the different knowledge constructs to be employed by the different methods in order to accomplish the tasks in different phases. Given the TSL architecture it needs to be determined now what kind of learning knowledge is to be employed in different phases of the architecture?, what kind of communication mechanisms are to be employed between different phases?, etc. All in all symbolic knowledge representation constructs, belief knowledge, learning knowledge, communication knowledge constructs, and other ontological constructs need to be outlined. Besides this, some constraints like knowledge sharing and maintenance from a user's perspective also need to be addressed. This is done at the computational level, which will be described in the next chapter.

6.5 SUMMARY

In this chapter the TSL intelligent associative hybrid architecture has been described for complex data intensive domains. The task structure level architecture is defined in terms of the information processing phases, tasks to be accomplished in each phase, constraints which need to be satisfied for accomplishing a task, and methods used to accomplish the tasks in different phases. The different aspects of the task structure level architecture have been derived from various perspectives including neurobiological control, cognitive science, levels of intelligence, physical systems, learning, forms of knowledge, user, and others. The TSL architecture is independent of the implementation details of the methods and the learning techniques employed. The vocabulary to be employed by the various intelligent methods in order to accomplish the tasks form a part of the computational level architecture. The computational level architecture and its vocabulary are described in the next chapter.

Notes

1. In fact in such circumstances a distinct control and decision phase may not be required and can be merged into one

References

Arbib, M. A. (1987, *Brains, Machines and Mathematics*, Springer Verlag, New York.

Beale, R. and Jackson, T. (1990), *Neural Computing: An Introduction*, Bristol: Hilger.

Bezdek, J.C. (1994), "What is Computational Intelligence?," Computational Intelligence Imitating Life, IEEE Press, New York.

Chandrasekaran, B. (1990), "What kind of information processing is intelligence," *Foundations of Artificial Intelligence*, Cambridge University Press, pp14-45.

Chandrasekaran, B, Johnson, T. R. and Smith, J. W. (1992), "Task-Structure Analysis for Knowledge Modeling," *Communications of the ACM*, September, vol. 35, no. 9., pp. 124-137.

Churchland, P.M. (1990), "Representation and High-Speed Computation in Neural Networks," *Foundations of Artificial Intelligence*, Cambridge University Press, USA.

Fieldman, J. (1988), "Neural Representation of Conceptual Knowledge," *Neural Connections, Mental Computation*, MIT Press, pp. 69-103.

Fodor, J. A. (1983), *The Modularity of Mind*, MIT Press.

Fodor, J.A., and Plyshyn, Z. W. (1988), "Connectionism and Cognitive Architecture: A Critical Analysis," *Cognition*, 28:2-71.

Khosla, R. and Dillon, T. (1992), "A Neuro-Expert System Architecture with Applications to Alarm Processing in a Power System Control Centre," *Fourth International IEEE conference on Tools for Artificial Intelligence*, Arlington, Virginia, USA.

Khosla, R. and Dillon, T. (1992b) "An Integrated Neuro-Expert System Model for Real-Time Systems" *International Joint Conference on Neural Networks, IJCNN'92*, Beijing, China, vol. 1, pp. 154-159.

Khosla, R. and Dillon, T. (1993) "Task Decomposition and Competing Expert System-Artificial Neural Net Objects for Reliable and Real Time Inference," *International IEEE Conference on Neural Networks*, San Francisco, USA, pp794-800, Mar/Apr.

Khosla, R. and Dillon, T. (1994a) "Learning Knowledge and Strategy of a Generic Neuro-Expert System Model," *AAAI International Symposium on Integrating Knowledge and Neural Heuristics*, Florida, USA.

Khosla, R. and Dillon, T. (1994b), "Cognitive and Computational Foundations of Symbolic-Connectionist Integration," *ECAI'94 Workshop Proceedings on Symbolic-Connectionist Processing*, Netherlands, Holland.

Khosla, R. and Dillon, T. (1994c) "Constructs for Building Complex Symbolic-Connectionist Systems," Sixth IEEE conference on Tools with Artificial Intelligence, New Orleans, Lousiana, USA.

Khosla, R. and Dillon, T. (1994d) "A Distributed Real-Time Alarm Processing System with Symbolic-Connectionist Computation," *Proceedings of IEEE Workshop in Real Time Systems*, July, Washington D.C., USA.

Khosla, R. and Dillon, T. (1995), "Task Structure and Computational Level: Architectural Issues in Symbolic-Connectionist Integration," *Working Notes of IJCAI95 Workshop on Connectionist-Symbolic Integration: From Unified to Hybrid Approaches*, Montreal,Canada.

Khosla, R. and Dillon, T. (1997), "Learning Knowledge and Strategy of a Generic Neuro-Expert System Architecture in Alarm Processing," to appear in *IEEE Transactions in Power Systems*.

Laird, J., Rosenbloom, P. and Newell, A. (1987), "SOAR: An Architecture for General Intelligence," *Artificial Intelligence*, vol. 33, pp. 1-64.

Laird, J. E., Yager, E.S., Hucka, M., and Tuck, M. (1991), "Robo-Soar: An integration of External Interaction, Planning, and Learning using Soar," *Robotics and Autonomous Systems*, vol. 8, pp. 113-129.

Lallement, Y., Hilario, M., and Alexandre, F. (1995), "Neurosymbolic Integration: Cognitive Grounds and Computational Strategies," *World Conference on the Fundamentals of Artificial Intelligence*, Paris, France.

Lallement, Y., and Alexandre, F. (1995), "Cognitive Aspects of Neurosymbolic Integration," *Working notes of the Workshop on Connectionist-Symbolic Integration: From Unified to Hybrid Approaches*, Montreal, Canada, pp. 7-11.

Malsburg, V. D. (1988), Goal and Architecture of Neural Computers, *Neural Computers*, Springer Verlag, Berlin, West Germany.

McClelland, J. L., Rumelhart, D. E. and Hinton, G.E. (1986), "The Appeal of Parallel Distributed Processing," *Parallel Distributed Processing*, vol. 1, pp. 3-40.

McClelland, J. L., Rumelhart, D. E. and Hinton, G.E. (1986), *Parallel Distributed Processing: Explorations in the Microstructure of Cognition*, vol. 2, Cambridge, MA: The MIT Press.

Medskar, L. A. (1995), "Hybrid Intelligent Systems," Kluwer Academic Publishers, Massachusetts, USA.

Minsky, M., (1991), "Logical Versus Analogical or Symbolic Versus Connectionist or Neat Versus Scruffy," *AI Magazine*, 12(2), pp. 34-51.

Mitchell, T. M. (1990), "Becoming Increasingly Reactive (Mobile Robots)," *Proceedings of the Eighth National Conference on Artificial Intelligence (AAAI90)*, vol. 2 pp. 1051-1058.

Newell, A., and Simon, H. A. (1972), "The Theory of Human Problem Solving," *Human Problem Solving*, Englewood Cliffs, NJ:Prentice Hall.

Newell, A. (1980), "Physical symbol Systems," *Cognitive Science*, vol. 4, pp. 135-183.

Newell, A. and Rosenbloom, P. S. (1981), "Mechanisms of Skill Acquisition and the Learning in Practice," *Cognitive Skills and their Acquisition*, J. R. Anderson (Ed.), Hillsdale, NJ:Erlbaum

Newell, A. (1981), "The Knowledge Level," *AI*, pp. 1-19.

Newell, A. (1990), *Unified Theories of Cognition*, Harvard University Press, Cambridge, Massachusetts, USA

Rao, R.P.N. and Ballard, D. (1995), "Dynamic Model of Visual Memory Predicts Neural Response Properties In The Visual Cortex," *Technical Report*

95.4, National Resource Laboratory for the study of Brain and Behavior, Department of Computer Science, University of Rochester, Rochester, USA.

Russell, S., and Norvig, P. (1995), *Artificial Intelligence - A Modern Approach,* Prentice Hall, New Jersey, USA.

Shepard, R.N., 1988, "Internal Representations of Universal Regularities: A Challenge for Connectionism," *Neural Connections, Mental Computation,* MIT Press, pp. 104-135.

Smolensky, P., (1990) "Connectionism and Foundations of AI," *Foundations of Artificial Intelligence,* Cambridge University Press, pp. 306-327.

Soucek, B., and the IRIS Group (eds) (1991), *Neural and Intelligent Systems Integration,* John Wiley and Sons, New York, USA.

Sun, R., (1991), "Integrating Rules and Connectionism for Robust Reasoning," Brandeis University, Computer Science Department, Waltham, MA, TR-CS-90-154, January, USA.

Tang, S. K., Dillon. T. and Khosla, R. (1995a), "Fuzzy Logic and Knowledge Representation in a Symbolic-Subsymbolic Architecture," *1995, IEEE International Conference on Neural Networks,* Perth, Australia

Tang, S. K., Dillon. T. and Khosla, R. (1995b) "Application of an Integrated Fuzzy, Knowledge-Based, Connectionist Architecture for Fault Diagnosis in Power Systems," *Sixth IEEE Intelligent Systems Applications Conference,* Florida, USA.

7 INTELLIGENT MULTI-AGENT HYBRID COMPUTATIONAL ARCHITECTURE - PART I

7.1 INTRODUCTION

An architecture is an information processing mechanism that can operate on the information represented in a form that is specific to the architecture (Chandrasekaran 1990). In the last chapter the TSL associative hybrid architecture has been described. In this chapter the primary goal is to outline the framework of the computational level architecture. This is done through the integration of the object-oriented model, agent model, and distributed operating system process model with the TSL architecture. Through this integration, an Intelligent Multi-Agent Hybrid Distributed Architecture (IMAHDA) with object-oriented properties is realized. IMAHDA at the computational level is built on four layers, namely object layer, software agent layer, intelligent agent layer, and problem solving agent layer. The PAGE descriptions of the software agent, intelligent agent, and problem solving agent layers are also outlined.

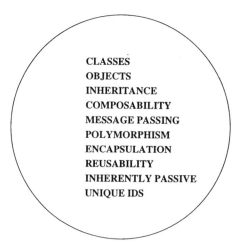

CLASSES
OBJECTS
INHERITANCE
COMPOSABILITY
MESSAGE PASSING
POLYMORPHISM
ENCAPSULATION
REUSABILITY
INHERENTLY PASSIVE
UNIQUE IDS

Figure 7.1.   The Object-Oriented Model

7.2 OBJECT-ORIENTED MODEL

Knowledge is generally organized into hierarchies with a twofold goal: devise a conceptually clean model of the represented world, and provide a compact storage organization which may also enable easier navigation through and search of knowledge. Various knowledge representation formalisms like frames, semantic networks, and objects have been developed over decades of symbolic AI research. These three formalisms as discussed in chapter 2 are based on hierarchical configurations of symbolic knowledge representation. An object-oriented model (see Figure 7.1) derives itself cognitively from human cognition related studies in symbolic knowledge representation and computationally from the research in object-oriented programming, object-oriented software engineering and databases (Rosch et al. 1976, 77; Whitfield et al. 1979; Murphy et al. 1985; Britannica 1986; Cox 1986; Booch 1986; Slatter 1987; Kim et al. 1988; Myer 1988; Unland et al. 1989; Coad et al. 1990, 92; Rumbaugh 1990; Dillon et al. 1993 and others).

In chapter 2 various characteristics of the object-oriented model have been described. These include inheritance, composability and other non-hierarchical relationships between various concepts from a modeling viewpoint. On the other hand, from a software implementation viewpoint object-oriented methodology provides attractive features like strong encapsulation, polymorphism and reusability. A common set of characteristics which define the general object-oriented model are shown in Figure 7.1 These characteristics are: unique object identifier, data[1] and operations, encapsulation, inheritance, composition, message passing, polymorphism and reusability. Here those characteristics of the object-oriented model are looked into which allow IMAHDA to identify problem domain concepts and their relationships.

7.3 AGENT MODEL

The agent model shown in Figure 7.2 is derived from the "agent and agent architectures" section in chapter 2. Figure 7.2 shows some of the characteristics of an agent model as distinct from an object-oriented model. An important point to note here is that the agent model is primarily driven by task abstraction and task-oriented behavior. On the other hand, the object-oriented model is primarily driven by structuring objects in the the real world and identifying relationships between them. Also an agent is dynamic by nature as against classes and objects which are passive by nature. Further, an agent facilitates task based communication and defines the nature of communication between different agents which can be intricate and complex (e.g. negotiation) depending upon the application domain.

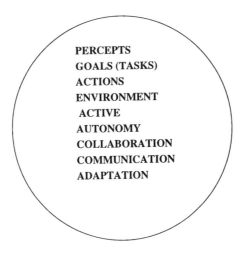

PERCEPTS
GOALS (TASKS)
ACTIONS
ENVIRONMENT
ACTIVE
AUTONOMY
COLLABORATION
COMMUNICATION
ADAPTATION

Figure 7.2.   The Agent Model

Besides the agent features outlined in this section, and object-oriented features outlined in the preceding section, a computational framework where the other characteristics like distributed control and parallelism can be realized is also needed. The computational framework should also enable effective realization of real time constraints like fast response time, reliability and temporal reasoning. A distributed operating system process model which satisfies some of these requirements is described in the next section.

7.4 DISTRIBUTED OPERATING SYSTEM PROCESS MODEL

Distributed control and parallelism are two activities which can be effectively simulated in a multi-process operating system environment. The term *process* in the operating system process model shown in Figure 7.3 denotes a program or component of a program which can be independently scheduled by an operating system in order to accomplish some task (Bourne 1983). In a single process design, all of the processing capabilities of a program are encapsulated into a single package. A multi-process application distributes processing services across several processes and may use operating system communication facilities to pass results and data among the processes. Communication facilities allow processes to execute and communicate across process or machine boundaries.

AUTONOMY
CONCURRENCY
INTER PROCESS COMMUNICATION
ENCAPSULATION
DISTRIBUTED PROCESSING
INHERENTLY ACTIVE
UNIQUE PROCESS IDS

Figure 7.3.   The Process Model

In multi-process environments the interaction among processes must be co-ordinated in order to prevent unsynchronized access and updates of data. One of the effective ways of coordinating interaction among processes in a multi-process environment is through inter-process communication.

Pipes are an effective means of communication in a multi-process environment (Christian 1988). A pipe is a communication channel which couples one process to another. A process can send data 'down' the pipe by using the *Write* system call, and another process can receive the data by using the *Read* system call at the other end.

Thus firstly, a multi-process architecture makes provision for concurrent and distributed processing of information. Secondly, the ability to simulate concurrent execution is conducive to the idea of competition for enhancing the computation reliability in a real time domain. Thirdly, inter-process communication provides an effective means for synchronous and asynchronous communication, cyclic and continuous operation in a real time and dynamic environment. Fourthly, distributed and parallel processing, synchronous and asynchronous communication facilitate hierarchical and autonomous control over inferencing done by various processes. The inter-process communication channels make provision for hierarchical control in terms of what type of data is going to be processed by the lower level process. The parallel and distributed processing feature contributes to the autonomous control each process has over its inferencing. These features also provide more flexibility for dealing with temporal reasoning constraints associated with real time systems.

Operating System Process (OSP) model like Object-Oriented (O-O) model encapsulates data and operations in the process paradigm.

Objects in an object-oriented system generally reside in a class hierarchy, while processes are not organized in this manner. Instead, a process is an independent entity which can initiate its own activity. Objects in the object-oriented model are typically passive, becoming active only when requested to do so (some classes may still remain passive as they exist only for facilitating inheritance). Processes are inherently active and become passive only when necessary.

Communication ports provide the vehicle for message passing in the process model. Ports may be typed in order to affect the manner in which messages may be sent and delivered. In contrast, typing exists throughout the object-oriented model, both internally with respect to the vehicles manipulated by the operations, and externally with respect to the messages and class structures supported by the model.

The OSP and O-O models differ in their overall architecture within a multi-processing environment. In the object-oriented model, all data and operations normally reside in a global object structure. The process model encapsulates only the operations and data used by a single process: no global structure which organizes a process exists.

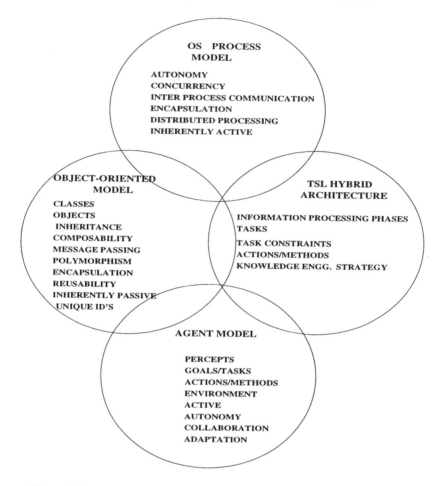

Figure 7.4.   Integration of O-O, Agent, OSP Models and TSL Architecture

7.5 COMPUTATIONAL LEVEL MULTI-AGENT HYBRID DISTRIBUTED ARCHITECTURE

In order to realize the associative hybrid architecture at the computational level, the TSL architecture is integrated with object-oriented, agent, and Operating System Process (OSP) models respectively as shown in Figure 7.4. In the integration process similar characteristics like communication, typing and encapsulation facilitate integration. On the other hand, complimentary characteristics like tasks and task-oriented behavior vs structured objects and relationships, inherently passive vs active, fixed class hierarchy vs autonomy and distribution bring about some of the enhancements as a result of integration. The result of this integration is an Intelligent Multi-Agent Associative Hybrid Distributed Architecture (IMAHDA) as shown in Figure 7.5. Figure 7.5 also provides a PAGE view of different problem solving agents of IMAHDA, and of IMAHDA as a whole. It provides a perspective of the percepts, goals, actions, and environment as perceived at the architecture level. In Figure 7.5 belief base agents have also been added in IMAHDA. These belief base agents represent the internal state of an agent (e.g. Preprocessing, Decision Agents) as they interact with the dynamic, and non-deterministic environment. The preprocessing, decomposition, control, decision, and postprocessing agents are also known as the problem solving agents of IMAHDA.

IMAHDA as shown in Figure 7.5 is built on four layers, namely, object, software agent, intelligent agent, and problem solving agent layers respectively (see Figure 7.6). The role of the object layer is to primarily capture the problem domain objects and define their structural relationships. The software agent, intelligent agent and problem solving agent layers are described in the next section.

7.6 AGENT BUILDING BLOCKS OF IMAHDA

Looked at more closely, IMAHDA can be seen as being constructed from software, intelligent, and problem solving agents as shown in Figure 7.7. The three tier agent structure consists of generic software agents, generic intelligent agents, and the problem solving agents shown in Figure 7.5. In this section, these software, intelligent, and problem solving agents are discussed.

7.6.1 Generic Software Agents

The PAGE description of the software agents, namely, Distributed Processing, Communication, Belief Base, and Relational is shown in Table 7.1. The agent structure gives these software agents a task orientation.

7.6.1.1   Generic Distributed Process Agent.   A process in an operating system environment is simply an instance of an executing program, corresponding to the notion of task in other environments. The list of percepts, actions, and goals shown in Table 7.1 may not be complete as the PAGE description relevant to IMAHDA has only been included. The generic action *Create New Process* is a basic process creation primitive. As the name suggests, a successful call to *Create New Process* causes the kernel in a Unix environment to fork a new process subject to available system resources. The newly created process and its parent process can both execute concurrently. The *Exit*, and *Kill* ac-

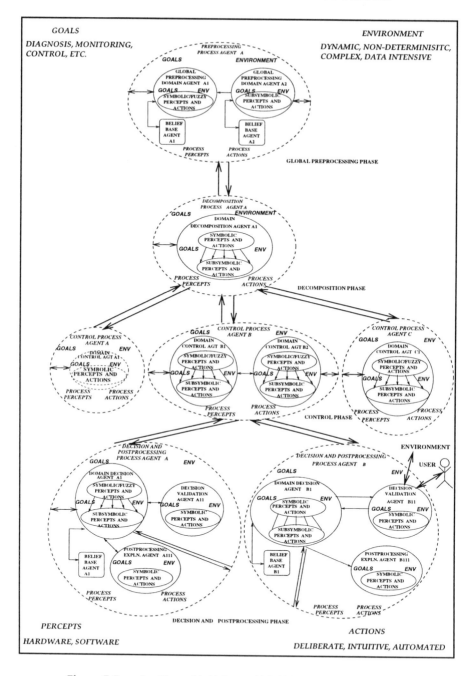

Figure 7.5.   Intelligent Multi-Agent Hybrid Distributed Architecture

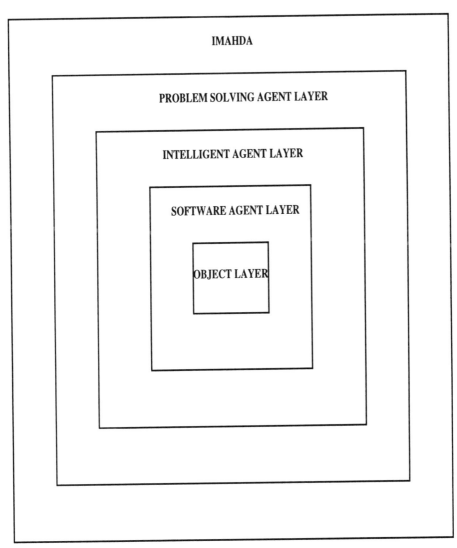

Figure 7.6.   Object, Software, Intelligent & Problem Solving Agent Layers of IMAHDA

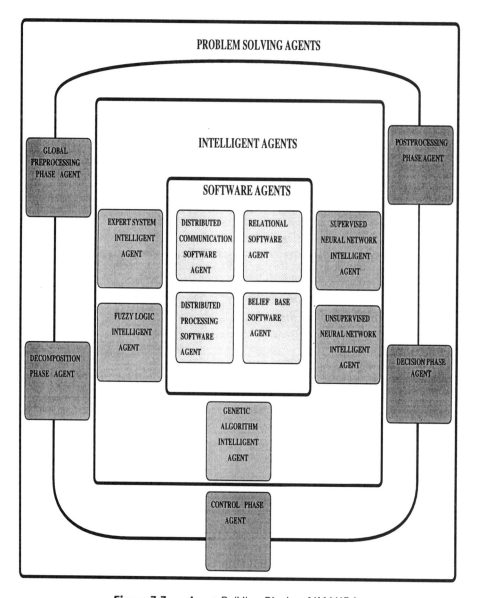

Figure 7.7.   Agent Building Blocks of IMAHDA

tions are used to terminate a process. The *Get Process ID* action is used to get the *Process ID* of the process to be terminated. Although, the distributed processing agent is explicitly derived from a Unix OS platform, its PAGE description and its implications are valid for other operating system platforms also.

7.6.1.2 Communication Software Agent. The goal of the communication agent is to facilitate distributed communication between problem solving agents in an hierarchical as well as lateral fashion. The *Write* and *Read* actions are used for inter-process communication. The distinctions in these methods between parent and child have been made for ease of implementation. The

Table 7.1. Generic Software Agents

AGENT	PERCEPTS	ACTIONS	GOALS	ENVIRONMENT
Distributed Processing	System Resources	Create New Process Exit, Kill, Get Process ID	Concurrency, Competition, Collaboration	MultiProcessor Dynamic
Distributed Communication	Pipes, Sockets, Process ID, Process State	Read, Write,	Hierarchical Communication, Lateral/Peer Communication	Distributed Processes Dynamic
Belief Base	Discrete, Fuzzy, Continuous inputs , Outputs, Inferences, Time.	Store, Retrieve Delete, Modify, Other User-defined Operations	Keep Track of Internal State	Dynamic
Relational	Agents, Sub-Agents Objects, Classes	Create ISA Link Create ISAPARTOF Link, Delete Link	Inheritance, Composition, Non-hierarchical relations	Dynamic

Process State signifies the different states of a process, namely, active, waiting/suspended state and inactive state. The pipes and/or sockets are used for writing and reading purposes.

7.6.1.3 Belief Base Software Agent. The belief base agent shown in Table 7.1 keeps track of the internal state of a problem solving agent in a dynamic and non-deterministic environment. The different types of inputs stored in the belief base agent represent the most recent snap shot of the environment. The outputs and inferences represent the results of the problem solving agents based on the recent snapshot of the environment. Actions like *Retrieve, Modify*, etc. are carried out to retrieve the previous state, and update it based on the new state. Besides the actions shown in Table 7.1, more user-specific actions can also be included.

7.6.1.4 Relational Software Agent.

The problem solving agents may inherit some of their knowledge and actions from different generic intelligent agents or other problem solving agents in the environment to accomplish their goals. The goal of the relational software agent shown in Table 7.1 is to facilitate this process. The actions *Create ISA Link*, and *Create ISAPARTOF Link* create inheritance and composition links between different agents and objects (in the object layer). Further, inheritance and composition links also enhance the knowledge representation capabilities of a problem solving agent. In a more general sense the relational agent can also be used to provide services for establishing hierarchical and non-hierarchical relations between objects. It can also be used for modeling the relational behavior as crisp or fuzzy depending upon the strength of the link.

7.6.2 Generic Intelligent Agents

The PAGE descriptions of generic intelligent agents, namely, Expert System, Supervised Artificial Neural Network, Unsupervised Neural Network, Fuzzy Logic, and Genetic Algorithm Agent are shown in Table 7.2. It may be noted that the genericity in the intelligent agents has been primarily defined in the context of IMAHDA (although it is likely to be equally applicable in the general case).

7.6.2.1 Generic Expert System Agent.

Expert systems have been used successfully for systems with symbolic reasoning. In Table 7.2, the PAGE description of the expert system agent captures its role in IMAHDA. Its goal is to enable the problem solving agents in IMAHDA to do high level reasoning like problem formulation and context validation. The generic action like *Rules* is used for inferencing on a given set of percepts. *Initialize Rule Variables* is used to initialize the rule variables based on the internal state of a problem solving agent. The Actions like *Loadkb* and *UnLoadkb*, as their name signifies, are used for loading (into working memory) and unloading (removing from working memory) the knowledge base of an *Expert System* agent. The percept *KBName* uniquely identifies the knowledge base name of a particular instance of the generic *Expert System* agent. Thus the generic Expert System agent provides an environment to a problem solving agent to build application dependent expert or knowledge based system agents.

It may be noted here that such an agent orientation can result in improving the modularity and reusability of the rule base. It can support the development of more consistent and homogeneous knowledge bases. Although, Table 7.2 only shows the PAGE description of a rule based architecture, model based, and other architectures can be similarly described and used.

7.6.2.2   Generic Supervised/Unsupervised Neural Network Agent.
The percepts of the supervised artificial neural network agent, and unsupervised artificial neural network agent shown in Table 7.2 highlight the basic difference between the two generic agents. The percepts of the supervised neural network agent involve both input and output symbols or patterns, whereas the percepts of the unsupervised neural network agent involve only the input patterns. The supervised neural network agent may engage in classification, whereas the unsupervised neural network agent may engage in clustering. The actions associated with the Supervised Neural Network Agent are applicable to multilayer perceptron using a backpropagation algorithm and other supervised learning algorithms. Similarly, the actions associated with the Unsupervised Neural Network Agent are applicable in neural network architectures like Kohonen nets (Kohonen 1990), and ART (Carpenter and Grossberg 1988). Other neural network architectures can also be modeled on a similar basis as described in this section.

7.6.2.3   Generic Fuzzy Logic Agent.
The actions in the fuzzy logic agent to a large extent represent the problem independent part of fuzzy logic. The problem dependent part is represented by the percepts like fuzzy membership functions, linguistic symbols, and structure of the fuzzy antecedent and consequent. The action *Fuzzy Inferencing* can represent min-max operations, compositional inference, and other forms of fuzzy inference. The *Defuzzify Data* can employ most commonly used centre of gravity method or other methods for defuzzifying data.

7.6.2.4   Generic GA Agent.
The generic GA agent shown in Table 7.2 is used for optimization of fuzzy rules, learning data sets, neural network structure, etc.. The action *Create Gene Type* is used for representing the percept/s as a bit, integer, floating point, or as a case chromosome. The actions *Initialize Population, Select Parents, Reproduce, Evaluate Offspring*, and *Replace Parents* represent the GA agent optimization life cycle. The *Select Parents* action selects parents who can possibly produce a fitter offspring. The *Reproduce* action involves crossover and mutation operators to produce the new offspring. The *Evaluate Offspring* action evaluates a offspring based on a given fitness function percept. Finally, *Replace Parents* replaces certain parents with the new or fitter offspring.

Table 7.2.   Generic Intelligent Agents

AGENT	PERCEPTS	(some)ACTIONS	GOALS	ENVIRONMENT
Expert/Knowledge Based System	Symbols, Antecedent Consequents, Hypothesis KBName	Rules, Load Knowledge Base, Unload Knowledge Base, Initialize Rule Variables	High Level Reasoning	Deterministic
Supervised Neural Network	Symbols(IO), Fuzzy/Linguistic Symbols(IO), Continuous Data(IO), Network Parameters	Normalize Input Data, Net Configuration, Train, Test & Validate	Learning, Automated Behavior, Modeling, Classification, Prediction, Pattern Recognition, Adaptation, Optimization, Rule Extraction	Non-Deterministic Dynamic
Unsupervised Neural Network	Symbols(I), Fuzzy/Linguistic Symbols(I), Continuous Data(I), Network Parameters	Normalize Input Data, Get Net Configuration, Train, Test & Validate	Learning, Clustering, Modeling, Prediction	Non-Deterministic Dynamic
Fuzzy Logic	Linguistic Symbols, Fuzzy Membership Functions, Fuzzy Antecedents, Fuzzy Consequent	Fuzzify Data, Fuzzy Inferencing Defuzzify Data	Human-like Approximate Reasoning, Fuzzy Knowledge, Representation, DeFuzzification	Non-Deterministic
Genetic Algorithm (GA)	Symbols, Fuzzy Rules, Neural Network Weights, Learning Data Set, Neural Network Structure, GA Parameters, Fitness Function	Create Gene Type, Select Parents, Intialize Population, Produce Offspring, Evaluate Offspring, Replace Parents	Optimization	Non-Deterministic

7.6.3   Problem Solving Agents

In order to provide a comprehensive view, the TSL associative hybrid architecture is looked upon as consisting of a set of problem solving agents. The PAGE

Table 7.3. PAGE Description of Problem Solving Agents of IMAHDA

AGENT	PERCEPTS	(SOME) ACTIONS	GOALS	ENVIRONMENT
Global Preprocessing	discrete, fuzzy, continuous data	Transform Data, Apply Global Noise Filtering Heuristics, Formulate Execution Sequence	Input Conditioning, Noise Filtering	Dynamic, Non-deterministic, Continuous, fuzzy
Decomposition	Conditioned discrete, fuzzy, continuous data	Apply Context Validation Rules, Select or Extract Input Features of Abstract Classes, Select or Extract Abstract Classes/Concepts, Apply Learning Algorithm, Formulate Execution Sequence	Input Validation, Context Validation, Problem Formulation Determine Abstract Classes	Dynamic, Non-deterministic, Continuous, Fuzzy
Control	Conditioned discrete, fuzzy, continuous data	Apply Local Context Validation Rules, Apply Local Noise Filtering Heuristics, Select Input Features for Decision Level Classes, Select Decision Level Classes, Learn New Decision Level Classes, Learn New Contexts, Calculate Time & Resources Fuzzify Input Features, Formulate Execution Sequence	Local Noise Filtering Input Validation, Context Validation, Problem Formulation Viability/Utility Determine Decision Level Classes, Optimization Rule Extraction	Dynamic, Non-deterministic, Continuous, Fuzzy

Table 7.4. PAGE Description of Problem Solving Agents of IMAHDA

AGENT	PERCEPTS	(SOME) ACTIONS	GOALS	ENVIRONMENT
Decision	Conditioned Fuzzy, Continuous data	Apply Local Noise Filtering Heuristics (optional), Apply Local Context Validation Rules, Select or Extract Input Features for Specific Decisions, Select/Determine Classes/Concepts Representing Specific Decisions Select Learning Algorithm, Encode Data or Structure into Genetic Strings, Apply Rules/Models for Resolving Conflicting Outcomes, Formulate Execution Sequence	Noise Filtering Input Validation, Context Validation, Problem Formulation Determine Specific Classification or Decisions Resolve Conflicting Outcomes	Dynamic, Non-deterministic, Continuous, Fuzzy
Post Processing	Fuzzy, Discrete, Continuous	Defuzzify Results/Decision/Specific Classification Determine Reliability of Results Apply Application DependentSpecific Decision Model Validate with User or Environment Learn Models Through Feedback Determine Explicit Knowledge for Explanation	Validate Results Explain Results	Dynamic, Non-deterministic, Continuous, Fuzzy

description of the five problem solving agents of IMAHDA which employ the services of the intelligent agents and software agents is shown in Table 7.3 and 7.4 respectively. These five problem solving agents are preprocessing agent, decomposition agent, control agent, decision agent, and postprocessing agent respectively. The problem solving agents are represented in terms of their percepts, actions, goals, and environment. The actions represent the behavior of the problem solving agents after any given sequence of percepts or sensors. The goals represent the tasks in different phases of the TSL associative hybrid architecture. The environment represents the software environment in which the problem solving agents operate.

IMAHDA shown in Figure 7.5 is a realization of the software, intelligent and problem solving agents described in this chapter. In the next chapter the communication knowledge, learning knowledge, and dynamic analysis knowledge of IMAHDA is described.

7.7 SUMMARY

In this chapter, firstly, the TSL associative hybrid architecture has been integrated with Object-Oriented (O-O) Agent and Operating System Process (OSP) models. In the integration process similar characteristics like communication, typing and encapsulation facilitate integration. On the other hand, complimentary characteristics like tasks and task-oriented behavior vs structured data and operations, and structural relationships, inherently passive vs active, fixed class hierarchy vs autonomy and distribution bring about some of the enhancements as a result of integration. An Intelligent Multi-Agent Associative Hybrid Distributed Architecture (IMAHDA) with object-oriented properties is realized through this integration. IMAHDA realized through this integration consists of four layers, namely, object, software agent, intelligent agent, and problem solving agent layers respectively. The software agent layer consists of Generic Distributed Process Agent, Communication Software Agent, Relational Software Agent, and Belief Base Software Agent. The intelligent agents consist of Generic Expert System Agent, Generic Supervised/Unsupervised Neural Network Agent, Generic Fuzzy Logic Agent, and Generic GA (Genetic Algorithm) Agent. Finally, the problem solving agent layer consists of preprocessing, decomposition, control, decision, and postprocessing agents respectively. PAGE description of all these agents is outlined to identify their task role in IMAHDA. These software, intelligent, and problem solving agents form the agent building blocks of IMAHDA.

Notes

  1. Data is used here interchangeably with the term "attribute".

References

Booch, G. (1986), "Object-Oriented Development," *IEEE Transactions software Engineering*, vol. SE-12, no. 2, February, pp. 212-220.

Bourne, S. R. (1983), *The Unix System*, Addison-Wesley, Massachusetts, USA.

Encyclopedia Britannica (1986), Articles on "Behavior, Animal," "Classification Theory," and "Mood," *Encyclopedia Britannica, Inc.*

Chandrasekaran, B (1990), "What Kind of Information Processing is Intelligence," in *The Foundations of AI: A Sourcebook*, Cambridge, UK: Cambridge University Press, pp. 14-46.

Christian, K., 1988, *The Unix Operating System*, Wiley-Interscience Publication, Ney York.

Coad, P. and Yourdon, E. (1990), *Object-Oriented Analysis*, Prentice Hall, Englewood Cliffs, NJ, USA.

Coad, P. and Yourdon, E. (1992), *Object-Oriented Design*, Prentice Hall, Englewood Cliffs, NJ, USA.

Cox, B. J. (1986), *Object-Oriented Programming*, Addison-Wesley.

Dillon, T. and Tan, P. L. (1993), *Object-Oriented Conceptual Modeling*, Prentice Hall, Sydney, Australia.

Khosla, R. and Dillon, T. (1995), "Integration of Task Structure Level Architecture with O-O Technology," in *Seventh International Conference on Software Engineering and Knowledge Engineering (SEKE'95)*, Maryland, USA. Published under the section *Advances in O-O Technology*.

Kim, Ballou, Chou, Garza and Woelk (1988), "Integrating an Object-Oriented Programming system with a Database System," in *ACM OOPSLA Proceedings*, October

Murphy, G. L., and Wright, J. C. (1985), "Changes in conceptual Structure with Expertise: Differences Between Real-World Experts and Novices," in *Journal of Experimental Psychology: Learning, Memory and cognition*, vol. 10, pp. 144-15 5.

Myer, B. (1988), *Object-Oriented Software Construction*, Prentice Hall.

Rosch, E., Mervis, C. B., Gray, W. D., Johnson, D. M. and Boyes-Braem, P. (1976), "Basic Objects in Natural Categories," *Cognitive Psychology*, vol. 8, pp. 382-439.

Rosch, E. (1977), "Classification of Real-World Objects: Origins and Representations in Cognition," *Thinking: Readings in Cognitive Science*, pp. 212-22.

Russell, S., and Norvig, P. (1995), *Artificial Intelligence - A Modern Approach*, Prentice Hall, New Jersey, USA.

Rumbaugh, J. et al. (1990), *Object-Oriented Modeling and Design*, Prentice Hall, Englewood Cliffs, NJ.

Slatter, P.E. (1987), *Building Expert Systems: Cognitive Emulation*, Ellis Horwood Limited, Chichester.

Unland, R. and Schlageter, G. (1989), "An Object-Oriented Programming Environment for Advanced Database Applications," in *Journal of Object-Oriented Programming*, May/June.

Whitfield, T. W. A. and Slatter, P. E. (1979), "The Effects of Categorization and Prototypicality on Aesthetic Choice in a Furniture Selection Task," in *British Journal of Psychology*, vol. 70, pp. 65-75.

8 INTELLIGENT MULTI-AGENT HYBRID COMPUTATIONAL ARCHITECTURE - PART II

8.1 INTRODUCTION

The last chapter described the four layers of IMAHDA. In this chapter the communication knowledge constructs, and learning knowledge and strategy are described (Khosla and Dillon 1995a, 95b, 95c, and 97). This is followed by a framework for doing dynamic analysis and verification of IMAHDA. Finally, the emergent characteristics of IMAHDA are described. All this put together provides a comprehensive view of IMAHDA and its capabilities.

8.2 COMMUNICATION IN IMAHDA

Communication is an essential prerequisite for control. The integration of operating system process model and the object-oriented model in IMAHDA facilitates a two level, three channel communication mechanism as shown in Figure 8.1. The two levels are the process agent level and the domain problem solving agent level. The three channels are process agent to process agent, process agent to domain problem solving agent, and domain problem solving agent to domain problem solving agent. The process level communication provides

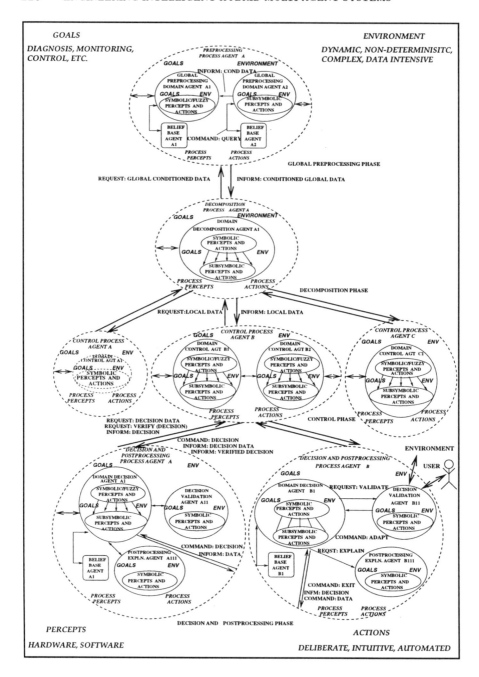

Figure 8.1.   IMAHDA based on Objects, Software, Intelligent & Problem Solving Agents

a mechanism where problem solving agents can respond to several messages at the same time and send several messages simultaneously. The two levels of communication are also representative of two levels of encapsulation, one at the process agent level and the other at the domain problem solving agent level.

The communication constructs between various problem solving agents are shown in Figure 8.1. These constructs have been based on the goals and the responsibilities of the problem solving agents and their role in information processing. It can be seen in Figure 8.1 that the nature of communication between decomposition agent and control agent is different from the communication between the control agent and the decision agent. This is primarily because of the difference in the nature of tasks undertaken by these agents.

The communication constructs in Figure 8.1 also establish the role of different problem solving agents and the stage of the problem solving process more explicitly. Communication construct like *INFORM:GLOBAL CONDITIONED DATA* represents a preliminary stage of the problem solving process, whereas, *REQUEST: VERIFY* or *REQUEST: VALIDATE* represent the final stages of the decision making process. The *REQUEST: VERIFY* is used to verify the decisions made by the decision level agents in terms of conflict resolution. It determines the correctness of a decision based on the internal domain structure representation. On the other hand, *REQUEST: VALIDATE* validates a decision in terms of its reliability, and usefulness. As shown in Figure 8.1, a user or other agents in the environment may form part of the *Decision Validation Agent*. The *COMMAND: ADAPT* is a communication construct used to adapt the decision agent to adapt in situations where the decision is invalid and is not acceptable to the user, or the decision validation agent. It may be noted that the decision validation agent can be the operational environment itself like in various control applications where a decision is validated through positive or negative feedback from the controlled process. On the other hand, the communication construct like *COMMAND: QUERY* is used to check the previous internal state of the problem solving agent for processing new data/percepts from the environment.

This section has outlined the communication constructs used in IMAHDA. Learning knowledge is another important aspect of IMAHDA. It is described in the next section.

8.3 CONCEPT LEARNING IN IMAHDA

There are various concept learning techniques in artificial neural networks. There are ones which involve interaction with environment like supervised learning (Rumelhart et al. 1986), reinforcement learning (Barto et al. 1983; Williams 1988) and unsupervised learning (Carpenter et al. 1988; Kohonen

1990). There are others which fall under stochastic vs. deterministic learning like the stochastic Boltzmann machine (i.e. generalized Hopfield network) and deterministic Boltzmann machine (Hinton 1989). Yet there are others which are based on input requirements, like similarity-based learning ("learning by examples") and instruction-based learning ("learning by being told").

The focus here is on supervised learning and similarity based learning. However, ideas espoused in the sections that follow applied in unsupervised learning, rule extraction, and data mining also. One of the most popularly used supervised learning model in real world problems is the multilayer perceptron with backpropagation learning (Rumelhart et. al 1986). However, when scaled up to complex data intensive problems there are numerous problems for training these neural networks as stated in the last chapter. The intention here is not to address problems like, size of the training data set, selection of network topology and learning parameters most of which form an intrinsic part of a neural network model. These are outside the scope of this work. It is intended to address the problem of training from a higher conceptual level which comes to grips with some of the inbuilt assumptions in neural network learning.

8.4   UNDERLYING TRAINING PROBLEMS WITH NEURAL NETWORKS

In order to understand the underlying problems an example from the animal kingdom domain with 75 input features and 104 classifications shown in Appendix A and B respectively is taken for illustrative purposes. The three basic steps involved for training a neural network are:

i. assigning all input features in the animal domain among different input nodes in the network

ii. assigning all output classes in the animal domain among different output nodes in the network

iii. including a partial data set for training and assessing the trained data set against the remaining data set for classification accuracy

However, there are inherent problems with this approach. The approach adopted by many supervised neural networks is to solve a problem in a domain as a pattern recognition problem. Associated with a pattern is the concept of holisticity. When it comes to large or data intensive domains, holisticity is a concept stretched too far. Let us take the 75 input features in the animal domain to illustrate the point.

■ The neural networks learn both similar and dissimilar features in order to come out with different classifications. However, with 75 input features

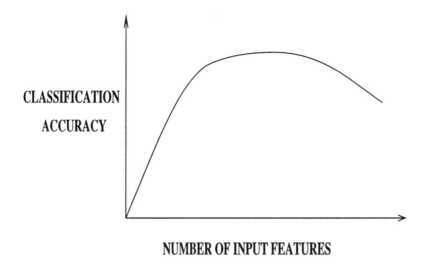

Figure 8.2.   Classification Accuracy vs Number of Input Features

the concept of relevant and irrelevant features is lost. That is, whether it is relevant for us to have 75 features for each classification in the sample animal kingdom domain is not explicitly determined. For example, 75 input features shown in Appendix A relate to different types of mammals and birds. We do not need all features for classification of a *Siberian Tiger* or a *Bluebird*. As a consequence it is quite likely that a lot of irrelevant features will be used for some classifications.

- Conceptually, generalization is associated with the notion of similarity. Comparisons (Chen 1992) between classification accuracy and the number of input features (see Figure 8.2) have shown that a consistent increase in number of input features results in decreasing classification accuracy. One of the reasons for this decline is that by using all the features at one level without any attention to ordering or relevance, the notion of similarity is likely to become more and more diffuse. As a result generalization is likely to suffer.

- Such a representation is indicative of the "Learning from scratch," philosophy as currently employed in almost all artificial neural network models. This is probably not adequate for complex domains, a point already made by Arbib (1987) and Malsburg (1988) in this area (see their quotes in chapter 6 in section 6.2.7).

In other words, the "Learning from scratch," philosophy leads to representations which generally lack structure and impose scalability constraints on
themselves.

- It also follows from the previous point that such a representation does not
  attempt to make use of techniques like observation, a priori or background
  knowledge and problem solving strategies for determining regularities in input data. Instead it only relies on neural network topology and the learning
  algorithm to form different clusters and learn the domain.

- From a computational point of view, a large number of input features increase the complexity of the network multifold. The multitude of connections
  only make the net less comprehensible. Besides, too many inputs, too many
  output classes (possibly many of them overlapping) only aggravate existing
  problems like selection and the size of training data set, local minima, and
  generalization associated with neural networks.

8.5 LEARNING AND IMAHDA

The central aspect of learning is constructing a representation of some reality.
The goal of training is to make a neural network learn or construct a representation of a domain being learnt. In IMAHDA a *priori* structure exists in which
learning can take place. This a *priori* structure consists of the decomposition,
control and decision phases of IMAHDA. This a *priori* structure which is motivated from philosophical, neurobiological, cognitive and practical viewpoints
facilitates concept learning. It facilitates concept learning at different levels of
abstraction which not only reduces the complexity of learning large amounts of
data but is also cognitively more comprehensible and intelligible. It represents
an abstract and hierarchical framework in which learning can take place. It
can be seen as an integration of macro level learning aspects (i.e. abstraction
and hierarchical structure) of symbolic learning systems with underlying neural
learning systems which accomplish the learning process.

By pursuing learning in this manner one is actually building up at each level
of abstraction background knowledge for the level/s below which provides scope
for performance improvement and incremental learning (Michalski et al. 1987.
Further, by learning at different levels of abstraction one removes redundancies
in the data and eliminates irrelevant features (Wolf 1987).

The a *priori* structure of IMAHDA also constrains the size and generalization of the neural networks due its localization effect (this will also become
apparent in section 8.7 of this chapter). Thus the a *priori* structure used by
IMAHDA not only benefits from the strengths of neural networks like fault tol-

erance and fast retrieval but also helps to realize a more cognitively plausible learning architecture.

However, what is not known in this a *priori* structure is how to select/extract input features and output classes in a complex data intensive domain? In order to accomplish this, learning knowledge and learning strategy has been devised for learning in the a *priori* structure of IMAHDA.

8.6 LEARNING KNOWLEDGE IN IMAHDA

Learning knowledge determines the type of knowledge or features and output classes to be used in different phases of IMAHDA. The three relevant phases are the decomposition phase, control phase and the decision phase. The learning knowledge is divided into three categories:

- decomposition knowledge

- control knowledge

- decision knowledge

8.6.1   Decomposition Knowledge

Elementary or decomposition knowledge is the one used for decomposing a global concept into abstract classes. This global concept can be a generalization or aggregation. For example in the animal domain a global concept like *Mammal* is a generalization or abstraction of abstract classes like *Ape, Cat, Whale, etc.*. These abstract classes themselves are generalizations, like for example *Lion, Tiger, Leopard, etc.* shown in Appendix B are subclasses of class *Cat.* Also, these abstract classes are globally independent classes. They exist at a level of abstraction lower than the global concept *Mammal.* At this level of abstraction they are independent classes because there are no overlapping boundaries between them (i.e. *Ape, Cat, Whale,* etc.). So the decomposition knowledge is used to select globally independent output classes and input features which classify them.

The input features used for classification of abstract classes are coarse grain features as mentioned in the last chapter. These coarse grain features are either general features or key features. A general feature is one which represents a generalization in terms of the features used for identifying an abstract class. For example, *Hair* and *Milk* are general features which are used for identifying an abstract class *Mammal.* A key feature is one which represents an important indicator of a class and can also represent a heuristic on that class. These general and key features have binary values (0 or 1) and/or structured values (one-sided, three-sided, etc.). The structured features have multiple discrete values and represent structure in a domain.

8.6.2   Control Knowledge

Control knowledge which functions within an abstract class is used for locally disaggregating a class into its constituent parts or locally specializing into its subclasses. For example, control knowledge will be used for determining the subclasses *Lion, Tiger, etc.* of an abstract class like *Cat*. Thus the output classes to be selected in the control phase are local constituent parts of an abstract class or its specializations. These constituent parts or specializations are what we call decision level classes of the domain. As defined in the last chapter, decision classes are those classes, inference on which is of importance to the user. For example *Tiger* is a decision level class in the animal domain as it is used for determining the instances like *Siberian Tiger, Bengal Tiger*, etc.. It can also be said that as a result of local disaggregation or specialization we move from relative heterogeneity to relative homogeneity (i.e. moving from *Cat* to *Tiger*). The extent of homogeneity achieved depends on the level of decision classification required in a domain. A decision class like *Tiger* is a solitary class and for that matter is purely homogeneous. However, decision classes can represent more than one class and thus may not be purely homogeneous.

Semi-coarse grain features which are used for doing such a classification are locally general features or key features. These features can be further broken up into binary, structured and linear/fuzzy features. Linear features are those which have a broad range from say *1...10* (e.g . range of a single analog measurement or a set of measurements or linguistic variables like large, small, etc.).

8.6.3   Decision Knowledge

Decision knowledge is used for doing specific classification. These are the classifications a system user is interested in. In the animal domain these can be instances of class *Tiger* like *Indochina Tiger, Bengal Tiger* and *Siberian Tiger.*

At this level of classification the distributed nature or the degree of similarity between features is high. In other words, there is a fine distinction between different input features. These fine distinctions are reflected through use of fine grain features at this level. They reflect refined distinctions between various fault classifications or instances of *Tiger*. They consist of refined linear or fuzzy features. That is, instead of broad range of *1..10*, the range is further refined to *2..4*, *4..7*, etc. or even a single value like less than *1* is used.

8.7 LEARNING STRATEGY IN IMAHDA

In the previous section the learning knowledge needed to train neural networks in the three phases of IMAHDA has been determined. That is, the type of input features and output classes which need to be selected and/or extracted in the decomposition, control and decision phases respectively have been determined. It now needs to be determined how to select and/or extract the output classes and input features in different phases of the architecture. Learning strategy, in other words, determines the method of selection of the output classes and connected with them the input features in different phases of IMAHDA.

The learning strategy in IMAHDA for selection and/or extraction of output classes involves:

1. Extraction of abstract aggregated or generalized classes in a manner that creates globally independent abstract classes in the decomposition phase.

2. Local disaggregation or specialization of each globally independent abstract class in the control phase. This entails selection of decision level classes in the domain, which are constituent parts or specializations of independent abstract classes.

3. Selection of instances of a decision level class or a specific classification related to the decision level class which is of interest to the user.

The learning strategy in IMAHDA for selection of input features involves:

1. Selecting/extracting the most discriminating features in different classes.

2. Selecting/extracting the common features among different classes.

3. Knowing when to stop selecting the common or discriminating features in a particular phase.

4. Retention and Elimination of redundant features.

5. Associated with '4', the provision of noise.

8.7.1 Learning in the Decomposition Phase

There are two aspects of learning in the decomposition phase. These include extraction of output classes and extraction/selection of input features which classify them. The extraction of output classes in the decomposition phase involves generating high level abstract classes in the animal domain. These abstract classes are not necessarily present in the domain and are extracted out of it through aggregation or generalization. In the case of animal domain the high level abstract classes are generalizations of the low level classes. Further, this abstraction of classes starts with the highest level of aggregated or generalized class distinction in the domain. For example, abstract classes like *Mammal* and *Bird* represent the highest level of class distinction in the animal domain under study. We then start moving down from this abstract level to levels below by selecting input features within these two abstract classes in two different ways. One type represents the overlapping boundaries between abstract classes and generates lower level abstract classes. The process ideally stops when we are left with no overlapping boundaries and have globally independent abstract classes. The other type represents non-overlapping parts of each abstract class. In the former and latter case the process stops before the disaggregation or specialization results into decision level classes.

For example, in the animal domain the disaggregation or specialization in this phase will stop, when classification into say *Ape, Cat, Whale, Monkey*, etc. classes has been achieved. Any further decomposition of say the *Cat* class will result into selecting *Lion, Tiger*, etc. classes which are decision level classes in the sample animal domain.

The process of extraction or selection of input features in the decomposition phase is determined by the fact that whether the coarse grain features are distinctively present/separated in the input data set or not. In case they are distinctively present in the input data set then they have only to be selected from the input data set. Otherwise, they have to be extracted from the input data set. In the sample animal domain the coarse grain input features are selected as they are distinctively present in the domain.

In the process of selecting input features in the decomposition phase not only discriminating features between abstract classes have to selected but also the common features between them. The combination of distinct and common features will generate largely uncorrelated abstract classes which can then be worked upon in the control phase.

8.7.1.1 Selecting Most Discriminating Features.
The 75 input features in the animal domain reflect a heterogeneous rather than a homogeneous mix. That is, there are coarse grain binary features like hair, milk, and feathers and egglaying which are common to all *Mammal* and *Bird* respectively. There are semi-coarse grain binary and broad linear (or fuzzy) features like *trunk* and *long-mane* respectively which are common to *Elephant* and *Lion* respectively, and also there are fine grain refined linear or fuzzy features like *very-narrow-dark-black-stripes* and *light-reddish-ochre-color* which are common only to *Javan Tiger* and *Indochina Tiger*.

The 75 features relate to two generalized classes, namely, *Mammal* and *Bird*. The features related to these generalized classes are shown in Figure 8.3.

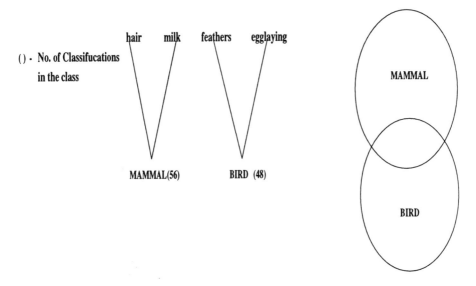

Figure 8.3.   Features for Mammal and Bird

Mammal(56) and *Bird*(48) mean that within the *Mammal* and *Bird* classes there are 56 and 48 sub-classifications respectively. From a computational point of view, this is not satisfactory as features required for 56 or 48 classifications are likely to be just too many.

One can also see that the features *hair* and *milk* do not have any eliminating ability. That is, the *Mammal* class cannot be subdivided by negating any of the two features, *hair* and *milk*. Similarly, the features *feathers* and *egglaying* cannot be used for subdividing the class *Bird*. Thus, these four features are too coarse for one to gain any significant progress in moving down from this abstract level. As a consequence more input features which can generate lower levels of abstractions or subdivide classes *Mammal* and *Bird* need to be selected.

8.7.1.2 Selecting Common Features Among Classes.

Common features among classes reflect their overlapping boundaries. It is useful to include the features which are common to more than one class for the purpose of elementary classification. By adding one common feature in two classes, there can be a multiple effect in terms of disaggregation or specialization. That is, by adding one common feature between two classes it is possible to create one or more distinct class/es through negation and affirmation. The number of new classes will depend upon the number of features selected prior to selection of the common feature.

In the animal domain, *oceanic* and *land* are features which are common to some types of *Mammal* and *Bird*. Also, by coincidence, the feature egglaying is also common to some types of *Mammal*. By using these common features among *Mammal* and *Bird* classes another six abstract classes namely, *Oceanic Mammal*, *Egglaying Land Mammal*, Non-Egglaying-Land Mammal, *Oceanic Bird*, *Land Bird* and *Lake-River and Tree Bird* are extracted as shown in Figures 8.4[1] and 8.5 respectively and move towards a lower level of abstraction.

8.7.1.3   Repeat Selection of Distinct and Common Features if Necessary.   In the previous two sections 5 coarse grain binary features have been chosen and 6 abstract classes have been extracted. However, the number of classifications in *Non-Egglaying Land Mammal* and *Non-Land Bird* (48 and 37 respectively) are still too large. Further, the goal in the decomposition phase to generate globally independent abstract classes has yet not been achieved as some abstract classes still have overlapping boundaries.

In order to further eliminate overlaps and specialize the classes the steps 1 and 2 are repeated. The most distinct features *placenta* and *tree* for the two classes respectively as shown in Figure 8.6 are selected. The *placenta* feature provides a clear demarcation between between *Land Mammal* which have *pouches* and which have not. It divides *Non-Egglaying Mammal* clearly into two abstract classes which facilitates autonomy at lower levels or phases. One could have chosen another apparently distinct feature *large* which occurs in classes like *Bear*, *Ape*, *Cat*, etc.. However, it does not provide a clear distinction. That is, say in the *Cat* family, *Lion* and *Tiger* are large mammals, whereas *Puma* and *Wild-Cat* are not large mammals. Same is true for the *Ape*. In other words, an *Ape* is not generically larger than *Bear* and a *Bear* is not generically larger than *Cat*. It is possibly a semi-coarse grain feature to be considered within an abstract class and not for distinguishing between abstract classes. Thus it is not a coarse grain feature to be considered at this level. Similarly, the feature *tree* has been chosen to distinguish between *Tree Bird* and *Lake-River Bird*.

This is followed with selection of a common feature *claws* among the newly created classes as shown in Figures 8.7 and 8.8 respectively.

Nine features have covered 11 classes but still there are 25 and 17 classifications in *Clawed-Land-Mammal* and *Non-Clawed-Land-Mammal* classes respectively as shown in Figure 8.9. In these circumstances, it is a good idea to stagger the elementary classification into more than one level. It is beneficial for three reasons:

- In this way more ground can be gained in elementary classification because features at two levels can be used

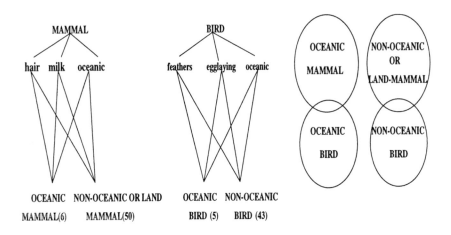

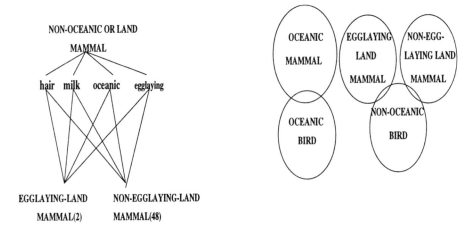

Figure 8.4. Features for Oceanic and Non-Oceanic Mammal and Oceanic Bird

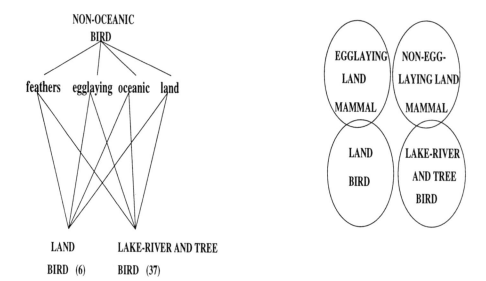

Figure 8.5.   Features for Non-Oceanic Bird

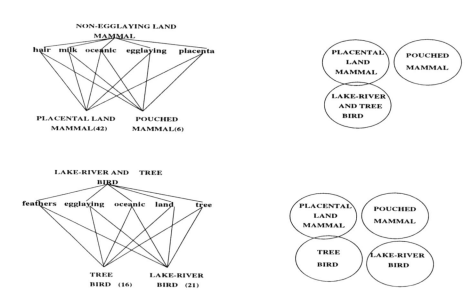

Figure 8.6.   Features for Placental Mammal, Land Bird, etc.

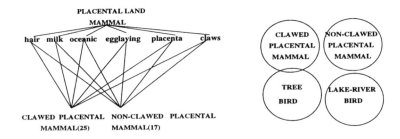

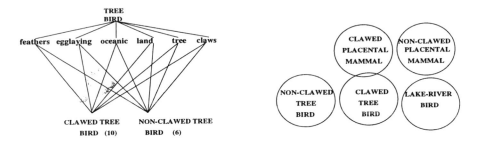

Figure 8.7.   Features for Clawed Mammal, Clawed Tree Bird, etc.

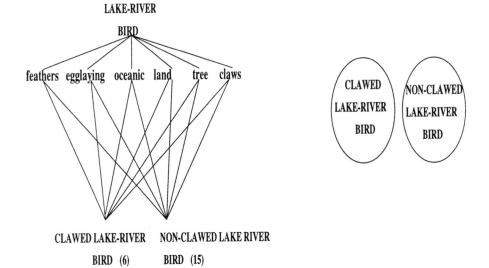

Figure 8.8.   Features for Clawed Lake-River and Non-Clawed Bird

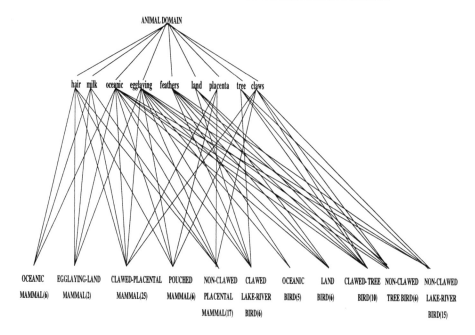

Figure 8.9.   Coarse Grain Features for Elementary Classification(level1)

- It provides additional flexibility for accommodating more features at each level in case of any future additions or scalability.

- If too many features are used at one level to start with, any scaling up will result in breaking up one level into two levels and will result in disturbing the class break up at lower levels. This can result in unnecessary retraining of neural networks at lower levels.

The features used to specialize *Clawed-Placental-Mammal* and *Non-Clawed-Placental-Mammal* are shown in Figure 8.10.

It may be noted here that features like *five-digit*, *odd-toed* and *even-toed* are structured features.

8.7.1.4   Stopping Criteria.   The process of selection of features in a particular phase stops when the selection infiltrates into the next phase or the phase below. In the transition from decomposition phase to control phase, the learning process stops when the effect of feature selection becomes highly localized rather than global. In other words, one is no longer considering generalizations or aggregated abstractions but specific classes like *Lion, Howler-Monkey,* etc..

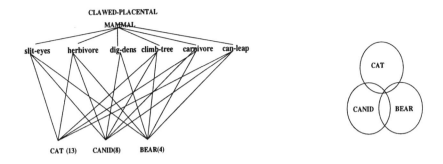

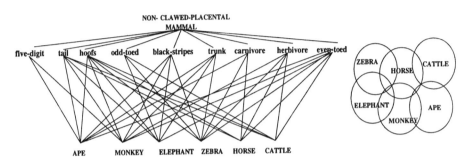

Figure 8.10.   Coarse Grain Features for Elementary Classification(level2)

The features which are selected are of a semi-coarse grain nature reflecting distinctions between *Tiger* and a *Leopard*. These distinctions can be reflected through broad range linear or fuzzy features (like *large*, *small* and *medium*), binary features and structured features within a localized class rather than any abstraction of it. Another indicator is that the abstract class is heterogeneous, whereas a specific class would be relatively homogeneous. Thus in the decomposition phase the learning process stops once the abstract independent classes *Cat, Canid, Bear, Ape*, etc. shown in Figure 8.10 have been determined. In this phase use of broad range linear or fuzzy features is avoided as they are not uniformly applicable at higher levels of abstraction. For example, *large* or *medium* size *Cat* is not necessarily a *large* or *medium* size *Bear*. These features are applicable within *Cat* and *Bear* class and hence are used in the intermediate classification phase.

8.7.1.5   Retention and Elimination of Redundant Features.   The different phases of IMAHDA makes provision for redundancy and thus noise in the domain. What is relevant in one phase is redundant in the phase below.

For example, features like *hair, milk, placenta, land,* etc. are useful distinctions for *Clawed-Placental Mammal.* However, for distinguishing between a *Cat* and a *Bear* or a *Leopard* and a *Tiger* these features although relevant become redundant because they have been ascertained by the *Clawed-Placental-Mammal* abstract class.

The identification of redundant features has two implications. One is to retain them for coping with noise in the domain. The other is eliminating them to reduce the number of features in a particular phase for training. Some features are declared redundant as they do not provide any additional classification in their aggregated class whereas some are declared redundant through a process of elimination. The redundant coarse grain features identified in the elementary classification phase are, *hair, egglaying* and *pouched*. Features like *hair* and *egglaying*[2] are redundant features which do not add any additional classification within their aggregated class. Features like *pouched* are declared redundant because of process of elimination.

Features like, *hair* and *egglaying* are common to all *Mammal* and *Bird* respectively. Thus they are retained as redundant features to assist in coping with **noise**. On the other hand, *pouched* is not and there are other supporting features like *hair, milk, non-placenta* and *claws* which can be used for *Pouched Mammal* classification. Hence, *pouched* feature is eliminated.

8.7.2   Learning in the Control Phase

The strategy to be followed for intermediate classification is similar to that of elementary classification except that now one is selecting features within a narrowed scope of a single abstract class rather than global scope of disparate abstract classes like *Oceanic Mammal* and *Oceanic Bird*. It may be noted here that the classification done in the elementary classification phase facilitates such a training process where one can work within in a single abstract class like *Cat, Canid* or *Bear*. An advantage of this strategy is that it facilitates elimination of irrelevant features. That is, in the *Cat* class one considers features related to the *Cat* family only and not the *Canid* family.

The intermediate classification done in this sample animal domain is for the *Cat* family as shown in Figure 8.11. Features like *black-stripes* and *rosette-shaped-spots* are distinct semi-coarse grain binary features found in *Lion, Tiger,* and *Leopard* respectively. The feature *can-roar* is a distinct binary feature found in *Lion* and *Tiger* whereas *tapered ears* is a distinct feature found in *Puma* and *wild-cat*.

In this phase it may be noted that graded commonalities have started to appear. That is, features like *long-mane, large, medium,* etc. are graded commonalities and suggest a range within them. As a consequence, more features

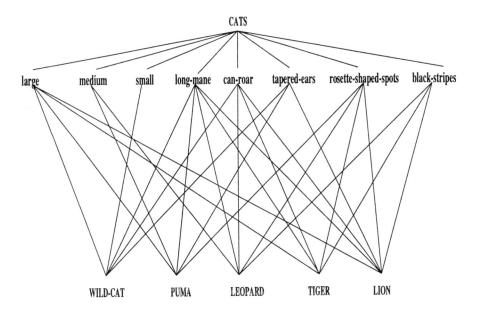

Figure 8.11.   Intermediate Classification of Cat

have to be used in order to determine the different classifications and the principal of elimination is also less effective. Redundancy is reflected in these graded commonalities. The graded commonalities also reflect a steady transition from semi-coarse grain features to fine grain features.

It can be observed that using linguistic features like *tapered-ears, large, medium, small* makes more sense in this case as they facilitate the strategy of determining the decision level classes and related to the class *Cat* only. These linguistic features can also be fuzzified. A description of their fuzzification can be found in Tang, Dillon and Khosla (1995).

In this phase the learning process stops once classification of the decision level classes of the *Cat* family which in our case are *Tiger, Leopard, Lion, Puma*, etc. have been determined. A decision level class like *Tiger* is a solitary class level and hence is purely homogeneous class. Now only the instances of the *Tiger* class in the decision phase are left to be determined.

8.7.3   Learning in the Decision Phase

In the decision phase fine grain features of the *Tiger* sub-class are used to classify the instances of *Tiger*. There are eight instances of tiger, namely, *Siberian Tiger, South-China Tiger, Indochina Tiger, Bengal Tiger, Javan Tiger, Caspian*

Tiger, Sumartan Tiger and *White Tiger*. The fine grain features used are shown in Figure 8.12.

It can be seen that fine graded commonalities are dominant indicating the heavily distributed nature of the features. The distinct features as such are almost entirely absent. Distinct feature like *black-stripes* is graded as *dark-narrow-stripes*, *dark-very-narrow-stripes* or *widely-spaced-stripes* and *closely-spaced-stripes*. This also indicates the breakdown in the discreteness of the features. In the decomposition phase, the features are coarse and discrete, whereas in the decision phase the features are fine, fuzzy and/or or continuous. This is true in many domains like alarm handling and process control where the high level discrete alarms are generated by low level continuous analog readings.

Here one can either reduce the number of input nodes by merging the nodes like *dark-narrow-stripes* and *dark-very-narrow-stripes* into a single *black-stripes* node and give it continuous values or retain input node configuration as shown in Figure 8.12. This decision depends upon the extent of continuity in the inputs in the decision phase. If the inputs are highly continuous, then it is best that input nodes with continuous values should be used. Further the input values like *dark-very-narrow-stripes* and *reddish-ochre-color* can also be fuzzified. A description of their fuzzification can be found in Tang, Dillon and Khosla (1995).

Another issue of interest here is generalization. It may be noted that the generalization in the decision phase is constrained to the decision level classes only. For example, in this phase generalization is constrained or localized to the *Tiger* class only. Besides the two previous phases provide background knowledge to this phase and tend to lend structural creditability to the generalization in this phase.

This completes the description of the learning knowledge and learning strategy of IMAHDA. This description provides a clear direction to a system designer that what needs to be learnt in the three phases and how to accomplish the learning process through use of neural networks.

Thus in the preceding paragraphs, firstly a *priori* structure for accomplishing neural network learning in data intensive domains has been outlined. Secondly, what are the different types of input features and output classes to be used in the three phases of the a *priori* structure has been determined. Thirdly, the learning strategy for selecting/extracting different types of input features and output classes in each of three phases of the a *priori* structure has been determined. This includes determination of the stopping criteria in each phase and provision for noise. The input features and output classes used in the a *priori* structure for the animal domain are shown in Figure 8.13. An overview of the learning knowledge and learning strategy is shown in Figure 8.14. This

FEATURE	ABBREVIATION
white-bellyside-extends-to-flanks	WBEF
white-tail	WT
yellowish-color	YC
reddish-ochre-color	ROC
light-reddish-ochre-color	LROC
closely-spaced-stripes	CSS
dark-narrow-stripes	DNS
widely-spaced-stripes	WSS
very-light-reddish-ochre-color	VLROC
white-bellyside-restricted	WBR
white-color	WC
blue-eyes	BE
short-neck-mane	SNM
heavy-cheek-hair	HCH
light-cheek-hair	LCH
very-heavy-cheek-hair	VHCH
belly mane	BM
dark-very-narrow-stripes	DVNS

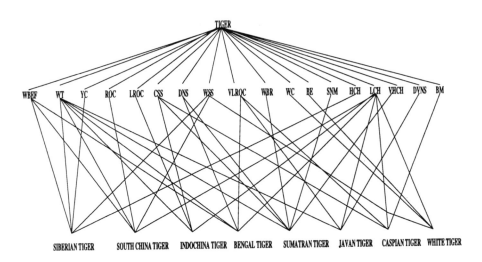

Figure 8.12.   Specific Classification of Tiger

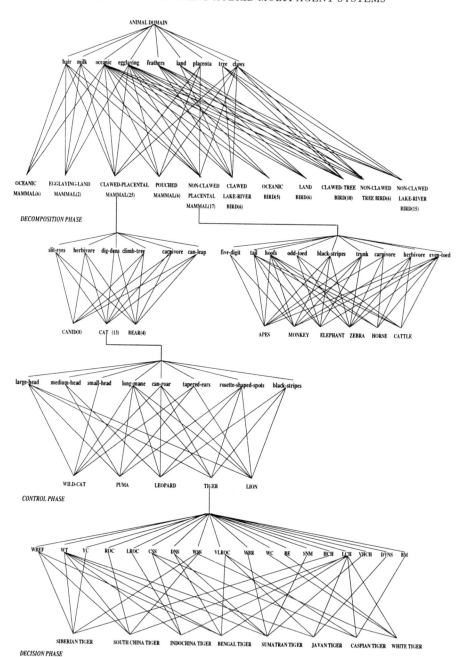

Figure 8.13.   Input Features and Output Classes in Three Phases

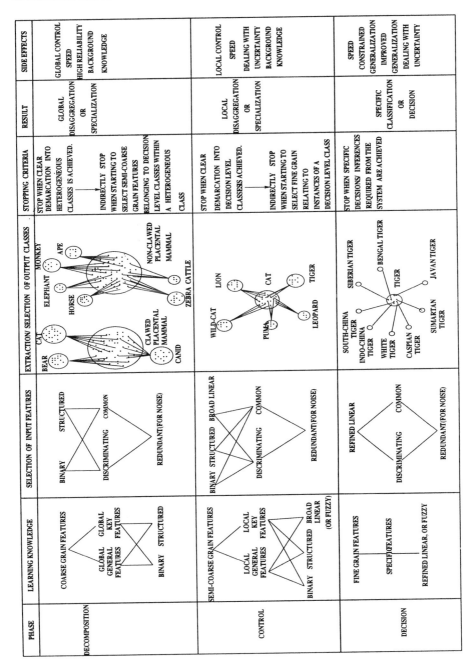

Figure 8.14.   Learning Knowledge and Strategy in   IMAHDA

learning knowledge and strategy has been applied in the data intensive alarm processing domain in chapter 11.

Further in the animal domain example discrimination between different classes has been achieved through use of general features in the decom position phase. Key features have not been used. As will be shown in chapter 11, key features can also be used to distinguish between classes in the decomposition phase.

This section completes the learning knowledge and strategy of IMAHDA. In the next section dynamic analysis of IMAHDA is undertaken in an effort to come to grips with some of its dynamic aspects, namely, agents states, cyclic and distributed operation, and synchronous and asynchronous communication.

8.8   DYNAMIC ANALYSIS OF IMAHDA

The objective of this section is to outline the constructs for dynamic analysis of IMAHDA keeping in view its distributed and real time character. Real time systems are complex because of their inherent parallelism, timing and temporal characteristics. They consist of distinct, distributed and concurrently running processes which communicate both synchronously and asynchronously. The failure to effectively realize these characteristics in real time operation can lead to catastrophic circumstances in hard real time systems and fault/error propagation in soft real time systems (Schiebe et al. 1992).

Thus it becomes necessary to come to grips with the dynamic architectures like IMAHDA. It is considered necessary not only for detecting catastrophes or programming errors/deadlocks but also to enable its future users to analyze consequences of changes made in design from time to time.

In the context of IMAHDA the dynamic analysis help us to analyze the process level (global) and neural network level (domain problem solving agent level) parallelism, synchronous and asynchronous communication, process states and domain agent states, and event of action execution. The dynamic analysis also help us in analyzing other aspects of real time behavior like continuous and cyclic operation.

The dynamic analysis of IMAHDA is done using State Controlled Petri Nets (SCPN). It is shown how features like global parallelism, agent states, action execution and synchronization/asynchronization can be made transparent using SCPN notations developed by Khosla and Dillon (1994).

8.8.1   State Controlled Petri Nets

SCPNs are extensions of the original Petri net developed by Petri (1966). Twenty-five years of theoretical work and practical experiences documented in several thousand journal papers and research reports have proved Petri nets to be one of the most useful languages for modeling of systems containing concurrent processes.

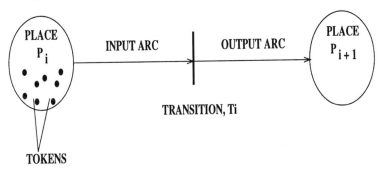

Figure 8.15.   A Petri Net Model

A Petri net (Petri 1966) shown in Figure 8.15 is a particular kind of directed graph with two types of nodes, namely the places (graphically depicted as circles) and transitions (graphically depicted as bars). The basic structure of a Petri Net consists of a set of places, a set of transitions and a set of directed arcs, which connect the transitions and places. The presence of token/s in the input place causes the transition to fire, leading to the removal of the token/s from input place and deposition of token/s in the output place of the transition. The pattern of placement of tokens through the net at a particular time is called the marking of the net. A given marking corresponds to a state of the net.

The above structure of Petri nets has been to extended to cover a wide range of discrete event situations (Landweber et al. 1978; Symons 1978; Genrich 1981; Wheeler 1985 ,86; Jensen 1983; Jenson 1985; Cardoso 1989; Lai and Dillon 1989a ; Zurawski and Dillon 1991).

SCPNs have been developed to widen the scope of application of the original Petri nets. SCPNs developed by Khosla and Dillon (1994) retain the basic principles, symbols and modes of operation of Petri nets, while adding a considerable amount of modeling expressiveness. A summary of their properties is as follows:

1. Each token has an identity, sometimes called color, and has been generalized to a tuple. In particular, the state token that represents the logic state of a predicate is separated from the control token required for execution of the rule.

2. A set of data variables is associated with the net.

3. The enabling and firing conditions of individual arcs are independent.

4. The transition enabling conditions refer to both tokens in input places and to data referenced by the rule.

5. The transition occurs when a) the transition is enabled, and b) the right types of both the state and control tokens are present.

   It includes firing of all tokens from input places, firing of tokens into output places, and operation on data referenced by the rule.

The places are taken to represent the states with variable extensions and transitions to represent categories of elementary changes of extensions. A token in a place denotes the fact that the predicate corresponding to that place is true for that particular instantiation of arguments contained in the token. Transitions are fired to represent categories of elementary changes of extensions. A token in a place denotes the fact that the predicate corresponding to that place is true for that particular instantiation of arguments contained in the token. Transitions are fired to represent events occurring. The net result is the exchange of token from places to places and a new marking, defined as the distribution of tokens over the places of the SCPNs, is obtained.

A list of symbols and their interpretations in SCPNs are shown in Figure 8.16. The definition of a transition in a SCPN is shown in Figure 8.17.

S_i : Place, which denotes the state of an object and is represented by a labelled circle,

t_i : Transition, which denotes an event and is represented by a labelled bar,

e : Simple token for control, represented by a small circle marked by e, e

y : State token, its presence at S_i denotes that predicate P_i is true and is represented by a small circle marked by y,

n : State token, its presence at S_i denotes that the negation of predicate P_i is true (i.e. Pi is false in a closed world situation) and is represented by a small circle marked by n,

Figure 8.16.   SCPN Symbols and their Interpretations

An example of the condition, cond(t), that must be satisfied before the transition is that there must be at a y state token in input place and at least one control token e in at least one of the input places.

A transition in a net N is defined as a 5-tuple as follows:

t = { pre(t), post(t), cond(t), fn(t), type(t) } and t $\leqslant$ T, where:

pre(t)   refers to the preconditions that need to be satisfied before the transition
         can be activated

post(t) refers to the post-conditions resulting from the firing of the transition

cond(t) specifies the condition on the predicate arguments that needs to be
         satisfied before t can be executed

fn(t)   defines the procedure carried out by the transition

type(t) classifies the transition into different types which are used to
         select the firing rule to be employed in executing the transition.

Figure 8.17.   SCPN Transition Definition

SCPNs can be considered as a structurally folded version of a regular Petri net for a finite number of the types of tokens. Thus, a SCPN can be unfolded into a regular Petri net by unfolding each predicate Pi into a set of predicates Pci, Psi, one for each type of token which the predicate place may hold, and by unfolding each transition t_j into a set of transitions t_{cj}, t_{sj}, one for each way that t may fire.

The timing associated with actions or states can be modeled using Timed Petri Nets. Thus, one can associate a time with a place for Timed Place Petri Nets (TPPNs) (Wong and Dillon 1984, 87). If one associates time with a place then the required time has to elapse from the instant the token arrives in a place before it becomes available for firing a transition. If time is associated with an event, then one has a Timed Event Petri Net (Ramamoorthy et al. 1980; Zuberek 1980). In this case, from the instant that a transition is activated the time required has to elapse before it fires.

8.8.2 Dynamic Behavior Analysis using SCPNs

Petri nets are often used for dynamic analysis of software programs. The SCPN model used for doing dynamic analysis of IMAHDA along with some modifications is shown in Figure 8.18. The definitions used in the model are outlined below:

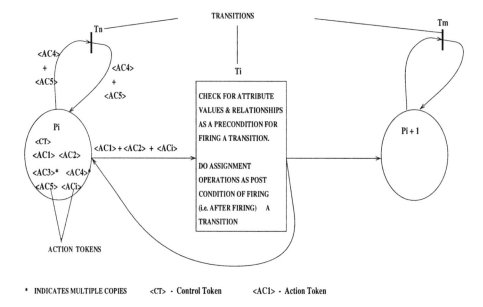

Figure 8.18.   A SCPN Model with Inscribed Transition Structure

- **Pi** - Places are denoted by circles which represent the state of an agent. The state of an agent is a consequence of action execution and a change in value of its attributes.

- **Ti** - Transitions denote the event of completion of action execution which could be an ES (Expert System) agent action (e.g. Rules -$< RUL >$), a Supervised Neural Network agent action (e.g. output-classifier - $< OUC >$), Distributed Process agent action (Create-child-process, $< CCP >$).

- **Tokens** - Actions in an agent are represented as tokens. Attributes are represented as variables. The action tokens are the state tokens. The presence of the action token in an agent state denotes that the given action in the agent is active in that state. The control token $< CT >$ denotes the flow of control among different agent states. Tokens can be created or destroyed on firing of a transition.

  Actions in an agent are executed in different agent states. The creation of action tokens represents those action tokens which are active in the new state and are being executed. The destruction of action tokens represents those action tokens which are invalid or inactive in the new state and thus need to be destroyed. Further, multiple copies of a action token can exist in a place.

- **IA** - Input Arc from place Pi to transition Ti represents the input condition. It denotes the number of action tokens to be executed to enable firing of the transition Ti. The firing of the transition results in transition of an agent from its previous state to new state. The tokens on the input arc are represented as a formal sum of different tokens.

- **OA** - Output Arc from transition Ti to place Pi+1 represents the output condition. It denotes the number of action tokens which should be present in the place Pi+1.

- **Loops** - The input and output arc can culminate and originate respectively from one place or state. That is, an execution of a single or set of action tokens may not result in a change of state or place. For example, a rule may loop around itself several times before its conditions are no longer satisfied.

Transitions can have inside them, an inscribed structure (see Figure 8.18), defining a collection of attributes with some relationships and operations applicable to them.

A transition is enabled whenever:

a. each place Pi contains the control token and at least as many action tokens as specified in the IA to Ti.

b. the attributes have values satisfying the conditions in the inscribed structure (if any) associated with a transition.

c. the capacity of each output place is not exceeded by firing the transition.

8.8.3 SCPN Modeling of the Decomposition Agent

The SCPN modeling of the decomposition agent is shown in Figure 8.19 for purpose of illustration. For purpose of simplicity, only a subset of actions related to the decomposition agent are shown in Figure 8.19.

The use of a formal description technique like SCPNs makes transparent transparent a number of aspects of IMAHDA. Some of the benefits which are derived from use of the SCPN model are outlined as follows:

a. It makes generic use of actions transparent from state to state as shown in Figure 8.19.

b. It makes transparent the different agent states associated with an agent. There are a total of six states associated with SCPN modeling. These are

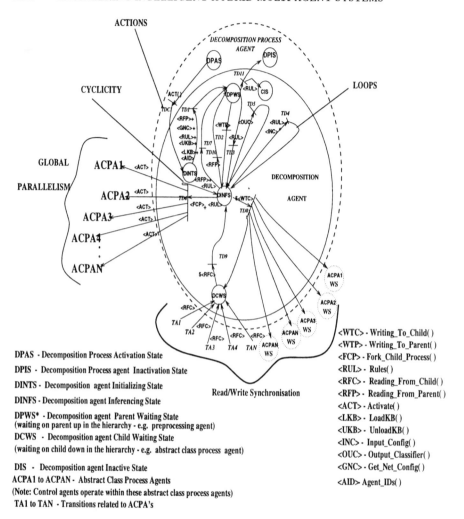

Figure 8.19.   SCPN Modeling of Dynamic Behavior of Decomposition Agent

process level agent inactivation and activation states and domain level initializing, inferencing, waiting and inactive states respectively. The process (e.g. Unix OS process) level agent (decomposition process agent in this case) and domain level agent (decomposition agent in this case) states are shown in Figure 8.19. The process level agent and domain level agent waiting states are common and thus have been merged into one.

c. The facility of creating and destroying actions enables a designer to know which actions are valid or invalid in a particular agent state and also which actions are not inherited by an agent from its parent class. For example, firing of transition TD1 in creates an action token 'AID'(Agent_IDs()) in place 'DINTS'. This token is not inherited from its parent. Also action token 'AID' is not executed in any other state except DINTS through transition TD1 (as it is invalid in other states). Hence it is destroyed.

d. It provides us with a facility to show dynamic inheritance of actions between two agents belonging to different hierarchical or domain structures. Places in two agents belonging to different structures can be connected through a transition, and an action token can be used for showing dynamic inheritance. Dynamic inheritance can also be shown between agents in different agent states. This allows us to model a condition when agents change their inheritance relationships dynamically in different agent states.

e. It makes transparent the process agent to process agent communication, process agent to domain agent communication, and domain agent to domain agent communication in different process and domain agent states. For example, transitions TD8 and TD9 of the decomposition agent show process agent to process agent communication. Transition TDC in Figure 8.19 shows process agent to domain agent communication.

f. The return arcs in transitions TD6, and TD9 in the decomposition agent make transparent the cyclicity associated in IMAHDA.

g. The transitions TD6 (creation of process agents) and TD8 (writing to children) in Figure 8.19 show the global parallelism associated with IMAHDA. Once created all process agents execute in parallel.

8.8.3.1   Synchronous and Asynchronous Communication Analysis.

Synchronization allows one to control the evolution of processes with respect to one another in a distributed environment. In a distributed architecture, like IMAHDA, the means of interaction between neighboring process agents takes place through exchange of messages. This is called synchronization on communication (Schiebe et al. 1992). The agents can be distributed on one machine or different machines.

In Figure 8.20 an attempt to analyze such a communication between the parent[3] (higher level) and child (lower level) process agents has been made. It shows the synchronizing mechanism used by the parent Decomposition Process Agent (DPA) with its children, say $ACPA1$ or $A1$ (Abstract Class Process Agent 1) and $ACPA2$ or $A2$ respectively for purpose of read and write operations associated with incoming data. Although, the communication takes place at process agent level the actions *Write to Child()* ($< WTC >$) and Read from Child() ($< RFC >$) are executed within the activated domain level agent,namely Decomposition Agent (DA) and thus the transitions are shown within DA. The preconditions for firing of transition TD8 in DA to accomplish

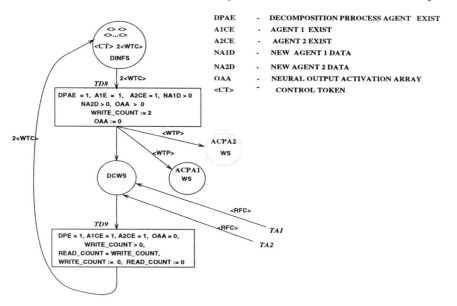

DPAE	-	DECOMPOSITION PRROCESS AGENT EXIST
A1CE	-	AGENT 1 EXIST
A2CE	-	AGENT 2 EXIST
NA1D	-	NEW AGENT 1 DATA
NA2D	-	NEW AGENT 2 DATA
OAA	-	NEURAL OUTPUT ACTIVATION ARRAY
<CT>	-	CONTROL TOKEN

Figure 8.20.   Read Synchronization between Parent and Child Process Objects

the write operation are as follows:

- The place DINFS must contain two $< WTC >$ method tokens.

- As indicated in the inscribed structure of TD8 in Figure 8.20, *DPA*, and *ACPA1* and *ACPA2* must exist (i.e., DPAE, A1E, and A2E respectively must be equal to 1) or be active for the write to be successfully accomplished.

- There must be new data (i.e., NA1D and NA2D should be greater than zero) for both A1 and A2 to justify the execution of method $< WTC >$.

The presence of two $< WTC >$ tokens is indicative of the fact that DPA can concurrently write to *A1* and *A2* respectively.

The post conditions of firing the transition *TD8* are as follows:

- The number of writes (i.e., write_count) must be assigned to 2 after the write is finished.

- The method tokens $< WTC >$ must be destroyed after the firing of transition $TD8$.

- New input method tokens $< WTP >$ (Write_to_Parent) must be created and placed in places ACPA1WS and ACPA2WS respectively.

Once parent DPA has successfully written to $A1$, and $A2$, it moves into the waiting state (DCWS). In order to emerge out of the waiting state (possibly for another write operation) the following preconditions and postconditions need to be satisfied.

The preconditions for firing of transition $TD9$ are:

- the place DCWS must contain two Read_from_Child ($< RFC >$) method tokens.

- all three, i.e., DPA, $A1$ and $A2$ must still be active, and

- the read_count must be equal to the write_count to ensure that $A1$, and $A2$ have already read the data.

The post condition of transition $TD9$ reinitializes the write and the read count to enable another cycle of read and write cycle to be executed. It may be noted here that although DPA has to synchronize its operations to emerge out of its waiting state. More details on SCPN modeling of applications based on IMAHDA and its simulation/verification in terms of deadlocks, and reachability graph can be found in Khosla and Dillon (1994).

The dynamic analysis of IMAHDA in this section provides us with dynamic constructs or knowledge like agent states, transitions, inscribed transition structure (preconditions, post conditions), and tokens. The present SCPN model although has accomplished the primary objective of developing a set of dynamic constructs for IMAHDA However, in the existing SCPN model we have made an assumption here that actions in an agent are executed instantaneously. This needs to be substituted with a notion of time in firing of the transitions. A time box can be incorporated along with the inscribed structure in each transition for keeping an account of time delay associated with execution of actions.

8.9 COMPREHENSIVE VIEW OF IMAHDA'S AGENTS

This chapter and the last one have outlined various computational aspects of IMAHDA. This includes the distributed hybrid multi-agent framework in which IMAHDA can operate, the learning constructs for the decomposition, control, and decision agents, the software, and intelligent agent building blocks of the problem solving agents, the communication levels and channels in IMAHDA, and the dynamic constructs for analysis of IMAHDA.

Table 8.1.   Problem Solving Agent Description

PROBLEM SOLVING AGENT	(SOME) PERCEPTS	(SOME) ACTIONS	GOALS	ENVIRONMENT
Global Preprocessing	discrete, fuzzy, continuous data	Transform Data, Apply Global Noise Filtering Heuristics, Formulate Execution Sequence	Input Conditioning, Noise Filtering	Dynamic, Non-deterministic, Continuous, fuzzy
Decomposition	Conditioned discrete, fuzzy, continuous data	Apply Context Validation Rules, Select or Extract Input Features of Abstract Classes, Select or Extract Abstract Classes/Concepts, Apply Learning Algorithm, Formulate Execution Sequence	Input Validation, Context Validation, Problem Formulation Determine Abstract Classes	Dynamic, Non-deterministic, Continuous, Fuzzy

Table 8.2. Problem Solving Agent Description Contd.

PROBLEM SOLVING AGENT	LEARNING KNOWLEDGE		KNOWLEDGE REPRESENTATION & REASONING CONSTRUCTS	SERVICE PROVIDER AGENT	DYNAMIC KNOWLEDGE CONSTRUCTS	COMMUNICATION KNOWLEDGE CONSTRUCTS
	INPUT FEATURES	OUTPUT CLASSES				
Global Preprocessing	Continuous Data	Noise Classes Symbolic/ Fuzzy Rules	Symbolic/ Fuzzy Rules, Distributed Neurons, Genes, Objects	Distributed Processing Communication Relational Belief Expert System Fuzzy System Genetic Algorithm Sup Neural Network	Agent States Events Tokens Transitions Inscriptions	Inform: Global Conditioned Data Request: Global Data Command: Query
Decomposition	Global General, Key, Binary Structured	Abstract Domain Classes	Symbolic Rules, Objects, Distributed Neurons	Distributed Processing Communication Relational Expert System Sup Neural Network	Agent States Events Tokens Transitions Inscriptions	Inform: Local Data Request: Global Conditioned Data

Table 8.3. Problem Solving Agent Description Contd.

PROBLEM SOLVING AGENT	(SOME) PERCEPTS	(SOME) ACTIONS	GOALS	ENVIRONMENT
Control	Conditioned discrete, fuzzy, continuous data	Apply Local Context Validation Rules, Apply Local Noise Filtering Heuristics, Select Input Features for Decision Level Classes, Select Decision Level Classes, Learn New Decision Level Classes, Learn New Contexts, Calculate Time & Resources Fuzzify Input Features, Formulate Execution Sequence Apply Intelligent Models for Resolving Conflicting Outcomes	Local Noise Filtering Input Validation, Context Validation, Problem Formulation Viability/Utility Determine Decision Level Classes, Optimization Rule Extraction Resolve Conflicting Outcomes	Dynamic, Non-deterministic, Continuous, Fuzzy
Decision	Conditioned Fuzzy, Continuous data	Apply Local Noise Filtering Heuristics (optional), Apply Local Context Validation Rules, Select or Extract Input Features for Specific Decisions, Select/Determine Classes/Concepts Representing Specific Decisions Select Learning Algorithm, Encode Data or Structure into Genetic Strings, Formulate Execution Sequence	Noise Filtering Input Validation, Context Validation, Problem Formulation Determine Specific Classification or Decisions	Dynamic, Non-deterministic, Continuous, Fuzzy
Post Processing	Fuzzy, Discrete, Continuous	Defuzzify Results/Decision/Specific Classification Determine Reliability of Results Apply Application DependentSpecific Decision Model Validate with User or Environment Learn Models Through Feedback Determine Explicit Knowledge for Explanation	Validate Results Explain Results	Dynamic, Non-deterministic, Continuous, Fuzzy

Table 8.4.   Problem Solving Agent Description Contd.

PROBLEM SOLVING AGENT	LEARNING — INPUT CLASSES	KNOWLEDGE — OUTPUT CLASSES	KNOWLEDGE REPRESENTATION & REASONING CONSTRUCTS	SERVICE PROVIDER AGENT	DYNAMIC KNOWLEDGE CONSTRUCTS	COMMUNICATION KNOWLEDGE CONSTRUCTS
Control	Local General, Key, Binary, Structured, Broad Range Linear or Fuzzy	Decision Level Classes	Symbolic/Fuzzy Rules Objects Genes Distributed Neurons Fuzzy Logic	Distributed Processing Communication Relational Belief Expert System Fuzzy System Genetic Algorithm Sup Neural Network Unsup Neural Network	Agent States Events Tokens Transition Inscription	Inform: Decision Data Command: Decision Verify: Decision Command: Query
Decision	Refined Linear or Fuzzy, Continuous	Instances of Decision Classes or Specific Decisions	Symbolic/Fuzzy Rules Objects Genes Distributed Neurons Fuzzy Logic	Distributed Processing Communication Relational Belief Expert System Fuzzy System Genetic Algorithm Sup Neural Network Unsup Neural Network	Agent States Events Tokens Transition Inscription	Request: Decision Data Inform: Decision Request: Validation Command: Query Request: Explain Display: Decision
Postprocessing			Symbolic/Fuzzy Rules Objects Genes Distributed Neurons Fuzzy Logic	Distributed Processing Communication Relational Expert System Fuzzy System Genetic Algorithm Sup Neural Network Unsup Neural Network	Agent States Events Tokens Transition Inscription	Command: Adapt Display: Explanation Inform: Validation

A comprehensive agent based description of the computational level is shown in Tables 8.1, 8.2, 8.3, and 8.4 respectively. The top half of these figures provides the PAGE descriptions and the bottom half provides the computational level extensions of the problem solving agents of IMAHDA described in this chapter. These figures summarize the vocabulary developed in the last chapter and this one to build intelligent hybrid multi-agent systems in complex data intensive domains.

8.10   EMERGENT CHARACTERISTICS OF IMAHDA

In this section, some behavioral characteristics of IMAHDA are outlined which provide it the richness and flexibility to cope with complexities of real world problems.

8.10.1   Goal Orientation & Autonomy

IMAHDA's problem solving agents operate based on goals/tasks assigned to them. They are distinguishable from each other based on goals undertaken by them. A problem solving agent in IMAHDA depends upon the external environment and on other problem solving agents for its information needs. They are largely autonomous in the sense that they accomplish the task on their own once provided with the required information. They are semi-autonomous in the sense that some of them (e.g. Decision agents) require information in a partially processed form from other agents or may require human intervention (or other external agents) for decision validation, and adaptation purposes.

8.10.2   Collaboration

IMAHDA as a whole is based on the concept of collaboration where all intelligent agents combine together in the transition from the problem state to solution state. Different agents discharge their responsibilities in different stages of the problem solving process. The decision agent depends upon the information and directions from the control agent, control agent depends upon the decomposition agent, and so on.

8.10.3   Flexibility and Versatility

IMAHDA's problem solving agents can seek services from a suite of intelligent agents and software agents in order to accomplish their goals. This feature enhances the flexibility of IMAHDA in terms of the range of the tasks and situations it can handle.

8.10.4   Global and Local Distribution/Parallelism

IMAHDA's agents exhibit global as well as local parallelism. Global parallelism is exhibited by the process agents, whereas local parallelism is exhibited by the artificial neural networks in the domain problem solving agents. The services of the *OSP* agent incorporate an arbitrary degree of complexity in IMAHDA's problem solving agents. For example, problem solving agent can be executed as a single process or the list of actions in the problem solving agent relating to two or more heterogeneous knowledge structures, can be executed through two or more processes in a competitive or a cooperative manner.

8.10.5   Learning and Adaptation

The decomposition, control and decision agents IMAHDA have learning and adaption properties based on the learning knowledge and strategy described in this chapter. This feature helps them to learn and adapt to new situations. The a *priori* learning structure of IMAHDA learns easy or simple data before it learns complex data.

8.10.6   Knowledge Representation

Knowledge in IMAHDA can be represented in various forms including objects, rules, fuzzy logic, distributed neurons, genes/chromosomes and others as shown in Tables 8.2 and 8.4 respectively. The object layer enriches knowledge representation in IMAHDA through inheritance, composition, and other non-hierarchical relationships.

Multiple inheritance from the generic intelligent agents like *Expert System (ES), Supervised Artificial Neural Network (SANN) or (ANN), Genetic Algorithm (GA),* and *Fuzzy Logic (FL)* agents enable creation of heterogeneous or hybrid problem solving agents.

8.10.7   Real-Time System Constraints

Temporal reasoning forms an important part of soft and hard real time systems. The term temporal reasoning here is interpreted in in two ways. Firstly, it is interpreted as reasoning under time constraints. Secondly, it is interpreted as reasoning with constantly changing data. In other words, data used for inferencing in time $t1$ is used along with new data in time $t2$ where $t2 > t1$. Further, the inferences made with data in time $t1$ can be used or discarded in light of new data in $t2$. Real time system requirements like temporal reasoning , reliability and fast response time impose stringent constraints on the design of such systems. Temporal reasoning constraints in real-time systems require increased predictability with respect to:

- states and behaviors required of the system (for example the execution precedence)

- states and behaviors exhibited by the system (for example process execution times)

The decomposition, control and decision problem solving agents of IMAHDA guarantee a macro level execution precedence for an application. The concepts of aggregation or composition, uniform communication interface, encapsulation reflected in IMAHDA increase the predictability level of execution precedence for large complex real-time systems.

8.10.8 Cyclic Operation

As highlighted in the last section IMAHDA as a whole can engage in cyclic operation without human intervention. This feature along with the persistency of the agents will be demonstrated in chapters 10, 11 and 12 respectively while describing a real time application of IMAHDA.

8.10.9 Two Level Three Channel Communication.

Multiple levels and channels of communication shown in Figure 8.1 provide additional flexibility to a problem solving agent in IMAHDA to achieve its goals. The communication constructs of the problem solving agents shown in Figures 8.1 define the nature of communication between different problem solving agents. Also, as explained in the previous section an IMAHDA's agent can engage in synchronous as well as asynchronous communication with other problem solving agents.

8.10.10 Maintainability and Knowledge Sharing

Maintainability is an important management issue for development of large scale systems. IMAHDA facilitates maintainability and knowledge sharing of large scale systems in a number of ways. Firstly, the problem solving agents of IMAHDA help the problem solver to deal with the complexity of the domain with a deliberative-cum-automated reasoning structure. Secondly, the object-oriented properties of IMAHDA help to encapsulate knowledge in different problem solving agents, thus facilitating change and maintenance. Thirdly, the various levels of information processing in IMAHDA facilitate horizontal and vertical scalability at each level. Finally, the genericity of the software, intelligent, and problem solving agents of IMAHDA facilitate knowledge sharing and reuse across applications. This will help to reduce the cost of application development.

Further, the process model with its capability of creating and deleting processes makes provision for dynamic memory management during run time. All these important properties facilitate use of IMAHDA for large and complex systems.

In chapters 9, 10, 11, and 12 a real time alarm processing application of IMAHDA is described.

8.11 SUMMARY

IMAHDA, at the computational level is made up of objects, software agents, intelligent agents, and problem solving agents. These components constitute the building blocks of IMAHDA. The knowledge content of IMAHDA includes among other aspects communication knowledge, learning knowledge and strategy and dynamic analysis knowledge. Communication, in IMAHDA occurs at two levels, namely, the process level, and the domain level. The communication constructs in IMAHDA explicitly establish the role of different problem solving agents and the nature of communication between the problem solving agents.

Learning knowledge determines the type of input features and output classes which need to be selected or extracted in the decomposition, control and decision agents of IMAHDA. Learning strategy on the other hand determines the method of selection of input features in different phases.

The dynamic knowledge of IMAHDA has been defined using State Controlled Petri Nets (SCPN). The dynamic knowledge consists of agent state, action tokens, transitions, transition firing rules, and transition inscriptions. These dynamic components, among other aspects, are used to model IMAHDA dynamic features like agent states, concurrency, cyclicity, distributed communication, action execution, and dynamic inheritance. Lastly, the emergent behavioral characteristics of IMAHDA in terms of goal orientation, autonomy, flexibility, learning and adaptation, global and local parallelism, synchronous/asynchronous communication, and maintainability and knowledge sharing are outlined. A comprehensive agent based description of the computational level which incorporates these characteristics for building intelligent hybrid multi-agent systems in complex data intensive domains is provided.

Notes

1. The lines from input features to output classes imply affirmation or negation.

2. Although, it is used in the Mammal class, it can be done away with if required through elimination.

3. The terms parent and child are associated with unix operating system function fork()
where a parent forks one or more children.

References

Arbib, M. A. (1987) *Brains, Machines and Mathematics*, Springer-Verlag, New York.

J. Cardoso, et al. (1989), "Petri nets with uncertain markings,," *10th International Conference on Applications of Petri Nets*, Bonn, Germany.

Carpenter, G. A. and Grossberg, S. (1988), "The ART of Adaptive Pattern Recognition by a Self-Organizing Neural Network," *IEEE Computer*, vol. 21, no. 3, March, pp. 77-88

Chen, L, 1992, "Integration of Neural Networks and Expert Systems for Condition Monitoring," *Proceedings of the 5th Australian Joint Conference on Artificial Intelligence*, pp. 95-100.

Dillon, T. and Tan, P. L. (1993), *Object-Oriented Conceptual Modeling*, Prentice Hall, Sydney, Australia.

Genrich, A. and Lautenbach, K. (1981), "System Modeling with High Level Petri Nets," *Theoretical Computer Science*, vol. 35, pp. 1-41.

Fischer, K., Muller, J.P., and Pischel, M. (1994), "Unifying Control in a Layered Agent Architecture," *Technical Report TM-94-05 from DFKI GmbH*, German Research Center for Artificial Intelligence.

Hanks, S., Pollack, M.E., and Cohen, P.R. (1993), "Benchmarks, Test Beds, Controlled Experimentation, and Design of Agent Architectures," *AI Magazine*, winter, pp. 17-39.

Jensen, K. (1983), "High Level Petri Nets," *Application and Theory of Petri nets* , Informatik Fachberichte 66, Springer Publishing.

Jensen, K., "Colored Petri Nets and the Invariant Method," *Theoretical Computer Science*, vol. 14, pp. 317-36.

Khosla, R., and Dillon, T. (1994), "Dynamic Analysis of a Real Time Object Oriented Integrated Symbolic-Connectionist System using State Controlled Petri Nets," *Fifth Intelligent Systems Applications Conference*, Avignon, France.

Khosla, R. and Dillon, T. (1995a),"Symbolic-Subsymbolic Agent Architecture for Configuring Power Network Faults," *International Conference on Multi-Agent Systems* , ICMAS'95, San Francisco, USA, June.

Khosla, R. and Dillon, T. (1995b),"Task Structure Level and Computational: Architectural Issues in Symbolic-Connectionist Integration," *International Joint Conference of Artificial Intelligence (IJCAI95), Working no.es of the IJCAI Workshop on Neuro-Symbolic Integration*, Montreal, Canada, August.

Khosla, R. and Dillon, T. (1995c), "Knowledge Modelling in Integrated Symbolic-Connectionist Systems," *IEEE Conference on Systems, Man, and Cybernetics*, Vancouver, Canada.

Khosla, R. and Dillon, T. (1997), "Intelligent Hybrid Multi-Agent Task Level Architecture for Knowledge Engineering of Complex Systems," to appear in *IEEE International Conference on Neural Networks*, Houston, USA.

Kohonen, T. (1990), *Self Organisation and Associative Memory*, Springer-Verlag.

Lai, R., Parker, K. R. and Dillon, T. S. (1989), "Application of Numerical Petri Nets to Specify ISO FTAM Protocol," *Proceedings of the 1989 Singapore International Conference on Networks*.

Landweber, L., and Robertson, E. (1978) "Properties of Conflict Free and Persistent Petri Nets," *Journal of the ACM*, vol. 25, pp. 352-64.

Liu, N. K. and Dillon, T. S. (1989), " An Approach towards Verification of Expert Syste ms Using Petri Nets," *International Journal of Intelligent Systems*, vol. 6 , no. 3, pp. 255-76.

Malsburg, V. D. (1988), Goal and Architecture of Neural Computers, *Neural Computers*, Springer Verlag, Berlin, West Germany.

Michalski, R. S. (1987), "Learning Strategies and Automated Knowledge Acquisition: An Overview," *Computational Models of Learning*, Springer-Verlag, Berlin, pp. 1-20.

Petri, C. (1966) "Communication in Automata" (translated into English), *Technical Report RADC-TR-65-377*, vol. 1, Rome Air Development Center, Griffths Air Base, USA, January.

Pressman, R. S. (1992), *Software Engineering: A Practioner's Approach* , McGraw-Hill International, Singapore.

Ramamoorthy, C. V. and Ho, G. S. (1980), "Performance Evaluation of Asynchronous Concurrency Systems Using Petri Nets," *IEEE Transactions on Software Engineering*, vol. SE-5, pp. 440-9.

Russell, S., and no.vig, P. (1995), *Artificial Intelligence - A Modern Approach*, Prentice Hall, New Jersey, USA.

Schiebe, M. and Pferrer, S. (1992), *Real-Time Systems Engineering and Applications*, Kluwer Academic Publishers, USA.

Symons, F. J. W. (1978), "Modeling and analysis of communication protocols using Numerical Petri Nets," *PhD thesis*, Department of Electrical Engg., University of Essex, Telecommunication System Group.

Tang, S. K., Dillon. T. and Khosla, R. (1995), "Fuzzy Logic and Knowledge Representation in a Symbolic-Subsymbolic Architecture," *1995, IEEE International Conference on Neural Networks*, Perth, Australia

Wheeler, G.R. (1985), "Numerical Petri Nets - A definition," *Technical Report 7780*, Telecom Australia, Research Laboratories, Clayton, Victoria, Australia.

Wheeler, G.R., Wilbur-Ham, M.C., Billington, J., Gilmour, J.A. (1986), "Protocol analysis using Numerical Petri Nets," *Lecture no.es in Computer Science*, 188:435–452.

Wolf, J. G. (1987), "Cognitive Development as Optimization," *Computational Models of Learning*, Springer-Verlag, Berlin, pp. 161-206.

Zurawski, R . and Dillon, T.S. (1991), "Systematic Construction of Functional Abstractions of Petri Net Models of Typical components of Flexible Manufacturing Systems," *IEEE Workshop on Petri Nets and Performance Models*, Melbourne, Australia.

9 ALARM PROCESSING - AN APPLICATION OF IMAHDA

9.1 INTRODUCTION

Real time applications to control industrial, medical, scientific, consumer, environmental and other processes is rapidly growing. Today such systems can be found in nuclear power stations, computer-controlled chemical plants, flight control, etc.. This growth however, has also brought to the forefront some of problems with the existing technologies. These problems have pushed for research into new techniques which could be used for solving these problems. 1991; Liu et al. 1992). The problems range from the enormous size of the power system to the fast response time constraints in emergency situations. This is the primary reason for selecting this domain as the domain of application.

This chapter looks into the various problems associated with alarm processing and the feasibility of IMAHDA in addressing these problems in a regional power system control center.

The chapter is organized into four parts. Firstly, the problems associated with alarm processing are described in general. Secondly, a brief survey of the existing alarm processing systems is carried out. Thirdly, the feasibility of IMAHDA for alarm processing systems is discussed. Finally, the focus is

narrowed down to description of alarms associated with Thomastown Terminal
Station (TTS), in Melbourne, Victoria, Australia (selected for application of
IMAHDA) and the objectives of the alarm processing system.

9.2 CHARACTERISTICS OF THE PROBLEM

The problems associated with alarm processing have been bothering the minds
of power system researchers since the 1977 New York blackout. The social
and economic consequences of a major interruption in the supply of electric
power are so great that every effort should be made to reduce the impact of a
disturbance (Electrical World 1990; Ewart 1978). In order to understand the
problems associated with alarm processing, it is useful to have a perspective of
the underlying structure in which the alarm processing systems function in a
power system.

A typical power system consists of three levels of control, namely system level
control or EMS (Energy Management System) control, grid or regional level
control, and distribution level control. Alarm processing systems are used at
all these three levels which can give some indication of the complexity involved.
The EMS systems are not only responsible for the overview of the power system grid but also the generation of power and interconnection to other power
systems. The grid or regional systems are responsible for transmission of power
between generation and distribution systems. Finally, the distribution based
systems are concerned with maintenance and delivery of power to the customers
from a distribution substation.

An alarm is a structured signal from the computerized supervisory control
and data acquisition (SCADA) system which is used in a power system control
center at system, regional or distribution level. A few example situations in
which an alarm message is produced are:

a. A line flow exceeds its normal limit.

b. A line flow exceeds its emergency limit.

c. A bus voltage drops below its minimum value or exceeds its maximum value.

d. A breaker is opened or closed.

e. A sensor detects an excessive temperature in a transformer.

f. The automatic generation control program detects an excessive area control
   error.

g. No data is obtained from a particular substation at the latest scan (perhaps
   due to a temporary failure in a communication link).

h. A peripheral unit is not functioning properly and the computer configuration control program has replaced it by its backup.

Alarm processing thus relies on status information or measurements gathered at a large number of points distributed throughout the system. The advances in computer and telecommunications technology in the last three decades have made it possible for a large SCADA system to scan 20,000 to 50,000 points every few seconds. Such a SCADA system will have the ability of displaying 500/1000 or even more alarm messages per minute (Kirschen et al. 1988; Munneke and Dillon 1989; Kirschen and Wollenberg 1992). These days three or more operators are needed to oversee such a complex system which scans dozens of terminal stations and substations spanning hundreds of miles.

The enormity of the system and rate at which messages are displayed on the monitor increases the complexity of the problem multifold. This complexity leaves a operator in the control center who has to analyze these messages in a highly constrained time frame suffering from high stress and cognitive overload as explained in the following section.

9.2.1   Operator Stress

Under normal circumstances, the operators carry out routine actions and adjustments to optimize the security and economics of the power system. Uncontrollable events such as sudden load fluctuations, equipment failures and atmospheric perturbations can propel the system from a stable and secure state to an insecure and unstable state. The operator in these circumstances has to take immediate action in order to pull back the system into an acceptable state. Failure to respond quickly, can lead to catastrophic circumstances like the state of the system may continue to deteriorate with some loads getting disconnected or in extreme cases, the entire system may collapse, leading to a blackout for hours.

An approximate estimate of number of alarms which could be triggered under such or similar events in a regional control center is (Durocher 1990):

- up to 150 alarms in 2 seconds for a transformer fault;

- up to 2000 alarms for a generation substation fault, the first 300 alarms being generated during the first five seconds;

- up to 20 alarms per second during a thunderstorm;

- up to 15000 alarms for each regional center during the the first five seconds of a complete system collapse.

Such an enormous rate of alarm messages makes a quick response from the operator difficult. First, by suggesting a catastrophe (because of large number

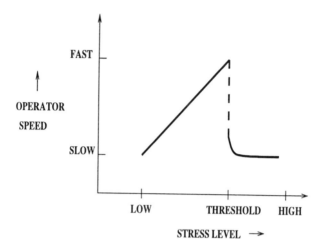

Figure 9.1.   Operator Speed vs Stress Level © IEEE 1992(Kirschen and Wollenberg 1992)

of messages) , it may increase the level of stress beyond the threshold level at which the performance of the operator drops down sharply as shown Figure 9.1. Having overcome the element of surprise, the operator must sift through a large number of messages to find the cause of the problem. This may even involve scanning through the design manuals or drawings as the operator finds it difficult to process the large number of messages on their own. A significant amount of time can be wasted in this search which might be crucial to prevent a deterioration of the situation. Finally, an operator working under stress and with an over abundance of data may easily be misled as to the true nature of the problem.

Besides the large number of alarms and the rate at which they are displayed on the monitor, there are certain other characteristics in power systems which get reflected in the alarms and have to be taken into account for analysis (Inoue et al. 1989). They are:

- Some alarms are needlessly repeated and distract from more important ones

- Some alarms are faulty alarms and need to be suppressed.

- Multiple sympathetic alarms can occur for a single event.

- Multiple alarms can occur for multiple events.

- Alarms are not annunciated in order of priority.

■ Alarms remain displayed after being acknowledged.

To further complicate matters, there is a temporal dimension associated with the alarms as there is with most real-time systems which increases the complexity manifold.

9.2.2   Temporal Reasoning

Temporal reasoning is an important aspect of alarm processing systems. The status of a power system constantly evolves over time. Alarms which monitor the constantly changing status of a power system have to be reasoned in a temporal fashion for the following reasons (Kirschen and Wollenberg 1992):

■ the time provided with alarm messages is the time when the corresponding information arrived at the control center and not the time when the corresponding fact really happened.

■ messages coming from different plants have different transmission. times

■ the meaning of a message depends on what happened before and, in certain cases, on what happened next.

Due to the above reasons, inference made on a particular group of alarms may become invalid in light of the new alarm messages. Thus, in an alarm processing system, reasoning cannot be based on static facts but has to be based on a constantly changing set of assertions. The conclusions made by the system have to be constantly updated to account for the changing set of assertions.

9.3   SURVEY OF EXISTING METHODS

The various approaches to solve the alarm processing problem have ranged from augmenting the conventional alarm processing programs to development of specialized expert systems and some suggestions as to the use of neural networks.

The augmentation of the conventional alarm processing programs through use of filtering mechanisms, priority and grouping schemes and message routing procedures does help in reduction of alarms to some extent (Amelink et al. 1986). However, considering the sophistication of present EMS and SCADA systems these are to some extent primitive measures as they cannot be used to synthesize messages with higher information content. Some researchers have used neural networks for alarm processing (e.g. Chan 1989, 90; Jongpier et al. 1991). However, they have experimented with their ideas only on small systems. Further, on their own admission, they have expressed the concerns about limitations of neural networks for use on full-scale systems (Chan 1989, 90).

The use of expert systems for alarm handling has been originally initiated by Wollenberg (1986). He describes in detail the various rules required for alarm processing. According to him, processing alarms using a rule-based expert system offers a lot more flexibility as against the conventional approach. Firstly, rule-based systems can be used used to filter and prioritize alarms. Secondly, they can be used to combine various alarms and correlate the information gathered from other sources.

The papers by Talukdar et al. (1986) and Kirschen et al. (1988, 89) are strategic papers which are applicable at any level of the power system hierarchy. Talukdar et al. propose a blackboard architecture and concurrent processing for development of alarm processing systems in control centers. They also describe a simulator - diagnostician model where the diagnostician creates several hypotheses on possible system topologies while the simulator produces the relay operations that will follow the present power system activity. The simulator is then used to highlight the most probable hypothesis. The main problem with the blackboard technique is the communication between various disparate knowledge sources, control of the global state of the system and interpretation of results for use by different knowledge sources.

The paper by Kirschen, et al. is an extension of Wollenberg's (1986) paper on Intelligent Alarm Processor (IAP). The discussion reiterates the stages of alarm processing as stated by EPRI (1983) that operator must:

- become aware of the alarm.

- determine the events that caused the alarm.

- analyze the consequences of that event.

- review the sequence of events leading to the alarm, and

- determine the course of action.

There are four criteria which govern the design of the IAP:

- Only those messages requiring the operator's immediate attention should be displayed.

- Keep the operator aware of problems as they occur.

- Reduce the alarm loading by eliminating redundant alarms and summarizing similar alarms.

- Perform deeper analysis. The operator often needs to make inferences based on the alarm messages received. Such inferencing it is claimed can be accomplished faster and more reliably by a rule-based system.

Inoue et al. (1989) describes an Intelligent Alarm Processing for EMS and SCADA systems which which gives a graphical picture to the operator of the exact location of the fault, the list of alarms related to the fault along with the protective equipment which has operated and which has not.

The expert system developed consists of the following functions:

- Determination of the faulted area.

- Conflicting information checking.

- Information priority ranking.

- Information correlation

The expert system processing is initiated if a circuit breaker has operated or under voltage has occurred in the network.

The knowledge is represented in the form of rules and inferred in the forward direction. The system only uses simple knowledge to increase reliability. The general location of the fault is isolated by using information priority ranking on the information of protective zones of relays. The conflicting information check is done by cross checking information on relay operation and circuit breaker tripping. A summary of the alarms related to the faulted area or equipment is presented to the operator as part of information correlation. This results in the suppression of low-priority alarms.

The expert system has combined two parts, that is, improving the quality of alarm information presented to the operator (by suppressing low priority alarms) and doing fault diagnosis. The method adopted for information priority ranking is good. However, as mentioned in the paper it uses simple knowledge for fault diagnosis. In certain fault conditions, the information may not be presented in a straight forward form and would thus require more complex analysis. Choosing only rules for knowledge representation will make the system inefficient and may seriously impede real time application.

Purucker (1989) in his paper discusses the design and operation of an Communication Alarm Processor (CAP), installed at a central operations control center which receives real time alarm data from two microwave monitoring and data transmission systems.

The real-time alarm data is read, compressed, filtered and buffered until the expert system is ready for new data. The expert system reads in the new alarms, gets rid of old ones, then diagnoses an alarm using the current alarm data. The expert system concludes its diagnosis by writing the results to a disk. The operator interface reads the diagnostic file and presents the diagnosis to the operator. It also allows operator to examine the past diagnosis, log observations, archive conclusions, and thus gives the operator the control to use his expertise when the expert system falls short.

There are four inferencing stages in CAP involving suppression of alarms, prioritizing alarms, mapping alarms to the topology and performing diagnosis. Confidence factors are used to limit the number of diagnostics or causes presented to the operator.

CAP has 230 rules. The real-time response time in CAP (which is the difference between the alarms arrival at the expert system and the expert system's diagnosis) of the expert system was measured as less than five seconds. However, under very heavy input the expert system's performance degrades appreciably.

The SAA (SOCCS Alarm Advisor) system built by Shoop, et. al. (1989) achieves the following three objectives:

- Alarm filtering - involves filtering of nuisance alarms from the input stream.

- Alarm Interpretation - involves recognizing and reporting out of service, alive or back feed and other specific events. The deduced status is compared against analogs for discrepancies.

- Recommended Actions - involves displaying recommended operator actions based on Con Edison company's (where it is installed) internal procedures.

The interesting aspect of SAA is its knowledge representation and temporal reasoning of alarms. SAA has a layered representation of the power network. The first layer is the physical model which describes the devices contained in the network as well as their connections. The physical devices are organized in objects called "clusters" and "switch groups" based on the relay protection zones. A functional model is constructed on top of the physical model to aggregate devices and facilitate causal reasoning. It is used to describe the condition of the cluster given the switch group status. The main heuristic to limit the search for cluster isolation is analog confirmation. The third layer is a temporal model which is used to maintain an accurate representation of the network at every moment of time and to reason about trends and previous history. It relies on a Doyle-style truth maintenance system for retracting conclusions which have become invalid in light of new data. Automatic retractions are allowed up to five minutes after the initiating alarm for a particular event. However, this conflicts with a response time expectancy of 20 seconds by operators.

A temporal reasoning strategy similar to SAA is also employed by the intelligent alarm processor developed by Bijoch, et al. (1991) at Northern States

Power (NSP) at Minneapolis, USA. The alarm processor maintains a situation stack of unresolved alarms. The unresolved alarms are a) combined with new ones for reinferencing, b) canceled if the expiration time has elapsed and c) canceled if the reported event no longer exists.

Tesch et al. (1990) describe a KBAP (Knowledge Based Alarm Processor) being built for Wisconsin Electric Power Company. The purpose of KBAP is to reduce numerous, ungrouped alarms to a small number of diagnostic messages in a timely manner. For achieving this objective, KBAP employs a "station-oriented" approach where the alarms are related to a particular station and not to the entire system topology. KBAP uses meta-rules and configuration or topology database to hypothesize the cause of the alarms. It then backward chains through so called object-like rules which attempt to prove the hypothesis. Any new alarms have to wait till the existing alarms have been inferenced. It the hypothesis is proven as true it is then passed to the conclusion processor. The conclusion processor then converts the valid hypothesis into a form suitable for presentation to the SCADA system. The conclusion processor also sends messages to the SCADA system to advise of messages which are no longer valid. If the hypothesis is false, next one is tried till the list is exhausted. If all hypothesis fail, the system waits until a new change of state is received. Here again, in emergency situations meeting the real time constraints can be difficult because of the search involved and hard wiredness of the rules.

More sophisticated knowledge based and model based systems for alarm handling and fault location have been developed recently by Eickhoff et al. (1991) and Pfau-Wagenbauer et al. (1991) respectively. However, they too suffer from problems of search and response time.

9.4   IMAHDA AND ALARM PROCESSING

The usefulness of IMAHDA for alarm processing and fault diagnosis can be evaluated based on the following weaknesses in the existing expert system approaches

9.4.1   Temporal Reasoning

Temporal reasoning as has been discussed in the section 9.2.2 is an important characteristic of the alarm processing systems. These temporal characteristics directly conflict with the response time needs of these systems which vary from a few seconds to 30 seconds. The existing expert system approaches of retracting conclusions in light of new data (like in SAA) or maintaining a stack (like in NSP) of unprocessed alarms (which will be processed when new data comes in) is both time consuming and inefficient. It is time consuming because there is search involved with every inference. As a result the time difference between the old inference which has become invalid in light of new data and new inference can be as long as few minutes or even more which conflicts with stringent response time requirements. It is inefficient because there is no surety that the unprocessed alarms can be processed on arrival of new data because of the hard wiredness associated rule based expert systems. That is, the information may still not be sufficient for processing and drawing conclusions.

IMAHDA can better cope with these temporal characteristics because of neural networks and its parallel distributed properties. There is practically no search involved with use of neural networks. The successive conclusions drawn by them will be pretty fast. The invalid conclusions can be retracted more quickly than their expert system system counterparts and will be closer to the response time requirements. Further, the parallel distributed properties are likely to have a catalytic effect on the processing time of IMAHDA in general.

9.4.2   Incomplete Information

Incomplete information is a characteristic associated with any time dependent large scale system. The power system with its complexity is no exception to it. In rule based expert systems processing incomplete information is a inherent problem whereas, in model based expert systems, the processing is slow.

The associative properties of neural networks facilitate processing of incomplete information and hence make them suitable for processing incomplete alarm messages.

9.4.3   Incorrect Information

Incorrect information can neither be processed by a rule based or a model based expert system. The fault tolerant properties of the neural network make it somewhat immune to noise or incorrect information. Additionally, the parallel and distributed properties of IMAHDA facilitate concurrency and competition among alarm data coming from different levels of protection. That is, alarm data related to different levels of protection (e.g. circuit breaker, relay and analog) can be processed in a fast, concurrent and competitive manner. This will enhance the reliability and also reduce the possibilities of catastrophic effect noise or malfunctions can have on inference made by the system. Thus, IMAHDA can cope better with noise or malfunctions in the power system.

9.4.4   Alarm Processing Including Fault Diagnosis

Search is a problem which restricts the incorporation of fault diagnosis in alarm processing systems. As far as the operator is concerned alarm processing has not only to do with what is happening but also why it is happening. The former largely involves filtering, reduction and summarization of alarm messages whereas, the latter involves identifying which component of the system is faulty and needs to kept in view while undertaking restoration.

Most existing alarm processing systems only provide information to the operator on what is happening and do not undertake fault diagnosis or detection because of problems of search associated with use of a deeper model required for fault diagnosis. Search related problems are minimal with neural networks. Search will only be involved in the case of a wrong classification by a neural network in the decision phase. In those circumstances a symbolic fault propagation model will have to be used as a standby for obtaining correct classification. However, the three phases (decomposition, control and decision), parallel, distributed and modular properties of IMAHDA as stated in chapters 6, and 8 respectively have a localizing effect on this search. That is, they will have a restrictive effect on the size of the fault propagation models associated with proving the results from different neural networks in the decision phase.

9.4.5   Processing Speed

Through the use of IMAHDA, there is going to be an overall improvement in the processing speed of the alarms. This means that more alarms can be handled in less time as compared to the existing approaches. Also with faster processing the functional capacity of the alarm processing system would improve.

9.4.6   Development Time

The development time in existing expert systems has varied from 6 months to a few years depending upon the complexity and objectives of the system (Dillon 1991; Liu et al. 1992). The existing expert system approaches involve writing hundreds if not thousands of rules to cope with the size of the power system. This makes writing of symbolic rules cumbersome and error prone. The problem is not only writing a large number of symbolic rules but also carrying out consistency checks. This puts a lot of strain on the human resources in a power system utility and is one of the problems of getting such developmental projects through the top management. By using IMAHDA the development time would be cut drastically. Where it is possible, an exhaustive use of neural networks will short circuit the need of explicitly writing hundreds of symbolic rules.

9.4.7   Maintenance

Maintenance of large expert systems in power systems is known to be time consuming and inefficient (Liu et al. 1992). The parallel, distributed and modular properties of IMAHDA facilitate easier maintenance as they restrict the size of the different components of the system. IMAHDA as a whole is conducive to vertical and horizontal scalability.

9.4.8   Hardware requirements

To compensate for the slowness, some of the existing expert system approaches rely on dedicated AI processors for meeting the response time requirements of the operators. This means buying separate hardware is a prerequisite for use of such systems.

Although, the use of IMAHDA will not rule out the need for separate hardware, it certainly would not rely on it so heavily as do the existing systems because of its reduced memory (e.g. trained neural nets occupy less memory as compared to expert systems) requirements and dynamic memory management (as stated in chapter 8), and for reasons mentioned in the previous sections.

9.5   APPLICATION OF IMAHDA

In previous sections we have discussed the problems associated with the existing approaches for processing alarms and fault diagnosis, and the feasibility of IMAHDA.

In the rest of this chapter the preliminary steps in the application of IMAHDA are undertaken. We discuss the problems associated with processing of alarms in a 220kv automated Thomastown Terminal Power Station (TTS) which is monitored by a regional control center under the jurisdiction of State Electricity Commission of Victoria (SECV). We conclude the discussion by outlining the objectives of the real time alarm processing system to be developed for TTS.

9.5.1   Alarm Monitoring in SECV

The regional control centers in SECV which are primarily responsible for commission's extensive transmission and subtransmission system are the most alarm intensive. The SCADA system in SECV's regional control center presents the data graphically as schematic or single line diagrams of the system topology and in tables most common of which are the alarm lists. These single line diagrams are complex schematics containing the system topology (with only the switching elements changing on any diagram), several analog values appearing in close proximity to the associated plant element and a single line of text at the bottom of diagram for alarm messages. In the regional control centers these complex schematics can be displayed on any one of the three monitors.

The alarms (220KV and 66KV) received by these control centers often originate from fully automated terminal stations (transmission substations) and distributed substations. As a consequence these alarms are often the only data that indicates equipment malfunction. During a single alarm event involving a circuit breaker trip, a single-line diagram of the affected part on one monitor and the alarm lists for the station concerned on other monitor are usually sufficient to isolate the cause of the event. However, the problem arises in case of a multiple alarm event when one single-line diagram and a single alarm list is not sufficient for interpretation and isolation of the cause of the event. The operator has to continuously swap through various displays in order to make proper sense out of the alarms. This problem becomes even more complicated when hundreds of alarms appear during a system emergency and the operator has to respond in a constrained time frame.

The operators in a SECV regional control center describe their work as "99% boredom and 1% sheer panic". This summarizes the single greatest problem of current SCADA technology at times of system emergency.

9.5.2   Characteristics of Alarms at TTS

We have chosen TTS which is monitored at a regional control center for developing the alarm processing system as it is one of the more complex automated terminal stations under the jurisdiction of SECV. It is responsible for transmitting and distributing power to 16 substations. There are 65 circuit breakers in TTS itself and its 16 associated substations. The topology of the TTS circuit breaker network is shown in Appendix C and D respectively.

The alarms emerging from TTS can be broadly categorized into three groups:

- Single Alarms

- Multiple Alarms

- Expected Event Alarms

Single alarms occur sufficiently often, particularly in relation to events outside the system topology or grid elements. These alarms require separate treatment and should be displayed with sufficient information.

Multiple alarms can be categorized as a) repeated alarm messages where the same message is received a number of times, b) sympathetic alarms that result from some form of equipment failure and can produce a disparate group of alarm messages, c) multiple alarm message groups that are indicative of the occurrence of a single event, and d) alarm messages that result from multiple plant failures.

Expected event alarms can be categorized as a) alarms that occur as a result of operations, correctly completed, and initiated by control room staff, b) alarm messages that result from maintenance activity or installation and commissioning of power system plant, and c) fleeting alarms produced by alarms that have been identified as faulty by the control room staff.

Description of various types of alarms which fall under these broad groups is as follows:

Routine Alarms: These are alarm messages that occur on regular basis. They indicate the proper functioning of equipment that has operated as a result of the activity of a timing device or a logic circuit. These alarms must be suppressed in the case of correct operation. In case of a maloperation, a maloperation message must be displayed.

Known Fault Alarms: These are alarms which have been identified by the operator as faulty. These alarms are usually single alarms and are related to faulty plant or telemetry equipment. These need to be suppressed.

Emergency Alarms: Single building fire alarms are indicative of a station or substation fire and need to be displayed immediately. However, if a fire equipment malfunction alarm is associated with building fire alarm then the alarm messages need to be suppressed.

DC Failure Alarms: These are a string of uncorrelated multiple alarms caused by a DC supply failure alarm. These uncorrelated alarms are usually communication alarms and ancillary equipment alarms. All other alarms except the 'ALARM-EQPT-DC-FAIL' need to be suppressed. A message is to be displayed indicating alarm equipment failure.

Repeated Alarms:

As the name suggests, the same alarm repeats itself one or more number of times. Each repetition of the alarm needs to be suppressed. The first occurrence of the alarm has to be processed and inferenced. If it cannot be inferenced, it is be displayed as faulty. Future occurrences of the alarm in a certain time range are to be suppressed. The first occurrence of the alarm is also recorded in a database for future use.

Sympathetic Alarms: The alarm or alarms indicating the actual event are sometimes swamped by alarms from ancillary or related equipment like motor-generator alarms, oscillograph alarms and communication alarms. These are called sympathetic alarms. These alarms indicate a voltage-dip or a communication equipment failure. If the communication alarms occur along with the motor-generator alarms and oscillograph alarms the event is classified as a voltage-dip and this message should be displayed. The communication alarms should be suppressed in the case the event has been classified as voltage-dip. Otherwise, depending upon the pattern of communication alarms, the event classification will be loss of a communication channel, microwave equipment failure, communication equipment power supply failure, etc.

Network Circuit Breaker (CB) Alarms: Multiple Network CB alarms may indicate multiple events (e.g., line, bus and transformer faults) or single events within an individual terminal station (in our case TTS). They could also indicate multiple station failure events which could be at a higher voltage level (i.e. 220kv terminal stations) or lower voltage level (i.e., 66kv substations). These events have to be identified reliably and displayed immediately for urgent operator action.

Supervisory Cable Alarms: There is another class called supervisory cable (the cable containing information gathered with the switchyard at a remote site) alarms. These are a group of alarms which spring up due to maintenance work on these cables. Such alarms have not been considered in the present alarm processing system because of insufficient alarm information available on these cables.

9.5.3   Objectives of the Alarm Processing System

The topology of the TTS circuit breaker network, different types of alarms in TTS and the problems associated with identifying and reasoning about them gives one some idea about the complexity of the problem involved. Three objectives were finalized after discussion with the SEC's engineers for development of real time alarm processing system for TTS. These objectives are to assist the operator in:

- isolation of the event.

- isolation of the cause of the event, and

- fast and reliable isolation of the event.

Isolation of the event involves reducing the number of alarms, providing summarized messages and eliminating noisy and faulty alarms.

Isolation of the cause of the event is limited to detection and reporting of the faulty component in the system.

Fast and reliable isolation of the event involves reliable isolation of the event within the desired response time constraints. The response time constraints associated with the alarm processing system are soft constraints. That is, the decisions or inferences made by the system will be be useful even if they do not fall into range of 4 to 30 seconds (it is the reason for separation of the first and second objectives). However, as a general guideline, the response time requirement of 4 seconds is for single alarms and simple (2/3) multiple alarms. On the other end of the scale the response time of 30 seconds is for severe cases/worst case scenarios (e.g. multiple faults with 25/30 circuit breaker alarms).

The enumeration of the objectives provides the focus based on which a detailed analysis and design of the problem can be carried out. This is undertaken in the next two chapters.

9.6 SUMMARY

Real time applications to control industrial, medical, scientific, consumer, env ironmental other processes is rapidly growing.etc.. This growth however, has also brought to the forefront some of problems with the existing technologies. Real time alarm processing in power systems is one such application where the existing AI technologies (e.g. expert systems) have not been able to adequately address the real time and other issues associated with alarm processing. These issues include temporal reasoning, incomplete information, incorrect information, search, speed, power system size, development time, etc.. Parallel and distributed processing, scalablity, learning, and other properties described in the last chapter make IMAHDA better suited for the alarm processing domain as compared to the existing approaches.

The 220kv automated Thomastown Terminal Power Station (TTS) selected for application of IMAHDA, with 65 circuit breakers and 16 connected substations is one of the most complex terminal power station monitored in the regional control center of the State Electricity Commission of Victoria. The alarms emanating from TTS fall under three groups, namely, single alarms, multiple and expected event alarms. The alarms in these groups are classified into routine, known fault, emergency, DC failure, repeated, sympathetic, network circuit breaker alarms and supervisory cable alarms based on the different characteristics exhibited by these alarms. The objectives of the real time alarm processing system to be developed are to assist the power system operator in isolation of the event, isolation of the cause of the event and fast and reliable isolation of the event. The enumeration of the objectives provides us the focus based on which a detailed analysis and design of the problem can be carried out. This is undertaken in the next two chapters.

References

Amelink, H., Forte, A. M. and Guberman, R. P. (1986), "Dispatcher Alarm and Messa ge Processing," *IEEE Transactions on Power Systems*, August vol. PWRS-1, pp. 188-194.

Bijoch, R. W., Harris, S. H. and vol.man, T. L. (1991), "Development and Implementation of the NSP Intelligent Alarm Processor," *IEEE Transactions on Power systems*, May, vol. 6. pp. 806-812.

Chan, E. H. P. (1989) "Application of Neural Network Computing in Intelligent Alarm Processing," *Proceedings of Power Industry Computer Applications Confere nce*, pp. 246-251.

Chan, E. H. P., (1990), "Using Neural Networks to Interpret Multiple Alarms," in *IEEE Computer Applications in Power*, April, pp. 33-37.

Dillon,T. (1991), "Survey on Expert Systems in Alarm Handling," *Electra* , no. 139, pp. 133-147.

Durocher, D. (1990), "Language: an Expert System for Alarm Processing," *Eleventh Biennial IEEE Workshop on Power Systems Control Centers*, September, pp. 19-21.

Eickhoff, F. Handschin, E. and Hoffmann, W. (1991), "Knowledge Based Alarm Handling and Fault location in Distributed Networks," *Proceedings of the Conference on Power Industry Computer Applications*, pp. 495-98.

Electrical World, (1990), "How Much can a Solar Storm Cost?," September, pp. 9-11.

EPRI - Electrical Power Research Institute, Palo Alto, CA, Report no. EL-1 960, vol. 1-6, 1983, "Human Factors Review of Electric Power Dispatch Control C enters".

Ewart, D. N., (1978), "Whys and wherefores of Power System Blackouts," *IEEE Spectrum*, April, pp. 36-41.

Inoue, N., Fujii, T., Shinohara, J., Mochizuki, K. and Kajiwara,Y. (1989), "An Expert System for Intelligent Alarm Processing in EMS and SCADA Systems," *Second Symposium On Expert System Applications to Power Systems*, July, pp. 89-95.

Jongepier, A. G., Dijk, H. E. and Sluis, L. V. D. (1991), "Neural Networks Applied to Alarm Processing," *Third International Symposium on Expert System Applications to Power Systems*, Tokyo, Japan, pp. 615-20.

Kirschen, D. S. (1988), "Artificial Intelligence Applications in Energy Management Systems Environment," *First Symposium on Expert Systems Applications in Power Systems*, pp. 17.1-17.6.

Kirschen, D. S., Wollenberg, B. F., Irisarri, G. D., Bann, J. J. and Miller, B. N. (1989), "Controlling Power Systems During Emergencies: The Role of Expert Syte ms," *IEEE Computer Applications in Power*, April, pp. 41-45.

Kirschen, D. S. and Wollenberg, B. F. (1992), "Intelligent Alarm Processing in Power Systems," *Proceedings of the IEEE*, vol. 80, no. 5, pp. 663-672.

Liu, C. C. (1992), "Knowledge Based Systems in Electric Power Industry," *Proceedings of the IEEE*, vol. 80, no. 5, pp. 659-662.

Munneke, M. C. and Dillon, T. S., 1989, "Alarms-An Object-Oriented Alarm Interpretation Expert System," *Second Symposium On Expert System Applications to Power Systems*, July, pp. 72-78.

Pfau-Wagenbauer, M. and Brugger, H., (1991), "Model and Rule Based Intelligent Alarm Processing," *Third Symposium on Expert System Application to Power Systems*, Tokyo, April, pp. 27-32.

Purucker, S. L., Tonn, B. E., Goeltz, R. T., Hemmelman, K. M. and Rasmussen, R.D. (1989), " Design and Operation of Communication Alarm Processor Expert System," *Second Symposium on Expert System Application to Power Systems*, July, pp. 40-45.

Shoop, A. H., Silverman, S. and Ramesh, B. (1989), "Consolidated Edison System Operation Computer Control System (SOCCS) alarm advisor SAA," *Second Symposium On Expert System Applications to Power Systems*, July, pp. 84-88.

Talukdar, S. N. and Cadozo, E. (1986), "Artificial Intelligence Technologies for Power System Operations," *EPRI Project Report no. EPRI EL-4324*, January, pp. 1-6.

Tesch, D. et al., (1990), "A Knowledge-Based Alarm Processor for an Energy Management System," *IEEE Transactions on Power Systems*, February, vol. 5, pp. 268-275.

Wollenberg, B.F, (1986), "Feasibility Study for an Energy Management System Intelligent Alarm Processor," *IEEE Transactions in Power System*, May, vol 1, pp. 241-247.

10 AGENT ORIENTED ANALYSIS AND DESIGN OF THE RTAPS - PART I

10.1 INTRODUCTION

Analysis and design precede machine implementation of any large scale problem. The analysis and design process begins with the problem statement (or requirement analysis) and ends with solution statement or design specifications. Requirement analysis involves complete and accurate representation of the problem domain. It involves providing a natural language description of the problem, determining the different levels of abstraction which divide the problem space and determining the organization of these levels.

The solution statement may be further realized through implementation in some tangible or intangible form like a Real Time Alarm Processing System (RTAPS). This chapter covers the first part of Agent oriented analysis and design of the RTAPS. The first part primarily includes the problem domain structure analysis, and association of the problem solving agents of IMAHDA with problem domain structure analysis of the RTAPS. The aspects related to the vocabulary of the problem solving agents in the context of the RTAPS objectives, and the Agent oriented design are described in the next chapter.

10.2   AGENT ORIENTED ANALYSIS (AOA)

The process of analysis determines what the system does as distinct from how it does it. From a software engineering perspective, Object-Oriented Analysis (OOA) involves a complete and accurate representation of a problem domain in an object-oriented framework. It has emerged as premier problem domain structure analysis methodology in the past few years (Coad et al. 1990; Rumbaugh 1990; Champeaux et al. 1991; Dillon et al. 1993; Jacobson et al. 1995). somewhat removed from the way objects exist in the real world.

The agent oriented analysis perspective adopted in this chapter includes OOA problem domain structure analysis. Some of the advantages of association between OOA and agent oriented analysis are:

- OOA, because it supports structural abstraction facilitates realization of a software system in terms of the objects as they exist in the real world, thus making the overall system more intelligible to the user.

- Objects are also one of the knowledge modeling formalisms of IMAHDA for capturing domain specific knowledge.

- The encapsulating properties of OOA facilitate capturing of domain-specific actions. These domain-specific actions enrich the set of the domain-independent actions of the problem solving agents of IMAHDA. These domain-specific actions are however, limited by the data encapsulated by an object and hence are data driven.

- Agents which by definition engage in task abstraction and task-oriented behavior can be used for capturing the problem solving behavior of the user. They also capture domain-specific actions which are task driven.

- The communication constructs, learning constructs, and the problem solving agents of IMAHDA enrich the OOA as well as the agent oriented problem solving behavior analysis and thus facilitate the realization of the design architecture of the software system.

Based on the above advantages the agent oriented analysis of the RTAPS is described in the next section.

10.3   AOA OF THE RTAPS

The Agent Oriented (AO) analysis undertaken in the RTAPS includes the following steps:

- Understanding the domain.

- Problem domain structure analysis.

- Agent Analysis of problem solving behavior.

- Association of problem domain structure & agent analysis with IMAHDA.

- Defining the vocabulary of the problem solving agents.

10.3.1   Understanding the Domain

Before embarking on any application development one has to become familiar with number of aspects including the general characteristics of the domain, the nature and complexity of problems in the domain, and existing methods of addressing the problems. The motivation behind this is to enable the problem solver or knowledge engineer to become acquainted with the domain and model the solution effectively.

This step has been covered in the previous chapter. The general characteristics of the alarm processing in power systems and the problems encountered by some of the existing methods while solving the problems in the domain have also been outlined. The suitability of IMAHDA to address these problems has been studied. Finally, the objectives of the RTAPS in SECV, and a natural language description of type of alarms emanating from TTS has been provided. This preliminary part of the AO analysis sets the stage for conceptual modeling or problem domain structure analysis.

10.3.2   Problem Domain Structure Analysis

The problem domain structure analysis of the RTAPS is done using OOA methodology for reasons explained in section 10.2. Objects are also one of the knowledge representation constructs in IMAHDA.

Another reason for choosing the OOA approach for doing problem domain structure analysis is its compatibility with the domain structure of the RTAPS. Some of the reasons for this compatibility are:

- The power system is built hierarchically. The four levels, generation, transmission, sub-transmission and distribution are arranged in a hierarchical manner. The notion of hierarchy is an intrinsic part of OOA and it is only appropriate to use OOA for a domain which is structurally composed in a hierarchical manner. The structural components in a power system are not only organized hierarchically but also as whole-part structures (e.g. Grids, Zones, sections) which also makes it suitable for OOA.

- In any large scale system like a power system, issues like modularity, maintainability and reusability are extremely important to sustain the system from time to time because of incremental changes. OOA is a methodology which enables one to realize such constraints through the concept of modularity and consequently maintainability and reusability are stronger than the conventional methodologies.

- The different levels of abstraction built through OOA facilitate multiple use of the classes and objects in these levels.

10.3.2.1   OOA of the Alarm Processing System.   Various object-oriented analysis methods (Rumbaugh et al. 1990; Gibson 1990; Page-Jones et al. 1990; Coad et al. 1990; Champeaux 1991; Dillon et al. 1993; Jacobson et al. 1995) are available for purpose of doing the structural and behavioral analysis of a problem domain. The OOA employed for doing the problem domain structure analysis incorporates some concepts from Coad et al. (1990) along with extensions developed by Khosla and Dillon (1993). These concepts and their extensions are described next.

10.3.2.2   Basic Object-Oriented Constructs.   Before embarking on OOA, it is important to clarify the basic concepts used by us for expressing relationships between objects. The Generalization (Gen) - Specialization (Spec) structures and the Whole-Part structures are shown in Figure 10.1. The class and sub-class in the Gen-Spec structure 'A' are passive classes. These classes are generalizations which only facilitate inheritance and do not participate in any dynamic activity except possibly creation of objects. On the other hand, class-&-object in the Gen-Spec structure 'B' is an active class with at least one object within itself and is completely or partially executed during run time. The complete execution takes place when it is used in run-time to compute its own dynamic instance. The partial execution takes place when it is used to execute class-&-object specific methods which relate to its class-&-object/s or instance objects at the lower level. An instance object as the name implies is the instance of a particular class structure and represents the leaf of a particular class structure. The Gen-Spec and Whole-Part structures shown in Figure 10.1 have been used in doing OOA in this chapter. In the rest of the chapter where ever the term 'object/s' is used, it is intended to mean class-&-object/s.

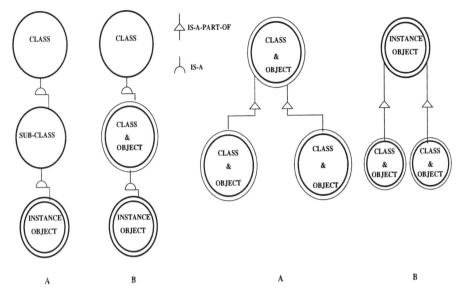

GENERALIZATION-SPECIALIZATION STRUCTURES   WHOLE-PART STRUCTURES

Figure 10.1.   Generalization-Specialization and Whole-Part Structures

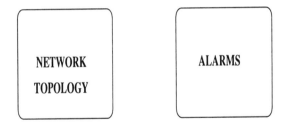

Figure 10.2.   Key Subjects in Alarm Processing

10.3.3 Steps Involved in the Problem Domain Structure Analysis

The OOA consists of the following steps:

- Identify the different subjects (or classes) from the natural language description of the problem domain. These subjects are the key determinants in the problem domain. They reflect the top most level of the problem. In our case there are two different subjects: alarms, and TTS power network topology as shown in Figure 10.2.

- Identify the classes and objects associated with these subjects. This involves identifying/creating different levels of abstraction within each of these subjects.

- Prune the identified classes and objects. The idea behind pruning is to restrict object proliferation. This generally involves deletion of those classes and objects which exist at higher levels and are there more for the understanding of the domain than representing actual physical structure or behavior of the domain.

- Identify the structural relationships between classes and objects in each of these subjects. This involves identifying the generalization-specialization and whole-part relationships between different classes and objects.

- Identify coarsely the attributes and methods associated with each class and object.

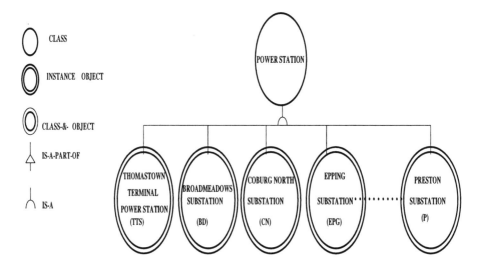

Figure 10.3.   Top or Station Level Structural Decomposition

- Identify intra-subject and inter-subject relationships and communication paths.

10.3.4   Identification of Classes and Objects

Based on the guidelines enumerated in the previous section, the OOA of the RTAPS is divided into two parts:

- OOA of TTS Power Network.

- OOA of Alarm Taxonomy.

10.3.4.1   OOA of TTS Power Network.   As the name suggests, this part of the analysis involves studying the static parts of the system, namely the TTS power network (see Appendix C and D respectively). The network structure can be analyzed at three different levels of abstraction. These are the top level, intermediate level, and lower or component level. These three different levels are shown in Figures 10.3, 10.5, 10.6 and 10.7 respectively in an object-oriented framework. The top level consists of the station level representation. The intermediate level structure shown in Figure 10.5 is derived from the sectional decomposition of the TTS power network shown in Appendix C and D respectively. The section level (e.g. 220KV, 66KV) decomposition is shown in Figure 10.4.

The lower or component level representation shown in Figures 10.6 and 10.7 respectively captures the way the different components, e.g. buses, lines, feeders, etc. are actually connected in the TTS power network shown in Appendix C and D respectively.

The OOA at different levels of abstraction is considered useful not only from the point of view of understanding the problem, but also different representations permit different problem solving strategies. By developing abstract representations one facilitates their easy, efficient and multiple use. For example, if the fault diagnosis strategy is based on the number of circuit breakers connected to different bus, transformer, etc. sections then section or intermediate level representation can be used. On the other hand, if the fault diagnosis strategy is based on the fault propagation model of each component then the lower level representation is more relevant. In the present version of the RTAPS the intermediate level representation has been used. Besides alarm interpretation and fault diagnosis, the three different representations of the power network can be used in other domains (e.g. in design) where information about network topology is considered useful.

10.3.4.2   OOA of the Alarm Taxonomy.   As described in the previous chapter, there are different types of alarms associated with alarm processing in TTS. These are shown in Figure 10.8 in an object-oriented framework. The different classes like *Alarm, Expected Alarm, Unexpected Alarm* in Figure 10.8 are

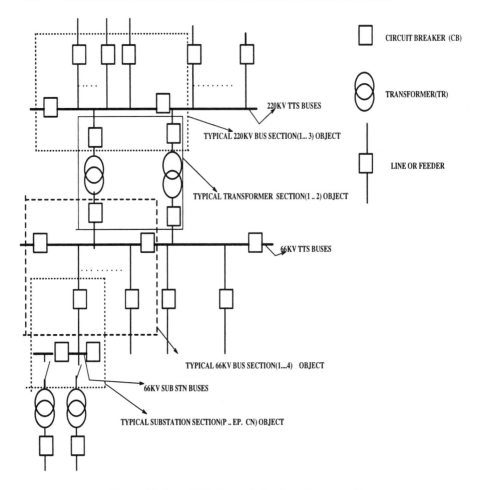

Figure 10.4.   TTS Network Sections Decomposition

generalizations or abstractions of alarms like *Network CB Alarm* and *Voltage Dip Alarm*, etc.. Some of these alarms are linked to the structure of the TTS power system network and telecommunication network. For example, *Network CB Alarm* and *Communication Alarm* are linked to the TTS power network and TTS communication network respectively. However, all these alarms reflect the dynamic behavior of power system network or the dynamic behavior alarms themselves in real time. For example, *Repeated Alarm* is a result of an alarm repeating itself on two or more occasions in real time. That is, this alarm is related to the dynamic behavior of the alarm itself rather than that of

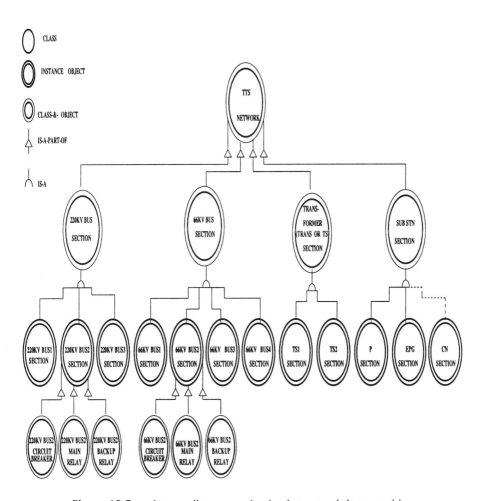

Figure 10.5.   Intermediate or section level structural decomposition

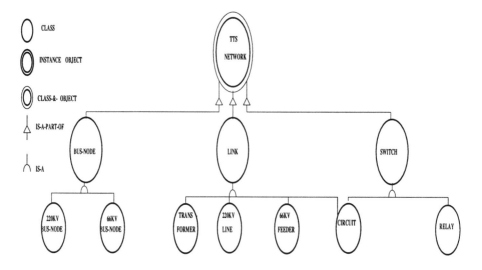

Figure 10.6.   Lower or Component Level Structural Decomposition

the network. The alarm terminology in Figure 10.8 is based on the everyday terminologies used by the operators and design engineers in the control center for classification of alarms.

10.3.4.3   Pruning.   The analysis done on the three subjects in section 10.3.4 has resulted in identification of various classes, class-&-objects and instance objects.

Pruning is imposed as part of OOA for two reasons. Firstly, pruning is undertaken to restrict object proliferation. Secondly, it is imposed by system constraints like fast response time, memory, etc. (although this aspect more appropriately belongs to the design stage).

Those class-&-objects are pruned or merged which contribute more to understanding of the domain rather than representing actual classes or class-&-objects in the domain.

Thus, the intermediate level class-&-objects identified in the OOA of the alarm taxonomy are merged with the higher level *Alarm* class as shown in Figure 10.9. That is, classes like *Expected Alarm, Unexpected Alarm, Multiple Alarm, Sympathetic Alarm*, etc. shown in Figure 10.8 are merged into the *Alarm* class. The pruned alarm class-&-objects in the alarm taxonomy are *Repeated Alarm, DC Failure Alarm, Known-Fault Alarm, Voltage-Dip Alarm, Network CB Alarm, Communication Alarm, Emergency Alarm*, and *Routine Alarm*.

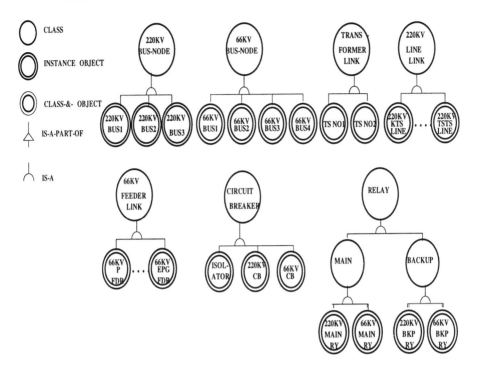

Figure 10.7.   Lower or Component Level Structural Decomposition Contd.

10.3.5   Agent Analysis of the Problem Solving Behavior

An important aspect of any complex software development process is to understand user's problem solving behavior. This helps to bring into focus aspects required to deal with the complexity in a domain. Secondly, such an analysis enhances the intelligibility of a computer system and facilitates more useful interaction between the user and the system. For example, in the neuro-fuzzy hybrid control application developed by Tani et al. (1996) (described in chapter 4) to control the tank level in a solvent dewaxing plant, the operator's problem solving behavior is used to determine the problem solving stages of the hybrid system. Likewise in case of the RTAPS it is essential to understand the problem solving behavior of an operator in the power system control center. Agents which engage in task and task-oriented behavior abstraction are used to model the problem solving behavior of the operator.

Briefly, in the context of the objectives of the RTAPS, the operator intially filters out *Repeated*, *Known Fault*, and *DC Failure* alarms which are considered

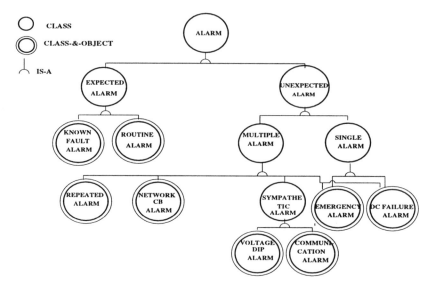

Figure 10.8.   OOA of Alarm Taxonomy

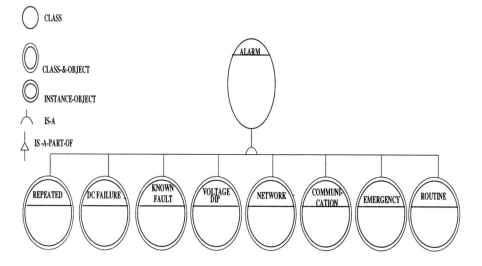

Figure 10.9.   Flattened Alarm Taxonomy

to be irrelevant and noisy. The remaining alarms are sub-divided into *Network CB (Circuit-Breaker)* or *Network, Voltage Dip, Communication, Emergency,*

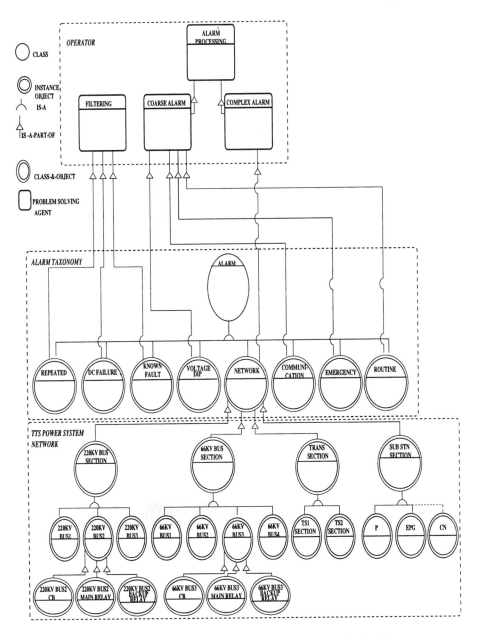

Figure 10.10.   RTAPS Inter-subject and Agents Relationships

and *Routine* alarms shown in Figure 10.10. The decision processes involved in *Voltage Dip*, *Communication*, *Emergency*, and *Routine* alarms are fairly coarse (in the present context) in that the problem solving is limited to identification of whether the alarm is a voltage dip alarm or say, a routine alarm. On the other hand more intricate decision making is involved with *Network CB* alarm. For example, in the *Network CB* alarm, the operator determines whether the alarms pertain to the 220kv bus section, 66kv bus section, transformer section, substation section, or a combination of two or more sections. Thus here, the operator is correlating the alarm with the structure of the power network. The operator goes on to further determine the cause of the alarm (or location of fault in the power network) based on the graphic display of the power network, power network drawings, and their prior experience with alarm patterns.

10.3.6   RTAPS Inter-subject and Agent Relationships

The relationship between classes and class-&-objects belonging to the two subjects Alarm, and TTS Power System Network along with the RTAPS agents modeling the operator's problem solving behavior are shown in Figure 10.10. The two agents *Filtering*, and *Alarm Processing* in Figure 10.10 are derived from the description of operator's problem solving behavior. The *Alarm Processing* agent relates to those alarms which are processed as against those which are filtered out. Another aspect of the operator's problem solving behavior relates to alarms which involve coarse decision making and alarms which involve intricate or complex decision making in the context of the RTAPS objectives. This part of the operator's behavior is reflected by the *Coarse Alarm*, and *Complex Alarm* agents. These two agents are a *PART-OF* the *Alarm Processing* agent.

The relationship between the *Network* (*Network CB*)) alarm and the TTS Power System Network class-&-objects reflects not only a structural relationship but also the fact that processing of *Network* alarms is related to the TTS Power System Network class-&-objects.

The partial list of domain-specific attributes, actions and communication paths between different class-&-objects is shown in Figure 10.11 and 10.12 respectively. These domain-specific attributes and domain-specific actions associated with objects and agents are primarily based on the problem domain structure, operator's problem solving behavior and the RTAPS objectives. Attributes like *Max_alarmid* and *Prev_alarmid* are inherited by the *filtering* and *Alarm Processing* agents from the domain *Alarm* class. For sake of simplicity, *Coarse Alarm*, and *Complex Alarm* agents have been merged with the *Alarm Processing* agent.

In the present version of the RTAPS only the circuit-breaker(CB) information is used for computing various network faults. Accordingly, the *Main*

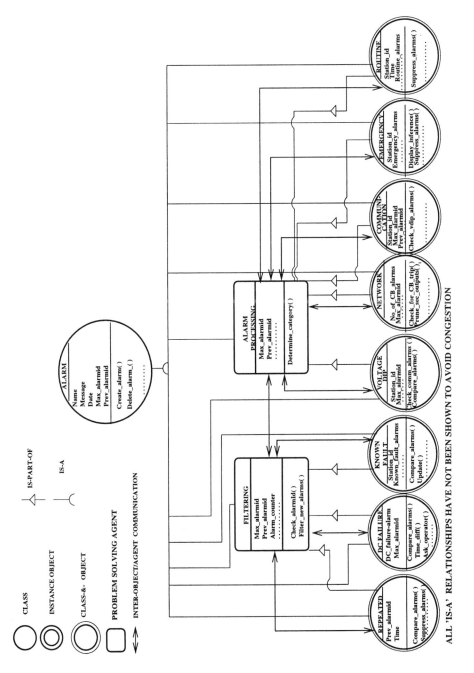

Figure 10.11.   Domain-specific Attributes, Actions and Communication Paths

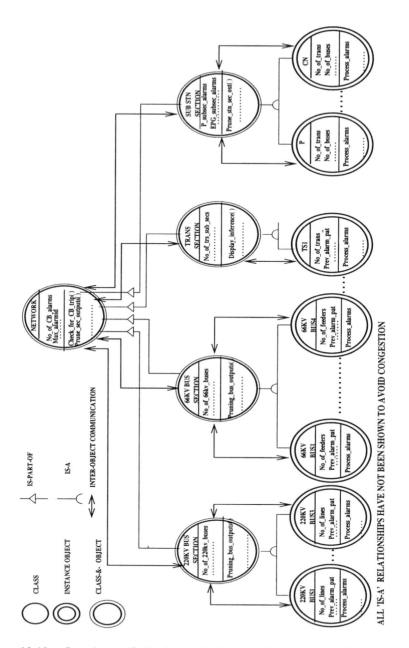

Figure 10.12.   Domain-specific Attributes, Actions and Communication Paths & Contd.

Relay and *Backup Relay* objects shown in Figure 10.5 and 10.10 have not been considered. The existing *Circuit Breaker* objects have been merged with their instance objects (e.g. *220kv Bus2 Section*) in Figure 10.12.

It may be noted here that *Emergency, Routine, Communication (Comms), Voltage-Dip, 220KV Bus2 Section*, etc. are completely active class-&-objects as they dynamically compute their alarm instances during run time.

10.3.7   Association of Problem Domain Structure & Agent Analysis with IMAHDA

As the title suggests, in this section the domain-independent problem solving agents of IMAHDA are mapped on to the results of the problem domain structure analysis of the RTAPS. The intelligent hybrid problem solving agents of IMAHDA include the global preprocessing, decomposition, control, and decision agents respectively. The PAGE description of global preprocessing (or simply preprocessing) agent (outlined in chapter 7) is associated with the *Filtering* agent as shown in Figure 10.13. The association is derived from operator's problem solving behavior and the PAGE description of the preprocessing agent. The preprocessing agent enriches the *Filtering* agent with its PAGE description, and other knowledge representation components. The preprocessing agent provides a prior domain-independent problem solving knowledge to the *Filtering* agent which has its domain specific attributes and actions.

As mentioned in the last chapter, the PAGE description and knowledge content of the decomposition agent corresponds to the abstract level (which does not provide immediate solution to the problem) of the problem domain. Here the concepts which do not provide immediate solution to the problem (in context of the RTAPS objectives) are alarm categories, namely, *Emergency, Routine, Communication (Comms), Voltage-Dip*, and *Network*. The *Alarm Processing* agent which has the responsibility of determining these categories is associated with the decomposition agent as shown in Figure 10.14.

Based on objectives of the RTAPS, intricate or complex decision making is only involved with the *Network* alarm object which gives it a high decision class priority. Hence, the control agent is associated with *Network* alarm object and its domain control structure. This association accounts for the complex decision making component of the *Alarm Processing* agent and operator's problem solving behavior.

The domain control structure is represented by the *220KV Bus Section, 66KV Bus Section, Transformer (Trans) Section*, and *Sub Stn Section* objects. These objects link the *Network* alarm object with power system network representation. This domain control structure as mentioned is context dependent. That is, if the component level power system network representation in Fig-

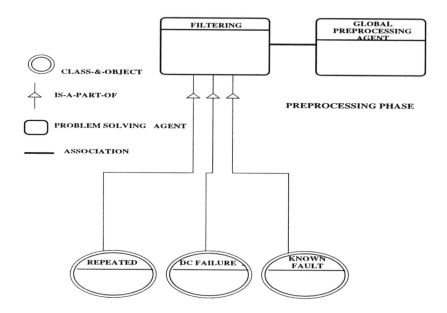

Figure 10.13.   Preprocessing Agent and Problem Domain Structure

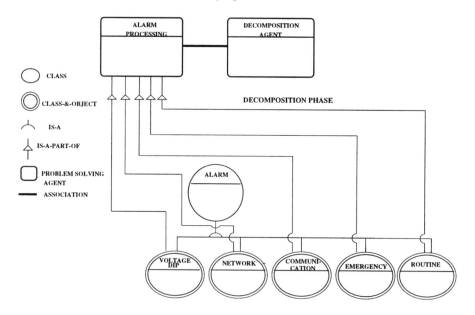

Figure 10.14.   Decomposition Agent and Problem Domain Structure

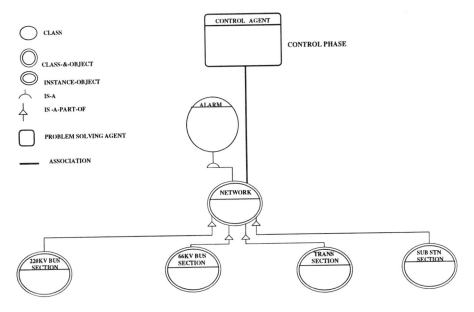

Figure 10.15.   Control Agent and Problem Domain Structure

ure 10.6 and 10.7 respectively, then the domain control structure represented by *Bus-Node*, *Link*, and *Switch* objects would be used. In the present context where fault diagnosis and isolation of the cause of the alarm event is to be done based on the number of circuit breakers connected to the different sections, the domain control structure shown in Figure 10.15 has been adopted. One of the reasons for adopting the domain control structure shown in Figure 10.15 is the availability of a number of intelligent methodologies associated with IMAHDA. For example, neural networks can handle a large number of circuit breakers associated with the section level representation. In other words, the availability of multiple constraint based neural networks (as one of the intelligent agents) facilitates the strategy to deal with the problem at the section level description, rather than being forced to reduce the number of variables and deal with it at the component level. Thus, IMAHDA provides more flexibility to the problem solver in terms of the fault diagnosis strategy.

The decision agent is associated with decision level objects of the power system network structure inferences on which are important for detection and isolation of single, and multiple line and bus faults. The association between the decision agent and these objects is shown in Figure 10.16.

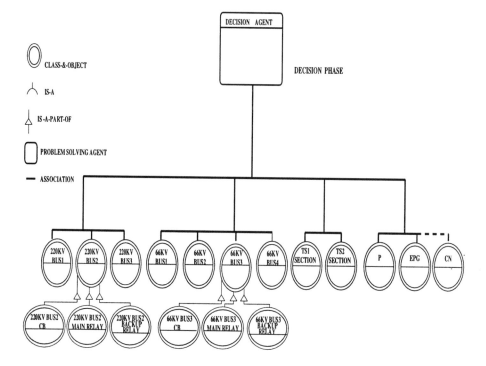

Figure 10.16.   Decision Agent and Problem Domain Structure

Figures 10.15 and 10.16 also show the distinction between the problem domain control structure knowledge and problem domain decision class level structure knowledge.

All put together, a complete view of the association of the preprocessing, decomposition, control, and decision problem solving agents with the problem domain structure is shown in Figure 10.17.

Before finishing this section, an important observation needs to be made regarding the implication of the problem domain structure analysis and its association with the problem solving agents of IMAHDA. The object-oriented problem domain structure analysis on its own (from a software engineering perspective) captures the domain specific data (attributes) and actions (methods) of the the problem domain structure objects. Problem solving agents of IMAHDA on the other hand, through their association, provide the problem solving knowledge to the RTAPS domain agents. This problem solving knowledge includes structuring the problem solving approach and mapping the problem domain structure in that approach (i.e the way different parts of the

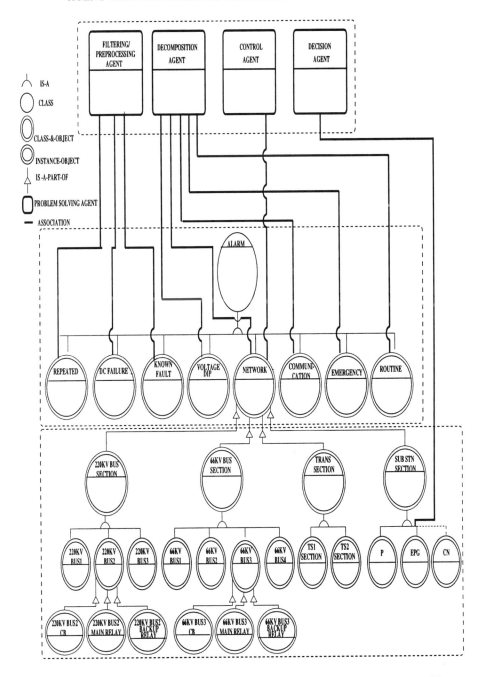

Figure 10.17.   Problem Solving Agents & Problem Domain Structure of the RTAPS

RTAPS structure have been mapped to the problem solving agents), goal and actions knowledge at different levels of the problem solving process, intelligent knowledge components like rules, neurons, and genes, learning knowledge and strategy, communication and distributed processing knowledge, and dynamic knowledge for doing a dynamic analysis. So the domain objects through their association with the problem solving agents not only inherit the problem solving approach but also inherit the entire vocabulary of these intelligent problem solving agents and become domain problem solving agents. In the next chapter the vocabulary and agent oriented design of the RTAPS is described.

10.4 SUMMARY

Analysis and design precede machine implementation of solution to any real world problem. Two aspects of agent oriented analysis of the RTAPS (Real Time Alarm Processing System) have been covered in this chapter. These are the problem domain structure analysis and association of the problem solving agents of IMAHDA with the problem domain structure analysis of the RTAPS. The problem domain structure analysis of the RTAPS has been done using the Object-Oriented Analysis (OOA) methodology. OOA enriches agent oriented analysis as it not only captures the intuitive structure of the domain but also captures the domain-specific attributes and actions. OOA includes the analysis of the power system network, alarm taxonomy. The agent analysis involves identification of the domain agents through analysis of the problem solving behavior of the operator in the control center. The problem solving agents of IMAHDA are mapped to domain agents and different parts of the domain structure analysis of the RTAPS. The process of mapping has integrated the problem solving approach into the domain agents and domain structure of the RTAPS.

References

Encyclopedia Britannica, (1986), Articles on "Behavior, Animal," "Classification Theory," and "Mood," Encyclopedia Britannica, Inc.

Champeaux, D. (1991), "Object-Oriented Analysis and Top-Down Software Development ," in *Lecture Notes in Computer Science 512*, Springer Verlag, NY, pp. 360-76.

Coad, P. and Yourdon, E. (1990), *Object-Oriented Analysis*, Prentice Hall, Englewood Cliffs, NJ.

Davlo, E., and Naim, P. (1991), *Neural Networks*, Macmillan computer science series, U.K.

DeMacro, T. (1978), *Structured Analysis and System Specification*, Prentice Hall, Englewood Cliffs, NJ, USA.

Dillon, T. and Tan, P. L. (1993), *Object-Oriented Conceptual Modeling*, Prentice Hall, Sydney, 1993.

Gibson, E. (1990), "Objects-born and bred," in *BYTE*, pp. 245-254, October.

Jacobson, I. et al. (1995), *Object-Oriented Software Engineering*, Addison Wesley, USA.

Khosla, R. and Dillon, T. (1993), "Application of Object-Oriented Technology in a Terminal Power Station," in *Technology of Object-Oriented Languages and Systems (TOOLS'93)*, December, Melbourne, Australia.

Martin, J. (1990), *Information Engineering*, Prentice Hall, Englewood Cliffs, NJ, USA.

Page-Jones, M. and Weiss, S. (1990), *Synthesis: An Object-Oriented Analysis and Design Method*, Wayland Systems Inc., Seattle, WA, USA.

Rumbaugh, J. et al. (1990), *Object-Oriented Modeling and Design*, Prentice Hall, Englewood Cliffs, NJ, USA.

Slatter, P.E. (1987), *Building Expert Systems: Cognitive Emulation*, Ellis Horwood Limited, Chichester.

Tani, T., Murakoshi, S., and Umano, M. (1996), "Neuro-Fuzzy Hybrid Control System of Tank Level in Petroleum Plant," *IEEE Transactions on Fuzzy Systems*, vol 4, No 3, pp. 360-368.

Yourdon, E. (1989), *Modern Structured Analysis*, Prentice Hall, Englewood Cliffs, NJ, USA.

11 AGENT ORIENTED ANALYSIS AND DESIGN OF THE RTAPS - PART II

11.1 INTRODUCTION

As the title suggests, this chapter is a continuation of agent oriented analysis of the RTAPS started in the last chapter. The remaining aspect of the agent oriented analysis of the RTAPS, namely, the vocabulary of the problem solving agents is described in this chapter. This is followed by the agent oriented design of the RTAPS which includes identification of software and intelligent agents to be used by the problem solving agents, mapping of the software and intelligent agents on to the problem solving agents, and establishing the order of execution precedence of the problem solving agents. Finally, emergent behavioral characteristics of this multi-agent distributed architecture of the RTAPS are outlined.

11.2 AGENT ORIENTED ANALYSIS CONTINUED

In this section the last part of the agent oriented analysis, namely, defining the vocabulary of the problem solving agents in the RTAPS is described.

11.2.1 Defining the Vocabulary of the Problem Solving Agents in the RTAPS

The vocabulary of the problem solving agents is determined by their knowledge representation and reasoning constructs, namely, objects, rules, fuzzy logic, genes/chromosomes/bit strings[1], and learning knowledge and strategy for distributed neurons. The object vocabulary of the problem solving agents has been defined in the preceding section. In this section, some rules related knowledge of the problem solving agents and the learning knowledge and strategy used by the decomposition, control, and decision agents of the RTAPS is described (Khosla and Dillon 1997).

Overall, the rule vocabulary consists of 60 rules used in global preprocessing, decomposition, control and decision agents of the RTAPS. In this section, some of the task related rules in these agents are outlined.

A noise filtering rule in the Preprocessing Agent

IF Number of New Alarm Messages is greater than Zero AND
Content of New Alarm Message is Identical to Previous Alarm Message AND
Time Difference Between New Alarm Message and Previous Alarm Message
is less than 5 secs THEN
Delete New Alarm Message

IF Number of New Alarm Messages is greater than Zero AND
New Alarm Message is Member of Known Fault Alarm Message List, THEN
Delete New Alarm Message

A Context Validation Rule in the Decomposition Agent
IF Conditioned New Alarm Message Does Not Belong to TTS THEN
Ignore It.

Context Validation Rules in the Control Agent
IF Context : Section Level THEN
Formulate Execution Sequence 1.

IF Context : Component Level THEN
Formulate Execution Sequence 2.

IF Context : Only Circuit Breaker Information Available THEN
Decision Level Agent Set A Valid .

IF Context : Circuit Breaker & Relay Information Available THEN Decision Level Agent Set B Valid .

11.2.1.1   Learning Knowledge of RTAPS.

Learning knowledge used in the RTAPS is based on the learning knowledge of the decomposition, control, and decision agents of IMAHDA described in chapter 8. In order to help the reader recapitulate, decomposition knowledge is used to classify abstract domain classes. These abstract classes are not of direct interest to the problem solver and are not the solution to the problem under study. In fact, these classes are high level generalizations or aggregations of the decision level classes. Coarse grain features are used for classifying the abstract classes. These coarse grain features are globally aggregated features and/or key features alarms. These aggregated features and/or key features have largely binary values(0 or 1) and some structured values e.g. *STATION, GRID, ZONE*) Like for example *STATION* is a set of discrete values consisting of different terminal station and substation names.

Control knowledge reflects the underlying control structure of a domain. In IMAHDA control knowledge is used for determining the decision level classes in each abstract class classified in the decomposition phase. Semi-coarse grain features are used for determining the decision level classes. The granularity of the features used in this phase is less than those used in the decomposition phase because features used in the control phase are applicable only within the class and not outside.

The semi-coarse grain features are locally aggregated and/or key features. These features can be further broken up into binary(e.g. *B_T* for Bus_Tie, etc.), structured(e.g. *220KV, 66KV*, etc.) and some linear features. Linear or fuzzy features are those which have a broad range from say *1...10* (e.g . range of a single analog measurement or a set of measurements).

Decision knowledge is used for determining the instances of the decision level classes classified in the control phase. These decision instances are the solution to the problem. These are the specific classifications which a power system operator/system user or a problem solver is interested in. For example, the RTAPS determines in this phase specific single line faults, multiple faults, bus faults and malfunctions which can occur under critical conditions in TTS. Fine grain features are used for classifying the decision instances.

The fine grain features reflect refined distinctions between various fault classifications. They consist of refined linear or fuzzy features (Tang, Dillon, and Khosla 1995). That is, instead of broad range of *1..10*, the range is further refined to *2..4, 4..7*, etc. or even a single value like less than or equal to *1*. This is reflected in the RTAPS by the alarms for each element/component in a bus section.

In general, the type of knowledge used in different phases works in a way as background knowledge for the phase below and helps to constrain the generalization in the decision phase. That is, if Artificial Neural network (ANN) is used in the 220KV Bus1 Section decision agent, then its generalization pertains to 220KV Bus1 Section only.

The learning strategy described in the next section explains how the learning knowledge of IMAHDA has been applied in the RTAPS domain.

11.2.2 Learning Strategy of RTAPS

Learning strategy in the RTAPS mainly involves selection of input features. The process of selection of input features is five steps:

1. Selecting the most distinguishing or discriminating features

2. Selecting the common features among various classes.

3. Knowing when to stop selecting the features in a particular phase.

4. Retention and elimination of redundant features, and

5. Associated with '4', the provision of noise.

The learning strategy of IMAHDA has been applied in the decomposition, control, and decision agents of the RTAPS.

11.2.2.1 Learning Strategy of the Decomposition Agent. In most complex domains like alarm processing, there are a large number (190 in the RTAPS) of input features (e.g. there are 65 CB alarms associated with different network faults) and large number of decision instances, and decision level classes. In the decomposition agent, as mentioned in the preceding section abstract classes are created by aggregating a large number of decision level classes.

The extent of aggregation varies from class to class. It depends upon the complexity of the domain and the level of classification(coarse or fine) required in a particular aggregated class. For example, level of classification required in the network CB alarm class is fine (component faults need to be identified). On the other hand, the level of classification required in the emergency alarm class is coarse in that whether an alarm classified as an emergency alarm indicates " building fire" or not.

The five different abstract output classes chosen by the agent are the Communication, Network Control, Voltage-Dip, Emergency and Routine. The next step is to select features which characterize these classes.

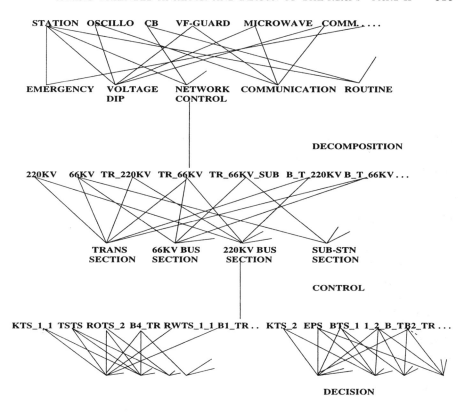

STATION   OSCILLO   CB   VF-GUARD   MICROWAVE   COMM., ...

EMERGENCY   VOLTAGE   NETWORK   COMMUNICATION   ROUTINE
             DIP   CONTROL

DECOMPOSITION

220KV   66KV TR_220KV TR_66KV TR_66KV_SUB B_T_220KV B_T_66KV ...

TRANS   66KV BUS   220KV BUS   SUB-STN
SECTION   SECTION   SECTION   SECTION

CONTROL

KTS_1_1 TSTS ROTS_2 B4_TR RWTS_1_1 B1_TR .. KTS_2 EPS BTS_1 1_2 B_TB2_TR ...

DECISION

Figure 11.1.   Sample Neural Network Stages in the RTAPS

In order to select binary and structured coarse grain features, the first step is to select the most discriminating features belonging to the abstract classes at this level like $COMM^2$, CB, $OSCILLO$, etc shown in Figure 11.1 belonging to communication, network and voltage dip alarms respectively. The second step is to determine the common features between the output classes in order to reflect the degree of overlap between the classes. A combination of the discriminating features and common features enables one to use the principal of elimination in order to determine validity of different abstract classes in different situations. Some such common features are *Station*, *Microwave* and *VF-Guard-Loss*. The last two features belong to the communication alarms class and voltage-dip alarms class. In the latter case they occur as sympathetic alarms and are not valid communication alarms.

In this phase, the selection of coarse grain input features should stop once the classification of the abstract classes in the domain (in this case alarm processing) has been achieved.

Redundancy is both an explicit and implicit phenomenon. Some features are declared redundant as they do not provide any additional classification. On the other hand, some are declared redundant through a process of elimination (i.e. absence of a certain feature is equivalent to the presence of another). For instance, in the decomposition phase the fact that it is a transformer CB alarm or a substation CB alarm is redundant as it does not give us any additional information besides that they are CB alarms the which belong to the *Network* class.

They are two implications of redundant features One is to retain them for coping with noise in the domain. The other is eliminating them to reduce the number of features in a particular phase for training. Thus, if there is Relay status information a feature (say, *RY*) along with the *CB* can be included as part of the input vector classifying the *Network* alarms. Although, the relay feature is redundant but it caters to noise. In other words, in case a CB malfunctions then the relay feature, *RY* can be effectively used for classifying the input vector.

Overall, in the RTAPS redundancy is inherently reduced by the three agents Decomposition, Control and Decision which operate at different levels of classification or abstraction. What is relevant in one phase is redundant in the phase below.

Thus through selection of coarse grain binary and structured features like *CB, COMM, OSCILLO, FIRE, STATION*, and others one is able to classify the abstract output classes in the decomposition agent.

11.2.2.2   Learning Strategy in the Control Phase.   In the control agent decision level classes are identified within an abstract class (e.g. Network/ Network Control agent). The decision level classes based on the decision level domain structure described in chapter 10 (section 10.3.4) are *220KV BS1* , *220KV BS2, 220kV BS3, 66KV BS1, 66KV BS2, 66KV BS3, 66KV BS4, TS1, TS2, Sub Stn Epping, ...,Sub Stn Broadmeadows*. However, in the present version of the RTAPS (where alarm information related to 65 CB alarms is available for configuring network faults) the decision level classes selected by the Network Control agent are *Trans Section, 66KV Bus Section, 220KV Bus Section*, and *Sub-Stn Section* respectively as shown in Figure 11.1.

That is, the need for determining whether the incoming alarms belong to a *220KV Bus1 Section (BS1)* or *220KV Bus2 Section (BS2)* has not been felt in the present version of the RTAPS [3].

In other words, the alarms are distributed to the individual 220KV bus sections based on the fact that whether they belong to the *220KV Bus Section* or not.

Then a set of input features to disaggregate the aggregated heterogeneous Network CB class into these four decision level output classes.

Local disaggregation is done using semi-coarse grain input features within a class. The discriminating features selected are *220KV and 66KV*. The discriminating features here are structured features as they reflect the structure of the network topology. The common features selected are *B-T(Bus-Tie)*, *TR(Transformer)* as they represent overlaps between different bus sections, substation sections, etc.. The selection of the input features stops once the decision level classes have been determined.

The redundancy in this phase can be reflected by the CB and relay data and will be explained in the next chapter while describing the implementation of the RTAPS.

11.2.2.3 Learning Strategy in the Decision Phase. In the decision phase, firstly the decision instances are determined which represent the solution to the problem. For example, single line fault and multiple line fault are the decision instances of the 220KV Bus1 Section class. In the decomposition and control phases the coarse grain and semi-coarse grain features have been filtered out. the decision phase largely consists of fine grain or linear features. These fine grain or linear features can also be fuzzified (Tang, Dillon, and Khosla 1995) as will be explained later in the next section. These fine grain features which reflect refined distinctions interact more closely than the semi-coarse and coarse grain features. In other words, they are much more distributed than the semi-coarse and coarse grain features. As a result most of the features are common features for the decisions made in this phase. The decisions are made on the different line or feeder faults, malfunctions and bus fault in a bus section(e.g. 66KV Bus2 Section, 220KV Bus1 Section, etc.,) Some of the features used in individual 220KV Bus Sections are shown in Figure 11.1. The extraction of discriminating features as such can be difficult in this phase. One way of having the discriminating effect is to stagger the number of decision classes. That is, instead of using the input features to specifically classify single/multiple line/bus fault/s and malfunction/s the output classification can be staggered using coarse classes. For example, in the 220KV Bus1 Section there are nine output fault classes for specifically classifying different single/multiple line faults, malfunctions and bus fault. These different output classes can be grouped together into Line Fault and CB Malfunction, Line Fault, Bus Fault and CB Malfunction classes. In other words, one ANN for deciding the coarse classes (e.g. Line Fault) can be used, and another ANN can be used if required

318   ENGINEERING INTELLIGENT HYBRID MULTI-AGENT SYSTEMS

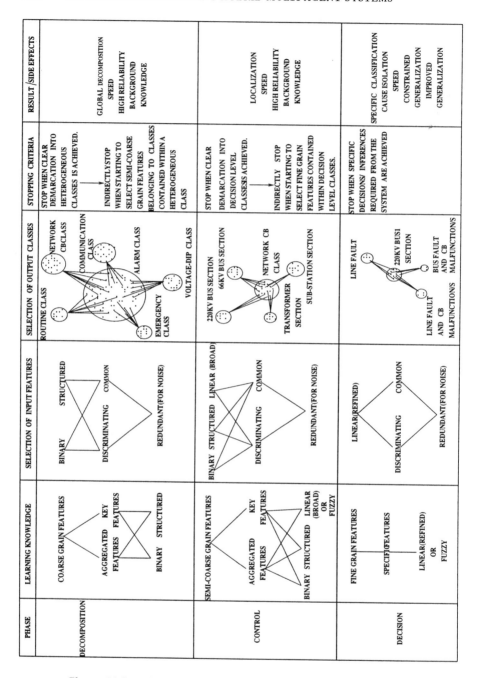

Figure 11.2. Learning Knowledge and Strategy in Different Phases

to decide which line/s is/are faulty and which CB/s has/have malfunctioned. Thus, the features used in both the ANNs are the same but the classifications are different.

The advantage of this strategy is that it works towards improving the generalization properties of the ANNs if they are used in the generalization mode in this phase. That is, by engaging in such a learning process an increasing number of patterns will belong fewer classes which leads to better representation in the partial data set and improved generalization.

In this phase redundancy can be incorporated in the presence of Relay and/or analog information. It is explained in the next chapter in the context of scalability properties of the RTAPS architecture. Figure 11.2 summarizes the learning knowledge and strategy of the RTAPS and also shows the side effects of using ANNs in different phases. In the next section, the application fuzzy logic in the decision phase for alarm processing application is described. This alternative decision knowledge modeling technique is shown for illustrative purposes only.

11.2.3 Application Fuzzy Logic in the Decision Phase

For the purpose of illustration, a model of a power system and protection system(consisting of CBs and Relays), as shown in Figure 11.3, is used (Sekine et al. 1989; Wang and Dillon 1992; Wang and Dillon 1991). This part of a power system represents the decision phase of the IMAHDA architecture. In this example, fuzzy logic is applied in the decision phase and the postprocessing phase for fault diagnosis in Bus 1 (B1) in Figure 11.3. There are eight possible paths to B1 in this example. The paths have been configured based on the operation/maloperation of the following circuit breakers and relays:

B1m CB2 CB3 CB4 CB5 CB6 CB7 CB9 CB10 T1s L1Cs T2s L2Cs

The eight paths are as follows:
Path-1: *B1m L2Cs CB10*
Path-2: *B1m CB6 L2Cs CB10*
Path-3: *B1m CB6 T2s CB3 CB5*
Path-4: *B1m T2s CB3 CB5*
Path-5: *B1m CB7 L1Cs CB9*
Path-6: *B1m L1Cs CB9*
Path-7: *B1m CB4 T1s CB4 CB2*
Path-8: *B1m T1s CB4 CB2*

Fault diagnosis at B1 is determined by the *activity level* of each path. The *activity level* is based on the number of active(1) / non-active(0) relays and

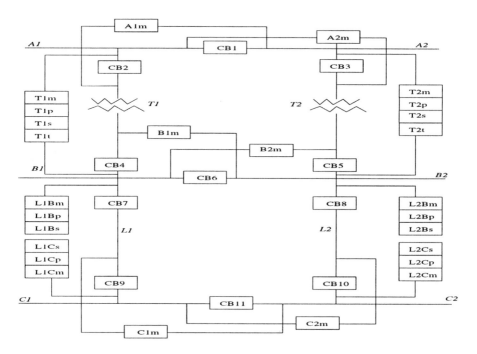

Figure 11.3.   Model of a Power System and a Protection System
Subscripts m, main protection; p,remote backup protection (distance 1); s ,
remote backup protection (distance 2); t, remote backup protection
(distance 2, opposite direction to subscript s)

circuit breakers in a path and their protection proximity to B1. The proximity
of the components to B1 enables us to determine their contribution values to
a fault in B1.

For example, the contribution ratio in Path-5 (B1m:CB7:L1Cs:CB9) is 2:2:1:1.
B1m and CB7 have higher contribution values as compared to L1Cs and CB9
because they are topologically closer to B1. If B1m and CB7 are the two components which have operated in Path-5, then the contribution value of this path
to a fault at B1 is 2/3. The contribution value of a path is mapped to the *active*
membership function in Figure 11.4 to determine its *activity level*. Thus, the
activity level of Path-5 with its operational components as above is 1/12.

The input patterns to the ANN can be of two forms. They can consist of the
operational status of the circuit breakers and relays mentioned above, or the
activity level of each path (which is a fuzzy feature) based on the contribution
value of each path to a fault in B1. The fuzzified/preprocessed input patterns

in the second case amounts to providing *a priori* knowledge to the ANN for training. Our results show that the output of ANN using preprocessed input patterns (inputs with prior knowledge) is closer to the truth value of their corresponding validation fuzzy rules.

The fuzzy rules used for validating the output of the ANN are of the form:

IF Path-1 is active THEN Bus 1 is faulty

where *active* is a fuzzy set. The membership function for *active* at each path is shown in Figure 11.4. The degree of certainty that B1 is at fault based on the *activity level* in each path leading to B1 has to be determined. The inferencing method chosen is the additive method (Earl 1994). In this method, the truth value of a consequence is determined by adding up the truth value of all fuzzy rules having the same consequence. For the power system diagnostic problem, there are many paths leading to a fault area. The additive inferencing method is used because the *activity level* at each path contributes to the possibility of fault in the diagnosed area. Therefore, the truth values of the rules leading to a diagnosed area are added up to determine the truth value of the area at fault. The final truth value is, however, bounded by 1. The decision made by the ANN, that an input pattern is faulty, is validated if its truth value assessed by fuzzy rules is above an alpha-cut threshold value. In this example, the alpha-cut threshold value for fuzzy rules is set at 0.5.

Test patterns 1 & 2 show the results with some unseen input patterns with the ANN outputs using both binary (a) and fuzzified inputs (b).

Test pattern 1:
a) Input:1 1 0 0 0 0 0 1 1 0 1 1 1
b) Fuzzified Inp:1.00 0.67 0.43 0.50 0.67 1.00 0.43 0.50
ANN output for B1 (a):0.900
ANN output for B1 (b):0.694
Truth value of the fuzzy validation rule:0.649

Test pattern 2:
a) Input:0 1 0 0 1 0 1 0 1 1 1 1 1
b) Fuzzified Inp:0.67 0.33 0.29 0.50 0.500 0.33 0.29 0.50
ANN output for B1 (a):0.200
ANN output for B1 (b):0.502
Truth value of the fuzzy validation rule:0.426

It can be seen from the test pattern results that the ANN outputs using fuzzified inputs (inputs with prior knowledge) are closer to the truth value of the fuzzy

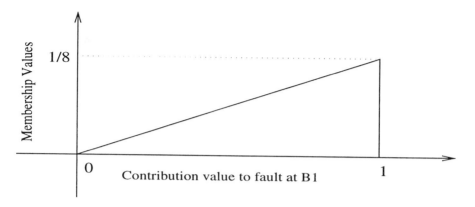

Figure 11.4.   Membership Function for active fuzzy set

validation rule than the ones with binary inputs. Although the decision made by the ANN may be validated or invalidated by fuzzy rules, the assessment of the certainty that B1 is at fault varies between the ANN and fuzzy rules. In the case of test pattern 2 where a fault at B1 is invalidated by the fuzzy rule because 0.426 is lower than the alpha-cut threshold value, the decision can be left to the user. rules are both close to their alpha-cut thresholds.

Test pattern 3:
Fuzzified Input Pattern: 0.33 0.33 0.43 0.50 0.83 0.67 0.43 0.50
ANN output for B1: 0.795
Decision made by ANN: B1 is at fault
Fuzzy rule truth value: 0.503

For this test pattern 3, the output of ANN is greater than the ANN alpha-cut threshold. Therefore, the decision made by ANN is that B1 is at fault. In this test pattern and test pattern 1, the ANN assessment of the certainty that B1 is at fault is higher than that of the fuzzy rule. The difference between the truth values of ANN and fuzzy rules that B1 is at fault could be explained as follows. ANN tends to generalize and the expert fuzzy rule determines the truth value of the generalization that B1 is at fault by the proximity of the different circuit breakers and relays protecting B1, and their operation status.

The agent oriented analysis of the RTAPS is complete with this section. In the next section aspects related to agent oriented design of the RTAPS are described.

11.3   AGENT ORIENTED DESIGN OF THE RTAPS

Design is a process of mapping system requirements defined during the analysis to an abstract representation of a specific system-based implementation. The abstract representation of a specific system-based implementation has in part been accomplishes through the association of IMAHDA's problem solving agents with domain structure analysis of the RTAPS. In the context of the RTAPS objectives and the agent oriented analysis a set of software and intelligent agents need to be identified which can help the RTAPS realize its objectives at the computational level. Once the required software and intelligent agents have been identified these need to be mapped to the problem solving agents of the RTAPS, and the RTAPS objects identified in the domain structure analysis. Further if the existing set of agents identified in the analysis stage are not sufficient to achieve the RTAPS objectives, new agents may have to be identified. The identification of such agents is generally a consequence of software solution to the problem under study (e.g. user interface agent) and not directly related to problem domain structure analysis. Finally, the order of execution precedence of the problem solving agents needs to established for implementation readiness. Keeping these aspects in view, the agent oriented design of the RTAPS involves the following steps:

- Identification of software and intelligent agents to be used by the problem solving agents.

- Mapping of the software and intelligent agents on to the problem solving agents.

- Establishing the order of execution precedence of the problem solving agents.

11.3.1   *Identification of Software and Intelligent Agents*

This step involves the identification of software and intelligent agents which facilitate realization of the goals of problem solving agents of IMAHDA in the context of the RTAPS. The prime goal of the the RTAPS is fast (between 4 and 30 seconds) and reliable alarm interpretation. The other two goals are isolation of the event and isolation of the cause of the event. Only Circuit Breaker (CB) information from TTS is available for cause isolation and fault diagnosis in the power system network. Also, intelligent agents like genetic algorithms, and fuzzy logic have not been used in the present version of the RTAPS [4].

Based on the agent oriented analysis of the RTAPS following software and intelligent agents have been identified:

- Distributed Processing (DP) Agent

- Communication COM Agent

- Relation (REL) Agent

- Belief Base BB Agent

- Expert System (ES) Agent

- Supervised Artificial Neural Net (SANN) Agent

- Data Sender Agent

- User Interface Agent

NAME:	DATA SENDER
PARENT AGENT:	DP, COM
COMMUNICATES WITH:	FILTERING AGENT
GOAL:	CREATE CONDITIONED ALARM DATABASE ESTABLISH COMMUNICATION WITH FILTERING AGENT
PERCEPTS:	ALARM FILENAME, RAW ALARM MESSAGES, DAY, MONTH, HOUR, MINUTES, SECONDS
ACTIONS:	READ ALARM FILE CREATE ALARM DATABASE COMMUNICATE CONDITIONED ALARM DATA TO FILTERING AGENT
ENVIRONMENT:	DYNAMIC, AND CYCLIC

Figure 11.5.   Data Sender Agent

The structure of software agents DP, COM, REL and BB, and intelligent agents like ES, and SANN is based on their definitions in chapter 7. These agents are identified based on the analysis phase of the RTAPS and the goals of the RTAPS. For example, in order to meet constraints like temporal reasoning, and fast response time distributed and parallel processing approach is essential, and can be achieved through use of DP, and COM software agents. Further, in a temporal reasoning environment, it is important to keep track of the internal state of a problem solving agent. This can be met through use of the BB agent. On the other hand, symbolic processing, and pattern recognition needs of the RTAPS can be met through the intelligent agents ES and SANN. However, the services of all these agents are realized within the context of the global preprocessing, decomposition, control, and decision agents. There are two more agents namely, Data Sender (DS), and User Interface (UI). The DS agent shown in Figure 11.5 is required for input conditioning. This task has not been exclusively covered by the Filtering agent in the last chapter. The DS agent reads alarms from an alarm data file and creates an alarm data base in the format acceptable for processing by the RTAPS. The UI agent is a software environment imposed agent of the RTAPS. It has not been explicitly developed in the present version of the RTAPS. At the moment the user interface is pretty simple and involves symbolic text display of the results of the RTAPS.

11.3.2 Mapping of the Software and Intelligent Agents

In this design step, the various software and intelligent agents are mapped on to the problem solving agents of the RTAPS. In the context of the RTAPS, the agent descriptions of the *Network Control* agent, and *66KV BS 2 Decision* agent after such a mapping is shown in Figures 11.6 and 11.7 respectively.

As mentioned earlier, the temporal nature of the RTAPS requires construction of temporal histories and maintaining the track of the internal state of the agents to enable correct inference. The description of the *66KV Bus Section 2 Belief Base* agent is shown in Figure 11.8. The *66KV Bus Section 2 Belief Base* agent contains knowledge in symbolic (e.g. results/inferences) and sub-symbolic (e.g. numeric input and output activations of neural networks). The Belief Base Agents at higher levels of the RTAPS (e.g. *Filtering* agent) contain only symbolic knowledge. Figures 11.9 and 11.10 establish the inheritance and compositional relationships between the software, intelligent, and problem solving agents of IMAHDA, and the RTAPS agents and objects. For example, *Filtering* and *Decomposition* agents shown in Figure 11.9 are representative of the domain independent as well as domain-specific (shown in Figure 11.9) attributes and actions associated with these problem solving agents. The inheritance link emnating from the software and intelligent agents determines the kind the software and intelligent services being employed by these problem solving agents. Similarly, the *Network Control* and *Decision* agents shown in Figure 11.9 and 11.10 are also representative of the domain independent as well as domain specific attributes and actions. The domain independent aspects of *Network Control* and *Decision* agents can be seen in Figures 11.6 and

NAME:	NETWORK CONTROL AGENT
PARENT AGENT:	DPA, REL, ES, SANN, IMAHDA CONTROL AGENT
PARENT CLASS:	ALARM
COMMUNICATES WITH:	DECOMPOSITION, BUS/TRANS/SUBSTN DECISION AGENTS
COMMUNICATION CONSTRUCTS:	REQUEST: NETWORK CB ALARM DATA
	COMMAND: 220KV/66KV BUS 1...3, SUBSTN DECISIONS
	INFORM: CB ALARM DECISION DATA
GOALS:	TIME VALIDATE ALARMS
	CONTEXT VALIDATION
	DETERMINE ACTIVE NETWORK CB DECISION AGENTS
	CREATE 220/66KV BUS1...3, TRANS, SUBSTN DECISION
	AGENTS
	ESTABLISH COMMUNICATION CHANNELS WITH
	DECISION AGENTS & DECOMPOSITION AGENT
	FORMULATE ACTION EXECUTION SEQUENCE
	RESOLVE CONFLICTS AMONG BUS, TRANS &
	SUB STN DECISION AGENTS
PERCEPTS:	CONDITIONED CB ALARM DATA, DECISION DATA
	FROM DECISION AGENTS
ACTIONS:	TIME VALIDATION RULES
	NETWORK CONTEXT VALIDATION RULES
	NETWORK DECISION AGENT CREATION RULES
	DISTRIBUTED COMMUNICATION RULES
	CONFLICT RESOLUTION RULES
	ACTION SEQUENCE RULES
	COMMUNICATE RELEVANT ALARM DATA TO ACTIVE
	BUS, TRANS & SUB STN DECISION AGENTS
	TRAIN, TEST & VALIDATE NEURAL NETWORK
	CLASSIFY ALARM PATTERNS
	CREATE/ACTIVATE DECISION AGENTS
	PROCESS NEW ALARMS
ENVIRONMENT:	DYNAMIC, CYCLIC, DISTRIBUTED, NON-DETERMINSTIC

Figure 11.6.   Network Control Agent

NAME:	66KV BUS SECTION (BS) 2 DECISION AGENT
PARENT AGENT:	DP, COM, REL, BB, ES, SANN, IMAHDA DECISION AGENT
PARENT CLASS:	ALARM, 66KV SECTION
COMMUNICATES WITH:	NETWORK CONTROL, OTHER DECISION AGENTS 66KV BS1 BELIEF BASE AGENT, 66KV BS1 DECISION VALIDATION AGENT,
COMMUNICATION CONSTRUCTS:	INFORM : 66KV BS 2 DECISION/S REQUEST: 66KV BS1 CB DECISION ALARM DATA, COMMAND: PREVIOUS RESULT, ACTIVATION STATE REQUEST : VALIDATE 66KV BS1 CB DECISION
GOALS:	CONTEXT VALIDATION DETERMINE SINGLE/MULTIPLE LINE/BUS FAULTS ESTABLISH COMMUNICATION CHANNELS WITH CONTROL AGENT & OTHER DECISION AGENTS FORMULATE ACTION EXECUTION SEQUENCE RECORD INTERNAL STATE
PERCEPTS:	CONDITIONED 66KV BUS SECTION CB ALARM DATA, RESULTS FROM OTHER DECISION AGENTS, VALIDATED RESULTS FROM VALIDATION AGENT, NEW/NOVEL 66KV BUS SECTION CB ALARM PATTERNS TO LEARN
ACTIONS:	66KV BS2 CONTEXT VALIDATION RULES DISTRIBUTED COMMUNICATION RULES TRAIN, TEST & VALIDATE NEURAL NETWORK CLASSIFY ALARM PATTERNS COMMUNICATE RESULTS TO OTHER DECISION AGENTS PROCESS NEW ALARMS
ENVIRONMENT:	DYNAMIC, CYCLIC, DISTRIBUTED, NON-DETERMINSTIC

Figure 11.7.   66KV BS 2 Decision Agent

NAME:	66KV BS1 BELIEF BASE
PARENT AGENT:	BB
COMMUNICATES WITH:	66KV BS1 DECISION AGENT, BELIEF BASE AGENTS OF OTHER DECISION AGENTS
COMMUNICATION CONSTRUCTS:	INFORM: PREVIOUS STATE, REQUEST: PREVIOUS/CURRENT STATE OF OTHER DECISION AGENTS
GOALS:	KEEP TRACK OF 66KV BS1 ALARM MESSAGES KEEP TRACK OF PREVIOUS RESULTS/INFERENCES/ ACTIVATIONS
PERCEPTS:	66KV BS1 ALARM MESSAGES, NUMERIC NEURAL NET INPUT & OUTPUT ACTIVATIONS, ALARM ID, DATE, TIME
ACTIONS:	RETRIEVE PREVIOUS RESULTS RETRIEVE NEURAL NET INPUT/OUTPUT ACTIVATIONS RETRIEVE ALARM MESSAGES MODIFY RESULTS/ACTIVATIONS ADD ALARM MESSAGES DELETE/RESULTS/MESSAGES.
ENVIRONMENT:	DYNAMIC, CYCLIC

Figure 11.8.   66KV Bus 2 Section Belief Base Agent

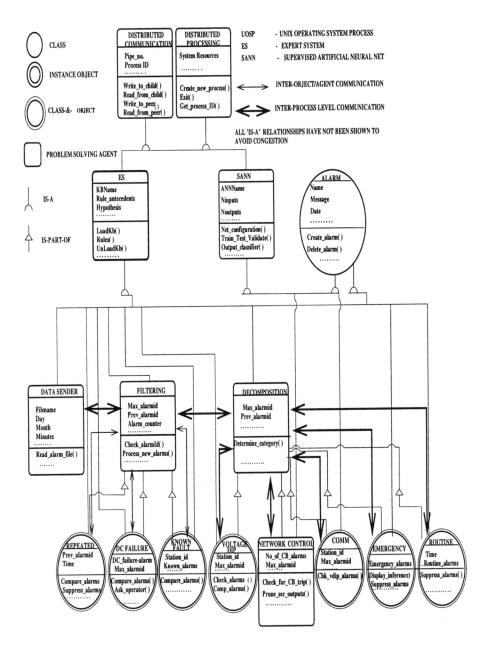

Figure 11.9. Mapping of Software, Intelligent, & Problem Solving Agents

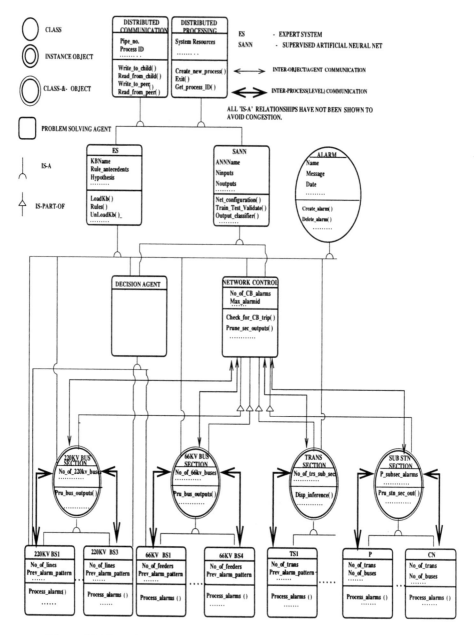

Figure 11.10.   Mapping of Software, Intelligent, & Problem Solving Agents

11.7 respectively. The *Voltage Dip* and other class-&-objects in Figures 11.9 and 11.10 are now identified as agents since they inherit the goals and actions of the software and intelligent agents of IMAHDA. Likewise *66KV BS 1...4, 220KV BS 1...3,* and other transformer and substation agents shown in Figure 11.10 are decision agents as they inherit the goals and actions of the IMAHDA decision agent. For example, *66KV BS 2 Decision* agent shown in Figure 11.7 inherits services from various parent agents including IMAHDA's problem solving decision agent. Also, *Voltage Dip, Repeated* and *Emergency* the RTAPS objects inherit services from software and intelligent agents like *COMmunication (COM), Distributed Processing (DP),* and *ES,* and alarm structure from the *Alarm* class.

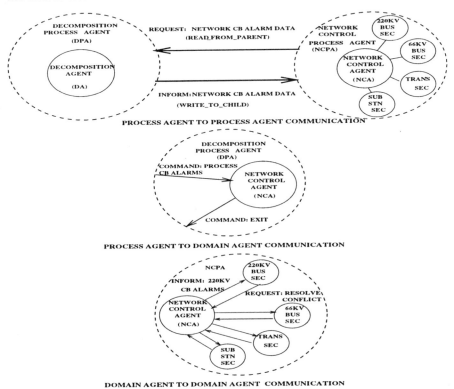

Figure 11.11.   Two Level Three Channel Communication

11.3.2.1   Two Level, Three Channel Communication.   The mapping of the distributed process agent introduces two levels of communication in the

RTAPS, namely, at the process agent level and domain agent level. Figure 11.9 shows the two level, three channel communication between the *Decomposition* agent and the *Network Control* agent, the *Network Control* agent and its domain agents. The three channels of communication shown in Figure 11.11 are process agent to process agent, process agent to domain agent, and domain agent to domain agent. The message *REQUEST: RESOLVE CONFLICT* as the name suggests is a request to the *Network Control* agent to resolve conflicting decisions made by different sections (e.g. 220kV Section and TRANS Section), whereas *INFORM: 220KV CB ALARMS* is a message to inform the *200KV Bus Section* agent to retrieve 220KV CB alarm data. It may be noted that the use communication construct *INFORM* between *Decomposition* agent and the *Network Control* agent involves inter-process communication through *WRITE_TO_CHILD* action, whereas the *INFORM* construct in the *Network Control* agent is used for communication between two domain agents and does not employ inter-process communication action like *WRITE_TO_CHILD*.

11.3.3 Establishing Order of Execution Precedence

An essential part of any design is to determine in what order the problem solving agents are going to be executed. It is called the order of execution precedence. It is determined by the criticality (with respect to the overall system objectives) and the purpose of each problem solving agent in the application domain. It helps in predicting the states and behaviors required of the system.

In Figure 11.12 the problem solving agents of the RTAPS along with their execution precedence are shown. In the following paragraphs a brief overview is provided of how this execution precedence facilitates the realization of the RTAPS objectives.

One of the critical objectives of the RTAPS is to provide a reliable inference. Therefore, it is essential that the data used for inference is in the correct format and free from noise as far is practically feasible. Thus, the *Data Sender* and *Filtering* process agents (which represent global preprocessing) shown in Figure 11.10 precede agents like *Decomposition* and *Network Control*. The *Decomposition* agent comes next in the order of precedence and makes the preliminary decision on the isolation of the alarm event. This is followed by the *Network Control* process agent which has to perform more intricate decision making related to the isolation of the cause of the event. The decision agents like *220KV Bus1 Section (BS1)* and *66KV Bus2 Section* which locally identify different causes follow the *Network Control* agent.

Another critical objective of the RTAPS is response time. The ingredients (i.e ANNs) in the problem solving agents are well suited for realizing fast re-

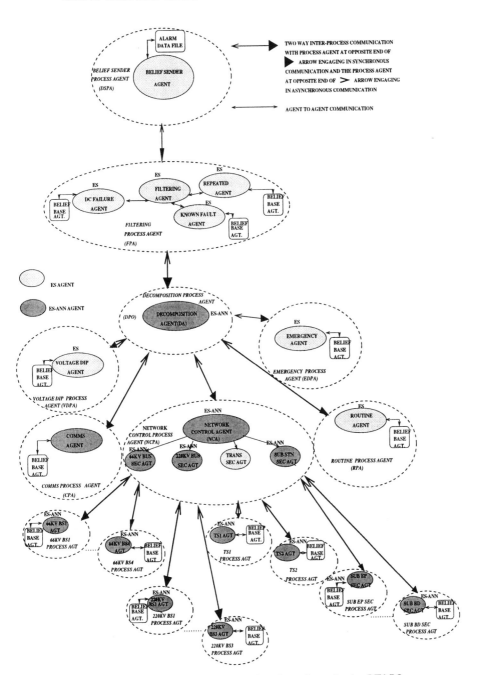

Figure 11.12.   Order of Execution Precedence in the RTAPS

sponse time (Khosla and Dillon, 1995). Last but not the least, the order of precedence also takes into account the problem solving behavior of the experts in the domain, which in our case are the operators in the regional power system computer control center. This is to promote comprehensibility and facilitate useful interaction between the user (operator) and the system if required.

11.4 EMERGENT CHARACTERISTICS OF THE RTAPS AGENTS

The agent oriented analysis and design of the RTAPS done in this chapter and the previous one includes static as well as dynamic aspects. The static aspects relate to the structural components or the physical structure of the TTS power system network. The dynamic aspect consists of two distinct parts. The first part relates to analysis of the alarm behavior in real time, i.e. the way alarms interact with each other. For example, concepts like *Repeated, Known Fault, Voltage-Dip*, etc. reflect the dynamic behavior of alarms. the second part relates to operator's problem solving behavior in real time.

The static and dynamic aspects of agent oriented analysis and design have been captured with help of IMAHDA in the RTAPS design structure shown in Figure 11.10. It may be interesting to determine the emergent behavior of intelligent hybrid distributed RTAPS based on the incorporation of these static and dynamic aspects In this section, these characteristics are briefly described.

- **Task Orientation & Autonomy:** The RTAPS agents operate based on tasks assigned to them. They are distinguishable from each other based on tasks undertaken by them. An agent in the RTAPS depends upon the external environments and on other agents in the RTAPS for their information needs. The information needs of an agent (e.g. *Filtering* agent and *66KV BS1 Decision* agent) are met from the external environment in a raw form (for *Filtering* agent) as well as in partially processed form (for *66KV BS1 Decision* agent) from other RTAPS agents. They are largely autonomous in the sense that they accomplish the task on their own once provided with the required information. They are semi-autonomous in the sense that some of them (e.g. *66KV BS1 Decision* agent) require alarm information in a partially processed form, or require human intervention in some situations (e.g. *Known Fault* filtering agent in some situations requests for operator intervention before making a decision for deleting or not a deleting an alarm). Although *Decision Validation* agents in the RTAPS (the reasons for this will become clear in the next chapter) have not been explicitly designed in present version of the RTAPS, they also require interaction with user or other non-system designed validation agents in the environment. Given these observations, the RTAPS agents can be said to be semi-autonomous.

- **Global and Local Distribution/Parallelism:** The RTAPS agents exhibit global as well as local parallelism. Global parallelism is exhibited by the process agents, whereas local parallelism is exhibited by the artificial neural networks in the domain problem solving agents.

- **Cyclic Operation and Persistency:** The RTAPS multi-agent architecture as a whole has a cyclic operation, wherein the new alarms are read in, processed, and read in again. The RTAPS agents have been designed in a manner they can engage in this cyclic operation for long periods without user intervention.

- **Synchronous and Asynchronous Communication:** The lower level agents in the RTAPS are designed in such a manner that their inferencing process is not constrained by inter-process communication with higher level agents. That is, the lower level agents like *Network Control* communicate asynchronously with their higher level agents like *Decomposition*. This is specifically related to the *Read( )* and *Write( )* system calls associated with inter-process communication. A lower level agent after reading the alarm data send a confirmatory message to the higher level agent asynchronously and goes ahead with its inferencing process with the new alarm data. However, the higher level agent has to ensure that the old alarm data has been read by the lower level agent before it sends new alarm data and so it has to synchronize with the read confirmatory message from its lower level agent before it proceeds with further processing. This manner of communication is shown through the solid and standard arrow ends of the double sided arrows in Figure 11.12.

- **Two Level Three Channel Communication:** Multiple levels and channels of communication shown in Figure 11.9 provide additional flexibility to a problem solving agent in the RTAPS to achieve its goals. The communication constructs of *Network Control*, and *66KV Bus 1 Section Decision* agents shown in Figures 11.7 and 11.8 respectively also highlight the difference in the nature of communication between different problem solving agents. It may also be noted, that although the RTAPS multi-agent architecture of homogeneous (e.g. the *Filtering* agent consists of high level rules only), and heterogeneous (e.g. *Network Control* agents consists of rules and artificial neural networks) agents, all agents maintain a uniform communication interface with their external environment and other agents.

- **Flexibility:** The existence of homogeneous and heterogeneous agents also demonstrated the flexible behavior of the RTAPS multi-agent architecture.

- **Learning and Adaptation:** The decomposition, control and decision agents in the RTAPS have learning and adaption properties which can help to learn from their mistakes and adapt to new situations. This process (as this book is being written) is being made autonomous or self-organizing. In its present state, learning from mistakes is undertaken through manually initializing the training regime of the artificial neural networks.

- **Collaboration:** The RTAPS multi-agent architecture as a whole is based on the concept of collaboration where all intelligent agents combine together in the transition from the problem state to solution state. Different agents discharge their responsibilities in different stages of the problem solving process.

- **Intelligibility:** The development process (e.g. capturing the problem solving behavior of the operator, capturing the problem domain structure through OOA) of the RTAPS agents makes them intelligible to operator. Thus, if required the user can interact with the various intelligent problem solving agents of the RTAPS.

- **Agent States** The behavioral characteristics like agent states, cyclicity, global and local parallelism, and synchronous/asynchronous communication can be analyzed using S tate Controlled Petri Nets (SCPN) described in the previous chapter. Figure 11.13 shows the dynamic modeling of the some parts of the RTAPS using SCPN model. More details in terms of reachability graph, deadlocks, etc. can be found in Khosla and Dillon (1994).

  genetic, fuzzy, explanation. The decision agents in the RTAPS can operate in cooperative or competitive mode. For example, *66KV Bus 1 Section, 66KV Bus 2 Section*, and *66KV Bus 3 Section*, and *66KV Bus 4 Section* decision agents can work cooperatively with one another to locate a fault in the 66KV section of the power system network. This cooperation may involve sharing information/results with each other since all these bus sections are connected with one another in the power system network.

  On the other hand, if CB, Relay, and Analog is available for fault diagnosis, then within, say *66KV Bus 1 Section* decision agent three sub-agents (i.e. CB, Relay, and Analog) can be activated simultaneous (through the services of the distributed processing agent) on the same diagnosis data, and compete with each other for locating the fault/s. This aspect of competition will be shown more explicitly in the next chapter.

- **Scalable Hybrid Agents** The RTAPS architecture also inherits the scalable characteristics of IMAHDA. The problem solving agents are horizontally, and vertically scalable. This characteristic is an important one as it

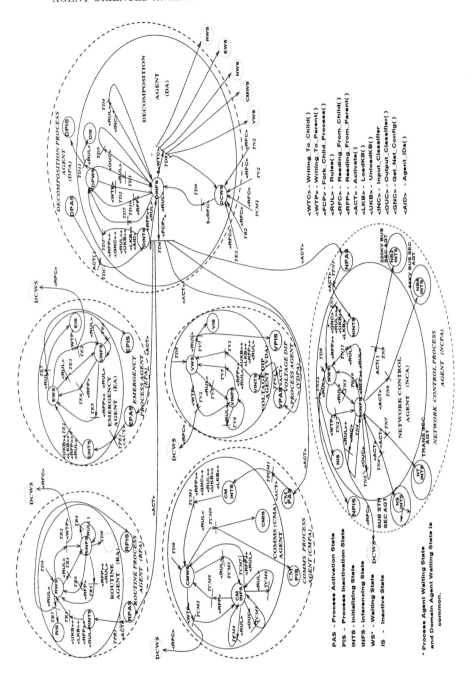

Figure 11.13.   Dynamic Model of the RTAPS

makes the architecture sustainable. For example, in the present version of the RTAPS only CB information is used for isolating the cause of the alarm event or fault diagnosis. However, when the Relay, and Analog information are made available, the existing RTAPS architecture will need to be scaled horizontally and vertically. This feature will be shown explicitly in the next chapter.

In the next chapter the implementation of the RTAPS architecture is described in terms of methodology related objectives, domain related objectives, and management related objectives.

11.5 SUMMARY

In this chapter the vocabulary of the problem solving agents of the RTAPS has been defined in terms of the time and context validation rules, and learning knowledge and strategy of the decomposition, control, and decision agents in the RTAPS. The agent oriented design of the RTAPS involves three steps. In the context of the RTAPS objectives and the agent oriented analysis a set of software and intelligent agents are firstly identified which can help RTAPS realize its objectives at the computational level. The second step involves the mapping of the software and intelligent agents to the RTAPS problem solving agents and objects. The mapping involves establishing inheritance and compositional relationships between the software agents, intelligent agents and problem solving agents and objects of the RTAPS. The problem solving agents of the RTAPS after the mapping contain domain independent and domain specific actions. The agent oriented design of the problem solving agents in specifying their parent agents, parent classes, goals, communication constructs, actions, percepts, and environment are outlined. The third design step establishes the order of execution precedence of the RTAPS agents which facilitates the realization of the RTAPS objectives. Finally, characteristics of the emergent behavior of the RTAPS agent oriented hybrid design architecture are outlined.

Notes

1. Genetic algorithms have not been used in the present version of the RTAPS.

2. These are keywords extracted from the real-time alarm data shown in Appendix E.

3.   One of the reasons is that the communication and processing overhead between *66KV Bus Section* and say *66KV Bus1 Section (BS1)* for not making the distinction between 66KV Bus Section CB alarms, and 66KV Bus1 Section CB alarms is not high.

4. The primary reason for this is that intelligent agents like (SANN, and (ES) have accomplished the present set of the RTAPS objectives sufficiently well.

References

Earl C., 1994 *The Fuzzy Systems Handbook: A Practitioner's Guide to Building, Using, and Maintaining Fuzzy Systems* , Academic Press, Boston, MA 1994.

Khosla, R., and Dillon, T. (1994), "Dynamic Analysis of a Real Time Object Oriented Integrated Symbolic-Connectionist System using State Controlled Petri Nets," *Fifth Intelligent Systems Applications Conference*, Avignon, France.

Khosla, R., and Dillon, T. (1995), "A Distributed Real-Time Alarm Processing System with Symbolic-Connectionist Computation," in *Proceedings of the IEEE Workshop in Real Time Applications*, Washington D.C., USA.

Khosla, R. and Dillon, T. (1997), "Learning Knowledge and Strategy of a Generic Neuro-Expert System Architecture in Alarm Processing," to appear in *IEEE Transactions in Power Systems*.

Sekine Y., H. Okamota, and T. Shibamoto, (1989), "Fault Section Estimation using cause-effect network," *Second Symp. on Expert Systems Application to Power Systems*, Seattle, USA, 17-20 July.

Wang, X. and T. Dillon, (1991), "A Second Generation Expert System for Fault Diagnosis," in *Proceedings of 3rd Symposium ESAPS*, Tokyo, Kobe, pp.751-756.

Wang, X. and T. Dillon, (1992), "A Second Generation Expert System for Fault Diagnosis," *International Journal of Electrical Power and Energy Systems*, vol. 14, no. 2-3, pp. 212-216.

Tang, S. K., Dillon. T. and Khosla, R. (1995) "Application of an Integrated Fuzzy, Knowledge-Based, Connectionist Architecture for Fault Diagnosis in Power Systems," *Sixth IEEE Intelligent Systems Applications Conference*, Florida, USA.

Zadeh, L.A. (1965), "Fuzzy Sets," *Information and Control*, vol. 8, pp. 338-352.

Zadeh, L.A. (1983), "The Role of Fuzzy Logic in the Management of Uncertainty in Expert Systems" *Fuzzy Sets and Systems* 11, pp. 199-227.

12 RTAPS IMPLEMENTATION

12.1 INTRODUCTION

Implementation of a large scale real time system like alarm processing involves realization of various objectives. These include methodology related objectives, domain related objectives, and management related objectives. The agent oriented analysis and design stages described in the last two chapters has provided the platform on which these objectives can be realized.

This chapter describes the realization of these objectives. For that matter, this chapter is divided into four parts. The first part and second part address the methodology related objectives. The third part and fourth part address the domain related objectives and the management related objectives respectively.

In the first part IMAHDA related issues are described. These are a) Multi-Agent Oriented Distributed Implementation, and b) Realization of various information processing phases in IMAHDA.

In the second part neural network related implementation issues like Supervised Artificial Neural Network (SANN) topologies, training cycles, diagnostic thresholds, countering noise, and fault prediction are discussed.

The third part deals with the power system aspects of the implementation, like a) Processing of alarms by different agents of RTAPS, b) Realization of RTAPS objectives set out in Chapter 8, and c) Real time Issues associated with RTAPS.

Finally, the scalability and cost effectiveness issues of the RTAPS architecture which facilitate its practical use and future enhancements are briefly discussed.

12.2   IMAHDA RELATED ISSUES

IMAHDA forms the backbone of the implementation strategy of RTAPS. In this section after outlining agent oriented implementation of RTAPS, the realization of various phases of IMAHDA in RTAPS is outlined.

12.2.1   Multi-Agent Oriented Distributed Implementation

The implementation hierarchy of the RTAPS is shown in Figure 12.1. It is a based on Figure 11.12 in chapter 11. RTAPS has been implemented as an agent-oriented, continuous, concurrently processed distributed system with object-oriented properties on a Sun4/280 workstation and Unix OS platform. RTAPS uses alarm data used by control room staff for real time processing. No relay status information is available at the regional control center at the moment for interpretation of network faults.

Nexpert Object, an expert system shell (Nexpert 1989) along with external C routines have been used to implement RTAPS. There are 60 rules, 28 SANNs, more than 100 objects (this includes intermediate as well as component level objects), and about 20,000 lines of external C code in the RTAPS. For the purpose of performance comparison, another version of RTAPS (using Nexpert Object) consisting of an expert system with rules only (executable in a parallel and distributed manner) has also been built.

The main reason for using Nexpert Object is that it provides the encapsulating and maintainable framework in which intelligent hybrid problem solving agent can be realized. Besides, it has the object oriented, and rule based reasoning framework in which agents object knowledge (e.g. power system network representation) can be represented as well as high level rule based reasoning can be performed. Soft computing aspects, like artificial neural networks are programmed externally and linked to the problem solving agents in the Nexpert environment. It makes provision for representing relational constructs like *is-a* and is-a-part-of and allows complete or restricted multiple inheritance. The principles of inheritance, modularity, reuse and maintenance are effectively realized with Nexpert Object. The problem solving agents shown in Figure 12.1

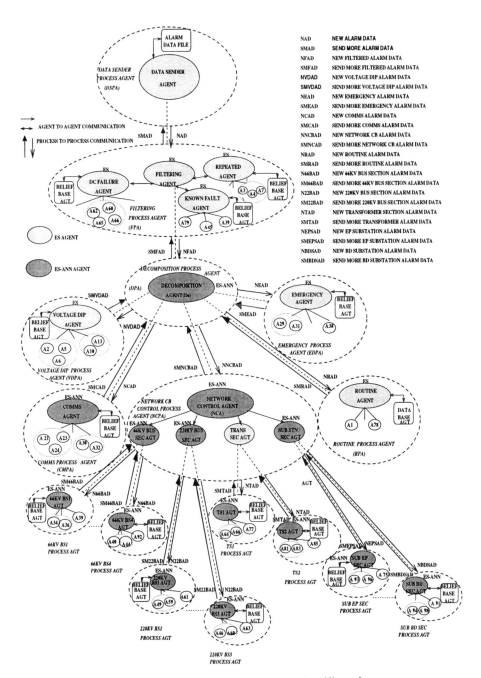

NAD	NEW ALARM DATA
SMAD	SEND MORE ALARM DATA
NFAD	NEW FILTERED ALARM DATA
SMFAD	SEND MORE FILTERED ALARM DATA
NVDAD	NEW VOLTAGE DIP ALARM DATA
SMVDAD	SEND MORE VOLTAGE DIP ALARM DATA
NEAD	NEW EMERGENCY ALARM DATA
SMEAD	SEND MORE EMERGENCY ALARM DATA
NCAD	NEW COMMS ALARM DATA
SMCAD	SEND MORE COMMS ALARM DATA
NNCBAD	NEW NETWORK CB ALARM DATA
SMNCAD	SEND MORE NETWORK CB ALARM DATA
NRAD	NEW ROUTINE ALARM DATA
SMRAD	SEND MORE ROUTINE ALARM DATA
N66BAD	NEW 66KV BUS SECTION ALARM DATA
SM66BAD	SEND MORE 66KV BUS SECTION ALARM DATA
N22BAD	NEW 220KV BUS SECTION ALARM DATA
SM22BAD	SEND MORE 220KV BUS SECTION ALARM DATA
NTAD	NEW TRANSFORMER SECTION ALARM DATA
SMTAD	SEND MORE TRANSFORMER ALARM DATA
NEPSAD	NEW EP SUBSTATION ALARM DATA
SMEPSAD	SEND MORE EP SUBSTATION ALARM DATA
NBDSAD	NEW BD SUBSTATION ALARM DATA
SMBDSAD	SEND MORE BD SUBSTATION ALARM DATA

Figure 12.1.   RTAPS Implementation Hierarchy

have been implemented as Nexpert agents (supported by external routines) which operate in a parallel and distributed manner.

The small solid line circles shown in Figure 12.1 which are not surrounded by light dashed line circles represent the alarm instances of various alarm categories (e.g. A38 represents an alarm instance of *Emergency* alarm category). The small solid line circles in Figure 12.1 surrounded by the light dashed line circles represent composite instances (e.g. as shown in *DC Failure* agent, the four alarms put together represent a *DC Failure* alarm category). "A65" represents an alarm with id 65. The dashed line/s between two process agents like DPA and CMPA shows the parent (DPA - one which creates the child) and child (CMPA) relationship between them.

At the lowest level in Figure 12.1, e.g., in the *220KV Bus1 Section* and *220KV Bus2 Section* decision agents, the different alarms small solid line circles or small solid line circles surrounded by light dashed line circles are alarm instances of the CB alarms associated with these bus sections respectively. Further, they are also indicative of the single/multiple line or bus faults.

12.2.2   Realization of the IMAHDA Architecture

This section elaborates on how the different phases of the IMAHDA architecture namely, global and local preprocessing, decomposition phase, control phase, decision phase and postprocessing phase are realized in the RTAPS.

12.2.2.1   Global Preprocessing.   As was explained in chapter 6, global preprocessing involves filtering out noise which is peculiar to the domain as a whole or common to many classes. The global preprocessing is done by the *Data Sender Process Agent (DSPA)* and *Filtering Process Agent (FPA)* shown in Figure 12.1. Global preprocessing is undertaken in RTAPS to condition the real time alarm data (see Appendix E) as well as filter out noise in the alarm data caused by *Repeated* alarms, *DC_Failure* alarms and *Known Fault* alarms. These alarms are filtered or suppressed. The above tasks are performed using software routines and symbolic heuristics or rules. The belief base agents in the FPA are used to keep track of the internal state of *Repeated Agent, DC_Failure Agent*, and *Known Fault Agent* respectively. The record of the internal state is used to preprocess new alarms. For example, if an alarm has been identified as faulty, it is recorded and this record is compared with any future occurrence of this alarm for filtering purposes.

In the context of RTAPS objectives, enough domain knowledge was available for preprocessing. Thus the need to adopt bottom up knowledge engineering strategy was not felt.

12.2.2.2 Decomposition Phase. Decomposition is done in the RTAPS by globally decomposing the filtered alarms into *Communication* alarms, Circuit Breaker (CB) (related to *Network Control* agent) alarms, *Voltage-Dip* alarms, *Emergency* alarms and *Routine* alarms. The input validation task involves context validation. The context validation rules ensure that only alarms belonging to the TTS (Thomastown Terminal Station) are processed and alarms relevant to the *Decomposition* agent are processed. The problem formulation task involves the conversion of alarms into input activation/vector of the SANN (used for determining abstract concepts), and initializing the SANN. The determination of abstract concepts like *Communication* alarms, Circuit Breaker (CB) (related to *Network Control* agent) alarms, *Voltage-Dip* alarms, *Emergency* alarms and *Routine* alarms.

Based on the input activations, the trained multilayer feedforward SANN determines which agents (i.e. *Communication*, *Network Control*, *Voltage-Dip*, *Emergency* and *Routine*) are to be activated in the next phase. Some of the actions executed in the *Decomposition* Agent (DA) in Figure 12.1 are LoadKb(), Rules(), Get_Net_Configuration(), Train_Test_Validate(), Output_Classifier(), Create_new_process(), and others.

The activated agents, i.e. *Communication, Routine, Emergency, Voltage-Dip* in Figure 12.1 do not have any explicit control phase. Infact, in these agents the control phase and decision phase have been merged together. The *Network Control* agent related to the *Network CB* alarms represents the control phase explicitly. Computationally, intelligent agents used are either ES agents or ES-ANN agents. The decision for constructing them as ES or ES-ANN agent is based primarily on level of decision making required (i.e. decision class priority) by these agents (except in the *Comm* agent. For example, in the *Voltage-Dip* agent at the moment RTAPS only needs to classify whether a particular set of alarms are voltage dip alarms or not. RTAPS does not need to find out the cause of these voltage dip alarms. The primary reason for using SANN in the *Comms* agent has not been because it has high decision class priority but to save development time. Computationally, an ES agent can also meet RTAPS objectives in the *Comms* agent.

12.2.2.3 Control Phase. Control is performed explicitly in the *Network Control Process Agent*. The *Network Control Agent (NCA)* controls the flow of incoming CB alarms from the *Decomposition* process agent into *220KV Bus Section*, *66KV Bus Section*, *Transformer Section* and *Substation Section* agent. It decides whether one or more agent/s is/are to be activated based on the received pattern of alarms from *Decomposition Process Agent (DPA)* in Figure 12.1.

The major tasks undertaken in the *Network Control Process Agent (NCPA)* are noise filtering, input validation, problem formulation, determining decision level concepts, and resolution of conflicts. A CB alarm is adjudged as noisy if it is not accompanied by a "CB-TRIP" alarm in a certain time range. The input validation tasks involves time validation and context validation. The time validation rules determine the time range of the "CB-TRIP" messages. The context validation rules determine problem formulation strategy based on whether intermediate or section level representation is employed or lower/component level representation is employed. Figure 12.1 shows the section level agents based on section level fault detection strategy. The problem formulation rules besides determining the action sequence of *NCA* for activating the section agents in NCPA, also determines the communication rules between *NCPA* and *DPA*, and between *NCPA* and the decision agents in the decision phase. *NCA* in *NCPA* uses a SANN agent to determine whether the incoming CB alarms belong to the *220KV Bus Section* , *66KV Bus Section* , *Transformer (Trans) Section* or the *Substation Section*. Based on the output activations of the SANN agent one or more of the four section agents in *NCPA* are activated. These section agents in *NCPA* are also used to resolve the conflicts among their respective decision agents in the decision phase. The *NCA* is used to resolve conflicts among the section agents as they are all linked with each other in the power system network. For example, the inferences made by the *220KV Bus1 Section*, *220KV Bus2 Section*, etc. agents are pruned by the *220KV Bus Section* agent before they are presented to the operator.

Based on the received CB alarm data from NCA, the section agents also decide whether one or more decision agent/s is/are to be activated or not.

12.2.2.4 Decision Phase. The *220KV Bus1 Section, 220KV Bus2 Section, 66KV Bus1 Section, TS1 Section, Sub EP Section* agents shown in Figure 12.1 represent the decision phase of IMAHDA. These decision agents make local decisions in terms of faults in different parts of the TTS power network. The input validation in these decision agents involves checking the time validity of the alarms. The problem formulation among other things involves mapping the previous internal state of the agent with its present state in light of new alarm data. In other words, previous input activations of the input vector of the SANN agent are combined with the present input activations in a valid time frame in order to form a more complete pattern of alarm events in a dynamic environment. This helps in revising the previous decisions of the agent which may become invalid in light of new alarm data. The belief base agent is used by the decision agent to keep track of all its decisions and the alarm events which led to those decisions. This aspect will be discussed in more detail later on in this chapter when temporal reasoning aspects of RTAPS are described.

The local decisions made by the decision agent are verified by the respective section agents in NCPA and NCA for any conflicts. The conflict resolution done by different section agents in the NCPA combines the local decisions made by the decision agents. At present peer to peer communication is being implemented between various decision agents. Once implemented, this will largely obviate the need for using vertical communication and conflict resolution facilities of NCPA.

12.2.2.5   Postprocessing Phase.   The post processing in the present version of RTAPS basically involves knowledge explanation. It is incorporated in *Communication, Voltage Dip, Routine, Emergency,* the decision agents (e.g. *220KV Bus 1 Section* in the decision phase. The knowledge explanation is provided by displaying to the operator the inference in natural language along with the list of alarm messages related to a particular inference made by RTAPS. Rules are used for providing knowledge explanation.

In the present stage of development of RTAPS, the need for providing a fault propagation model of each component for checking or validating the decisions made by the SANNs in the decision phase has not arisen. This is so because the neural networks at the moment are generalizing to 100% for the tasks assigned to them.

12.3   TRAINING OF NEURAL NETWORKS IN RTAPS

SANNs form an integral part of RTAPS. In the last chapter the underlying conceptual framework which has been used in the extraction/selection of input features and output classes for SANNs in different phases of the RTAPS has been outlined. In this section SANN training related issues and their fault prediction capabilities in the RTAPS are described.

12.3.0.6   SANN Training.   Training of SANNs in the RTAPS involves among other issues selection of input features and output classes in different phases, determination of network topology and diagnostic thresholds. SANNs are trained with discrete inputs and outputs. All SANNs used in RTAPS at the moment are trained with a single hidden layer using the backpropagation algorithm or a combination of Kohonen's LVQ (Kohonen 1990) algorithm and backpropagation algorithm (especially the ones in the decision phase). The training time has varied between 100 to 20000 cycles for different SANNs. Some of the SANNs are shown in Figure 12.2.

The maximum size of the SANN in the present version of the RTAPS is with nine inputs and eight outputs. The size of the SANNs has been deliberately kept small for scalability reasons in the future as will be explained in section

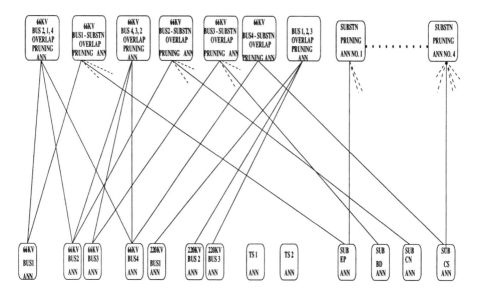

Figure 12.2.   SANN Network in the RTAPS

12.5. The SANNs have been trained with a partial data set in a manner so as to generalize to 100% on the remaining untrained data set. For example, a SANN with 256 patterns has been trained by us with a 8-7-3 configuration/topology and a representative training data set of 31 patterns only. We have used a combination of Kohonen's LVQ algorithm and the backpropagation algorithm as shown in Figure 12.3. The training of the SANN is done in two stages. In the first stage, Kohonen's LVQ algorithm is used to determine the clusters in the training data set. These clusters are then replaced by an equal number of hidden units in the hidden layer of the multilayer perceptron. The code book weight vectors which represent these clusters are now used as weight links between the input units and the hidden units of the multilayer perceptron as shown in Figure 12.3. The SANN is then trained with the backpropagation algorithm to achieve a generalization of 100%.

Figure 12.3 shows one such case of training on data associated with *220KV Bus1 Section* SANN in the decision phase of the RTAPS. It also shows the recognition comparison between stand alone Kohonen LVQ algorithm, backpropagation algorithm and their combination.

The diagnostic threshold for classifying a alarm pattern into a particular class with use of backpropagation algorithm is a function of overlap between patterns belonging to different classes and the characteristics of the data set

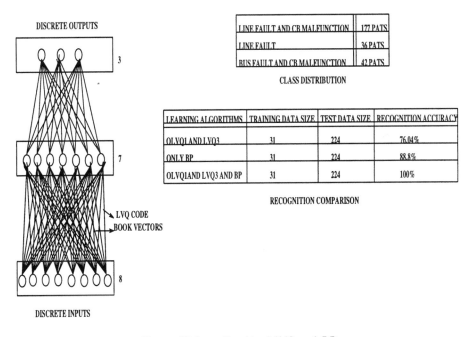

LINE FAULT AND CB MALFUNCTION	177 PATS
LINE FAULT	36 PATS
BUS FAULT AND CB MALFUNCTION	42 PATS

CLASS DISTRIBUTION

LEARNING ALGORITHMS	TRAINING DATA SIZE	TEST DATA SIZE	RECOGNITION ACCURACY
OLVQ1 AND LVQ3	31	224	76.04%
ONLY BP	31	224	88.8%
OLVQ1AND LVQ3 AND BP	31	224	100%

RECOGNITION COMPARISON

DISCRETE OUTPUTS

3

7

LVQ CODE
BOOK VECTORS

8

DISCRETE INPUTS

Figure 12.3.   Combined LVQ and BP

in general. In the SANNs trained by us, the threshold has varied from 0.5 in some SANNs to between 0.2 to 0.8 in some others.

12.3.0.7   Countering Noise and Fault Prediction.   In case of network faults provision for noise can be made not only in terms of extraction/selection of input features but also by incorporating some of the heuristics employed by the operator in inferencing different faults. One such heuristic has been incorporated for inferencing bus faults. Depending upon the number of CBs in a particular section (i.e, Bus1 or Bus2) the effect of number of CB malfunctions is taken into account. Like for example, in case of *220KV Bus1 Section* where there are 8 CBs, the SANN is trained for any 2 CB malfunctions to classify a bus fault. Another advantage of incorporating these malfunctions is that in case of severe faults, if for any reason the 1/2 CB alarm messages is/are missed out by RTAPS in real time operation while inferencing with the existing alarm snap shot or actually have malfunctioned it will still infer a bus fault correctly. However, in the absence of relay information, comprehensive CB malfunction detection is not possible. Since the SANNs are trained to generalize 100%, all

possible TTS network faults excluding a proportion of CB malfunctions are
covered.

12.4   POWER SYSTEM ASPECTS OF THE RTAPS

In chapter 9 different types of alarms associated with TTS have been described.
In this section, how the different alarms are processed in RTAPS and how this
processing leads to the realization of the RTAPS objectives is described.

12.4.1   Alarm Processing

Alarm processing as it is performed in the regional power system control centers of SECV (State Electricity Commission Victoria) is a function of operator
knowledge and experience, analysis of the previous real time alarm data, network topology, time, and the peculiarities associated with the alarms from
different terminal power stations and substations. The operator experience is
useful in developing shallow rules which are a function of frequency of the event,
falsity of the event, complexity of the event and duration of the event. For example, the alarms associated with fire in a power station or a substation may
be false because of a fault in the fire equipment and as a result inference needs
to be made keeping in view the false nature of these alarms. On the other hand,
the miscellaneous alarms associated with alarm equipment failure (as a result
of DC supply failure) can only be ascertained from operator's experience and
analysis of the real time data. The alarms associated with network topology
are inferenced using neural networks as already explained in the previous two
sections. The rules used for determining the time validity of alarms are derived
from the operator's experience and analysis of the real time data. For example,
an alarm which repeats itself within a minute is considered as a repeated alarm,
whereas if an alarm repeats after half an hour it is not considered as a repeated
alarm.

Some of the peculiarities associated with alarms from TTS relate to the
proneness of certain alarms to go on and off at regular intervals because of some
inherent problems in their design and physical location in the TTS. For example, OLTC (Overload Tap Changer) related transformer alarms and 220KV CB
Air alarms in TTS frequently go on and off because of some inherent problems
associated with design of these alarms and the OLTC.

The rules based on the above parameters have been used for global and local preprocessing and inferencing in different agents (e.g. *Filtering, Emergency,
Routine* and *Voltage-Dip*) of the RTAPS. The intention here is not to exhaustively address the design of these rules in each agent of the RTAPS because of
the sheer volume of the description. The intention here is to describe the end

result of using intelligent agents, software agents, and problem solving agents in RTAPS for processing of real time alarm data.

12.4.2   Data Sender Process Agent (DSPA)

For the purpose of simulation, the DSPA reads 10 alarms in each inferencing cycle from a alarm data file on a Sun 4/280 workstation. The real time alarm data used for this simulation is shown in Appendix E. The operation of an alarm is indicated by ON or OPER while OFF or RESET indicate that alarm message has reset. Reset messages are ignored because they are often automatically reset and the delay between operation and reset has no meaning. Thus all OFF and RESET messages are deleted by the DSPA. All ON or OPER alarm messages apply to status changes. OPEN followed by CLOSE associated with the same CB alarm message or only CLOSE associated with a CB alarm message indicates operator initiated actions and are not to be used for inferencing. These are deleted by the DSPA.

DSPA creates an alarm belief base for recording data like alarm Id, Time, Date, Location and, Message. It writes new data in a data file or creates the Filtering Process Agent (FPA).

12.4.3   Filtering Process Agent (FPA)

The filtering of the alarms is required in order to avoid improper classification by the Decomposition Process Agent (DPA) and other agents in RTAPS. The limitation on reading 10 alarms is to avoid excessive delay in the inferencing of critical alarms in emergency situations. If an unrestricted number of alarms are processed then a big spurt of alarms can cause a huge processing overload on DSPA, FPA, and DPA. As a result inferencing of important alarms like emergency, network CB alarms can be unnecessarily delayed.

The FPA encompasses four agents as shown in Figure 12.1. Before starting to filter or process the existing data, the FPA asks the DSPA if there is any more new data. If there is none it goes ahead and processes the existing data. In fact to ensure continuity and cyclicity of operation, all process agents in RTAPS look for new alarm data/messages after processing existing messages through inter process communication channels with their parent process agent. The Filtering agent activates the three agents, i.e. *Repeated, DC Failure* and *Known Fault* sequentially. The repeated agent filters out the repeated versions of a alarm message if any and stores the first occurrence of that alarm in a belief base agent. This is done in order to filter out the future occurrences in the next lot within a time difference of one minute. The first occurrence is processed for inference. The suppression of repeated versions of an alarm reduces the number of alarms to be processed and the confusion they cause to the operator in the control center.

The *DC Failure* agent filters out the uncorrelated alarms which encompass an alarm equipment malfunction alarm by matching them against a list of alarms in the *DC Failure* belief base agent. These listed alarms are known to go off from experience in case of alarm equipment malfunction. This matching, however, is only done if there is a ALARM_EQPT_DC_FAIL alarm in the group of new messages being processed. This ALARM_EQPT_DC_FAIL is stored in a temporal belief base agent and retrieved for the next alarm snap shot. The alarms in the next alarm snapshot are matched against the known list of alarms if the time difference is less than 15 minutes. Otherwise, the operator is prompted for advice on the alarm equipment status before deleting any future alarms. The prompting becomes necessary because sometimes such a situation has been known to prevail for as long as 45 minutes. However, analysis of the previous real time data shows that the DC Failure event does not occur frequently.

The *Known Fault* agent filters out the alarms which have been previously identified by the operator as faulty and are recorded in the *Known Fault* belief base agent. If the alarm is not one of those recorded in the belief base as faulty, it is processed as normal. The *Known Fault* agents caters in general for those alarms which frequently go on and off because of some inherent problems in their design.

After processing existing alarm data FPA again checks with DSPA for new data. If there is none, it goes to sleep and checks for new data every 1 second.

12.4.4   Decomposition Process Agent (DPA)

DPA uses a SANN to decide which child process agents (i.e. Routine, Emergency, Voltage Dip, Communication and Network Control) it has to write to or create. The rules are used here to check for new alarm data and convert the symbolic alarm data into discrete SANN input data. After the DPA determines which child process agents are to be activated it ensures that only the alarm data related to a particular child process agent is sent to it. It further ensures that data sent by it has been read by its children before engaging in another write operation.

12.4.5   Communication (Comms) Process Agent (CMPA)

CMPA is activated only if there are no voltage dip alarms in the new alarm data. This is because (as stated in chapter 9) in the event of a voltage dip, communication alarms appear as sympathetic alarms. They do not reflect any communication failure and need to be suppressed. The decision on presence or absence of voltage dip alarms is made by DPA.

A SANN is used in the CMPA for classification of communications alarms into different communication failures like loss of a communication channel, microwave equipment failure, etc.. This agent has further scope of expansion for identifying specifically the cause of different communication failures and for that matter more alarms and additional information will be incorporated into it in the near future (Besides saving on development time, this also one of the reasons of using neural network in this agent). The CMPA maintains a temporal history belief base agent for recording and retrieving numeric output and input activation values of the SANN, date and time of the previous alarm message/s. This is required for revising previous inferences in light of the new alarm data.

Like other agents in RTAPS, CMPA checks for new data before and after processing the existing alarm data.

12.4.6 Voltage Dip Process Agent (VDPA)

Voltage Dip alarms represent a common event in a power system. Alarms like OSCILLO_OPERATED and STATION_GENERAL indicate a voltage dip and loss of station power supplies in general. These alarms are linked to the backup emergency power facilities which become operational as consequence of voltage dip or loss of power. The responsibility of the *Voltage Dip* agent is to collect these alarm messages and present them to the operator in a concise form. Also, it must suppress other sympathetic communication alarms which may accompany such an event. Rules are used to accomplish this process. The *Voltage Dip* agent maintains a belief base agent to record and retrieve the existing voltage dip alarms and any new ones which may be added from time to time.

12.4.7 Emergency Process Agent (EPA)

Emergency alarms are isolated events and appear as single or a group of alarms. The EPA distinguishes between the emergency alarms related to building fire and the ones related to fire equipment fault. It uses rules to display the the emergency alarms in case of a building fire and suppresses all the emergency alarms if they appear as a consequence of a fire equipment fault. At the moment three emergency alarms have been identified in TTS which are processed by this agent. The alarms are stored and retrieved from a temporal history belief base agent to account for a false emergency situation when a building fire alarm and fire equipment faulty alarms come in two successive alarm snapshots.

12.4.8 Routine Process Agent (RPA)

The RPA suppresses the alarms associated with introduction and release of inductive industrial loads. These alarms indicate the introduction and removal of shunt capacitors in the power system in order to balance the inductive load.

12.4.9 Network Control Process Agent (NCPA)

NCPA uses a SANN agent to categorize the network circuit breaker alarms into Transformer (Trans) section, 220kv and 66kv Bus sections, and Sub Station section alarm agents. The rules used for local preprocessing also check the validity of the incoming circuit breaker alarms. That is, in case a CB alarm message is not preceded or succeeded by a "CB-TRIP" message within a time duration of 2 seconds (a heuristic derived from the operators) it is considered as an action initiated by the operator and is suppressed.

The section agents simply write to or create their child process instance agents (e.g. 220KV Bus1 Section, 66KV Bus2 Section agents) for processing different CB alarms. The single line, feeder and transformer faults are computed by the instance agents of different network sections and inferences are displayed to the operator. For computing single/multiple bus fault/s, and CB malfunctions the inferences made by the instance agents are pruned by their respective section agents before being displayed to the operator. This is done in order to take care of the overlaps between different individual bus, transformer and substation sections. Like for example, suppose *66KV Bus3 Section* agent inferences that "66kv Bus3 is faulty with no CB malfunctions," the *66kv Bus2 Section* and *66kv Bus4 Section* infer that interconnecting bus-tie CBs have malfunctioned. Then the conflict resolution (pruning) SANNs in the *66kv Bus Section* agent (shown in Figure 12.2) prune out these conflicting bus tie CB malfunctions reported by *66KV Bus2 Section* and *66KV Bus4 Section* decision process agents and display only "TTS Bus3 is Faulty" to the operator.

All the individual bus and transformer decision section agents maintain a historical record of input and output activations associated with the SANNs. This information is used in these agents for re-inferencing. Further, for dynamic memory management, unless there is new alarm data, all the process agents in the decision phase exit after processing existing alarm data.

In chapter 9 three objectives, namely, isolation of the event, isolation of the cause of the event or fault detection, and fast and reliable isolation of the event. Firstly, the event isolation and fault detection objectives are addressed (Khosla and Dillon 1993). These are followed with the description of how the real time objectives associated with RTAPS have been addressed (Khosla and Dillon 1997).

12.4.10   Event Isolation and Fault Detection

Figure 12.4 shows the results of a simulation based on real time alarm data collected from Keilor Control Centre of SEC, Victoria for the period 1990-91. The simulation has been done with 181 alarms shown in Appendix E. These 181 alarms consist of *Repeated, DC Failure, Known Fault, Emergency , Communication, Voltage Dip,* and *Network* bus section alarms.The bus section alarms in Appendix E (the ones with 'CB' at the end) also represent some of the input features used for training SANNs in the RTAPS. All these alarms have been accounted for in the simulation. The top rightmost window in Figure 12.4 shows the parallel execution of different agents in RTAPS.

The isolation of the event is accomplished in the RTAPS by reducing the number of alarms and providing summarized messages on different types of alarms.

12.4.10.1   Alarm Reduction. The *Repeated, DC Failure* and *Known Fault* agents exclusively engage in reducing the number of alarms to be processed and/or displayed to the operator.

19DEC90 1532:32 TTS ON COMM EQPT
19DEC90 1532:34 TTS OFF COMM EQPT
19DEC90 1532:46 TTS ON COMM EQPT
19DEC90 1532:51 TTS OFF COMM EQPT
19DEC90 1533:07 TTS ON COMM EQPT
19DEC90 1533:16 TTS OFF COMM EQPT
19DEC90 1533:16 TTS OFF COMM EQPT

Sample A: Repeated Alarms in TTS

Sample A shows a list of repeated communication alarms extracted from the overall list in Appendix E. All these repeated alarms are suppressed except the first instance of 'COMM_EQPT' which is processed as normal. Also note that the time difference of these alarms is taken into account before deleting them.

10DEC90 1327:11 TTS OFF 220/66KV TR TEMP
10DEC90 1328:32 TTS OPER 2 220/66KV TR GAS
10DEC90 1328:32 TTS ON 220/66KV TR TEMP
10DEC90 1330:08 TTS ON 415V SELECTED SUPPLY FAIL
10DEC90 1330:08 TTS ON COMM EQPT
10DEC90 1330:08 TTS ON ALARM EQPT DC FAIL
10DEC90 1331:00 TTS RESET 2 220/66KV TR GAS
10DEC90 1331:00 TTS OFF 220/66KV TR TEMP
10DEC90 1332:03 TTS OPER 4 220/66KV TR GAS
10DEC90 1332:03 TTS ON 220/66KV TR TEMP

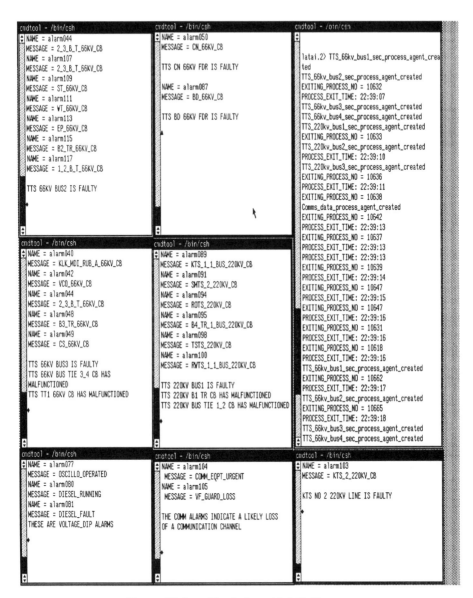

Figure 12.4.   Simulation with 181 Alarms

Sample B: DC Failure Alarms in TTS

9DEC90 1731:15 TTS ON 220/66KV TR OLTC EQPT
9DEC90 1731:24 TTS OFF 220/66KV TR OLTC EQPT

Sample C: Known Fault Alarms in TTS

8DEC90 1438:16 TTS ON BUILDING FIRE
8DEC90 1438:18 TTS OFF BUILDING FIRE
8DEC90 1438:18 TTS ON FIRE EQPT FAULT
8DEC90 1438:21 TTS OFF FIRE EQPT FAULT

Sample D: Emergency Alarms in TTS

Sample B shows a list of uncorrelated transformer and communication alarms encompassing the DC supply failure alarms like *ALARM_EQPT_DC_FAIL*. These alarms are suppressed by matching them against a list of alarms in the belief base of *DC Failure* agent. Likewise, the known fault alarms related to the OLTC equipment and shown in Sample C are filtered out after checking with the *Known Fault* belief base agent.

In the present simulation the emergency alarms shown in Sample D are suppressed because the 'BUILDING FIRE' alarm is associated with a 'FIRE EQPT FAULT' alarm. Such a suppression is helpful to the operator who is likely to overlook the 'FIRE EQPT FAULT' alarm in presence of a critical 'BUILDING FIRE' alarm in an emergency situation (where a large number of alarms appear on the monitor).

It can be seen from Figure 12.4 that the number of relevant alarms in the other windows is only 26. That is, there is a significant reduction (from 181 to 26) in the number of alarms which are displayed to the operator. This helps to reduce the information overload on the operator in real time.

12.4.10.2 Summarized Messages. Summarized messages related to *Communication* and *Voltage Dip* alarms are shown in Figure 12.4. We call them summarized messages because no fault detection (in terms of which component is faulty) or fault diagnosis (i.e. why a component is faulty) is being performed. The summarized messages provide pointers to the likely nature of the fault. Actual fault detection and diagnosis on the *Communication* alarms and *Voltage Dip* alarms can be performed with additional information on the communication network and analog information on the lines and feeders in TTS power network. This however, is not a priority with respect to (w.r.t) *Communication* and *Voltage Dip* alarms in the present set of objectives of the RTAPS.

10DEC90 1716:56 TTS OPER CB TRIP
10DEC90 1716:57 TTS OPER CB TRIP

```
10DEC90 1716:58 TTS OPEN KLK-MDI-RUB A 66KV CB
10DEC90 1716:58 TTS OPER CB TRIP
10DEC90 1716:59 TTS OPEN VCO 66KV CB
10DEC90 1717:00 TTS OPER CB TRIP
10DEC90 1717:01 TTS OPEN 2-3 B/T 66KV CB
10DEC90 1717:01 TTS OPER CB TRIP
10DEC90 1717:01 TTS OPER CB TRIP
10DEC90 1717:01 TTS OPER CB TRIP
10DEC90 1717:02 TTS OPEN B3 TR 66KV CB
10DEC90 1717:02 TTS OPEN CS 66KV CB
10DEC90 1717:03 TTS OPEN CN 66KV CB
```

Sample E: CB TRIP Alarms in TTS

```
13DEC90 1339:36 TTS OPEN CN 66KV CB
13DEC90 1339:53 TTS CLOSE CN 66KV CB
13DEC90 1344:41 TTS OPEN BD 66KV CB
13DEC90 1344:44 TTS CLOSE BD 66KV CB
13DEC90 1344:47 TTS OPEN SMR 66KV CB
13DEC90 1344:49 TTS CLOSE SMR 66KV CB
```

Sample F: Operator Initiated Actions in TTS

12.4.10.3   Fault Detection.   Samples E and F represent a small list (actual list can be seen in Appendix E) of CB alarm messages with and without 'CB TRIP' messages respectively. The ones without the 'CB TRIP' message are operator initiated actions and are suppressed. The ones with the 'CB TRIP' are used for fault detection in the TTS power network. These CB alarm messages are used to identify which component/s (e.g. bus, line, feeder, CB) of the TTS power network is/are faulty.

The *Network Control Process* agent along with different bus section instance agents (e.g. *220KV Bus1 Section*) are used to identify the faults in the power network. The single/multiple line faults shown in Figure 12.4 are computed and displayed by the bus section decision agents. However, the bus faults and CB malfunctions are computed by the instance agents but pruned and displayed by the *66KV Bus Section* and *220KV Bus Section* agents. As already mentioned in section 12.3 the system inferencing reliability is close to 100%. That is, it covers all the network faults except for a proportion of CB malfunctions.

12.4.11   Real-Time Issues

The implementation of hybrid RTAPS involves realization of various real-time constraints like continuous operation, cyclicity, synchronous and asynchronous communication, temporal reasoning and response time. Continuous operation, cyclicity, synchronous and asynchronous communication have been realized through parallel and distributed implementation (as described in section 12.1).

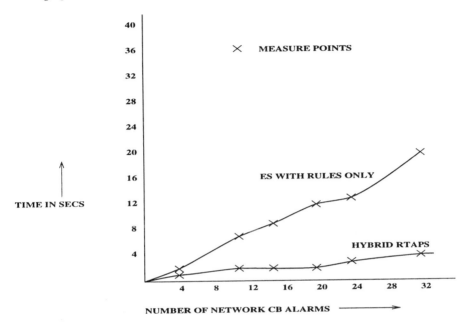

Figure 12.5.   Comparison of Response Times Excluding Agent Loading Times

Here the data variability and response time issues as addressed by the RTAPS are elaborated (Khosla and Dillon, 1995,97).

12.4.11.1   Temporal Reasoning. It is an important aspect in alarm processing that inference made on an existing alarm snap shot can become invalid in the light of the new data. In other words, the inferences or conclusions made by the system have to be constantly updated. RTAPS adopts a two fold strategy to deal with the constantly changing alarm data:

a. Before making an inference on a particular alarm snap shot (which is fixed at the moment to 10 alarms at the DSPA level for simulation purposes), all process agents look for more alarm data (before inferencing with existing data) , if any, from their parent process agent. This is possible, because all process agents which are executed concurrently can communicate independently with their parent process agent. Thus, FPA can take 20 alarms (or the 20th alarm) before starting inference, DPA can take 40 alarms (or the

40th alarm), NCPA, VDPA, CMPA, EPA and, RPA can take 80 alarms (or the 80th alarm).

b. All inferenced alarms are stored in a temporal history data base by different domain agents, thus making a provision for re-inferencing and retracting previous conclusions. In case of multiple network CB alarms where SANNs are used, we only need to store/retrieve the previous alarm snapshot input/output activation/s and time of the last alarm id in the previous alarm snapshot. This allows for very quick re-inference because of the massive parallelism inherent in SANNs. The present and previous alarm snapshot activations are used for re-inferencing. The previous and present alarm data has a time bound validity. That is, the previous alarm snapshot activations are considered valid if time difference between the last alarm in the previous alarm snapshot and the first alarm of the present alarm shot is less than or equal to one minute. (The concept of time bound inferential validity is also true in *Repeated* alarm agent. In this agent the alarm is classified as repeated only if the alarm repeats itself within a time period of one minute).

In the case of conflict between two inferences a revised message is displayed and the previous inference/conclusion is retracted. Say for example, based on a previous alarm snapshot, inference made was that 3 feeders connected to a *66kv Bus2 Section* are faulty. Based, on existing alarm snapshot and the previous alarm snapshot, now the SANN concludes that '66kv Bus2 is faulty'. This revised message will be displayed.

12.4.11.2 Reasoning Under Time Constraints. A comparison of response time (for fault detection with multiple CB alarms) achieved on Sun 4/280 with a previously built ES with rules only (an old version of the RTAPS) and the intelligent hybrid RTAPS is shown in Figure 12.5. The comparison shown in Figure 12.5 excludes the loading time of the two versions (as indicated by the 0 seconds starting point of the response tine curve). In fact in the worst case scenario the loading time of the ES version is 15 seconds more than hybrid RTAPS. In order to save on system resources, the problem solving agents in hybrid RTAPS are deactivated or even killed (especially the decision agents) if they have not been active for some time.

The total inferencing and processing time (from reading the alarms, activating/reactivating some distributed process agents, inferencing and displaying) for 181 alarms is 39 seconds on a Sun 4/280 workstation. The new Sun Ultra 2 workstation is more than 40 times faster than Sun 4/280 workstation. It is expected that the response time of 39 seconds would be drastically reduced on Sun Ultra 2 workstation. The inferences on 26 relevant alarms out of the total of 181 alarms as displayed to the operator are shown in Figure 12.6.

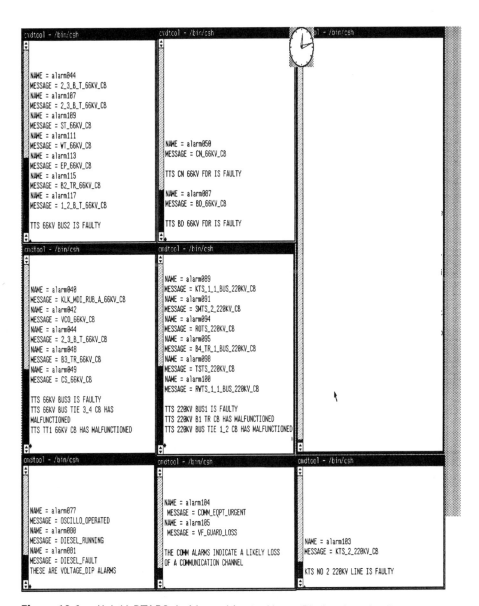

Figure 12.6. Hybrid RTAPS decisions with 181 Alarms Displayed to the Operator

Figure 12.5 shows that there is a significant improvement in the response time achieved by the hybrid RTAPS as compared to the one with rules only. In the worst case scenario, that is in case all the problem solving agents of the RTAPS are freshly loaded, the response time for severe multiple network faults is still within the acceptable response time requirement of 30 secs for severe faults.

12.5   SCALABILITY AND COST EFFECTIVENESS

In the development of large systems, it is essential to keep in perspective management related issues like scalability and cost effectiveness if one intends to put the system into practical use. The hybrid RTAPS has been built up with an open ended design approach keeping in view simultaneous constraints like fast response time, reduced memory requirements, distributed parallel processing, scalability and maintainability.

Although, one could have combined small sized SANNs to form fewer large sized SANNs, this has deliberately not been done for reasons of scalability and cost effectiveness.

At the moment only circuit breaker information has been used for inference in *Network Control Process* Agent (NCPA) and and its related agents. When relay and/or sensor measurements are used the size of small sized SANNs would increase manifold. These changes will also will be reflected upwards in the *Decomposition Process* Agent (DPA). The fact is that the hybrid RTAPS will be able to accommodate these changes with much more ease than it would have, had it been designed with few large sized SANNs and/or large sized ESs. By allowing the changes to flow through the architecture of RTAPS, provisions for scalability are being made. An example of vertical and horizontal scalability is shown in Figure 12.7. It shows the enhancements in the *Network Control, 220KV Bus Section* and *220KV Bus1 Section* agents in case main relay and backup relay information is incorporated in RTAPS. The *CB, Main Relay* and *Back Up Relay* agents will be executed in parallel and competing manner to configure single and multiple faults in the *220KV Bus1 Section* agent. A symbolic voting facility will be used to resolve conflicts and enhance the reliability of the inference. It can also be seen from Figure 12.7 that the *220KV Bus1 Section* and other agents have been reused in the enhancement process.

The hybrid RTAPS has been built without any need for dedicated AI processors as is the case with some purely symbolic systems developed in this domain (Moriguchi et al. 1989). The hierarchical, distributed and modular features of the hybrid RTAPS have contributed to the reduced size and training time of SANNs, reduced development time of different distributed and modular intelligent agents, and reduced dynamic memory requirements. All these features make the hybrid RTAPS architecture cost effective and facilitate its easier maintenance.

12.6 SUMMARY

Implementation of a large scale real time system like alarm processing involves realization of various objectives. These include methodology related objectives, domain related objectives, real time objectives and management related objectives.

The methodological objectives relate to the realization of the IMAHDA. IMAHDA is realized in RTAPS through agent-oriented, parallel and distributed implementation, clear demarcation of five phases of the IMAHDA. Nexpert object (an object-oriented expert system shell) with external C routines is used for implementing RTAPS. Learning knowledge and strategy of IMAHDA is applied in training neural networks in RTAPS. A combination of Kohonen's LVQ algorithm and backpropagation algorithm is used to improve generalization in neural networks.

The domain objectives relate to the power system aspects of the implementation namely, isolation of the event, isolation of the cause of the event or fault detection, and fast and reliable isolation of the event. RTAPS accomplishes isolation of the event through alarm reduction and providing summarized messages. Alarm reduction is accomplished through suppression of repeated alarms, DC failure alarms and known fault alarms. Summarized messages are displayed to the operator for isolation of voltage dip alarms, communication alarms and emergency alarms. Fault detection is performed in terms of identifying single and multiple bus,transformer, line and feeder faults. These faults are detected using neural networks and displayed to operator along with the related CB alarm messages. All TTS network faults are covered by the neural networks except for a proportion of CB malfunctions. Only CB alarms are used at the moment for inferencing network faults. The fast isolation of the event is accomplished by achieving a desired response time range of 4 to 30 seconds for single/simple and severe multiple alarms. Real time alarm data is used for alarm processing and inferencing. Rules used in RTAPS are a function of operator knowledge and experience, analysis of previous real time alarm data, network topology, time, and the peculiarities associated with the alarms from TTS.

Real time issues like data variability or temporal reasoning are well addressed in RTAPS. In order to account for data variability, RTAPS has two fold strategy. Firstly, all agents prior to inferencing with the existing data look for more data if any related to them from their parent agent through inter process

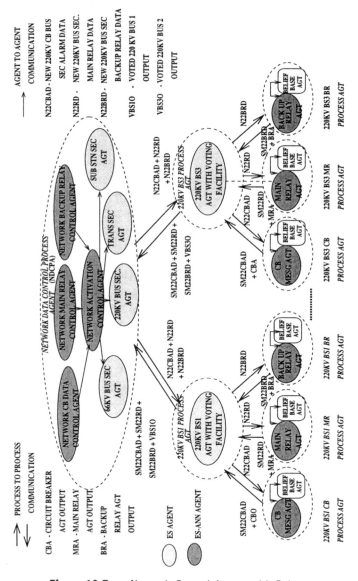

Figure 12.7.   Network Control Agents with Relays

communication channels. Thus it is possible to double the existing number of alarms before starting to inference with them. Secondly, the exhaustive use of SANN agents in the decision phase drastically reduces the re-inferencing time because of massive parallelism property of neural networks.

A comparison of response time (for fault detection with multiple CB alarms) achieved on Sun 4/280 with a previously built ES with rules/agents only (an old version of RTAPS) and the hybrid RTAPS shows a significant reduction in the response time achieved by the hybrid RTAPS.

Finally, dynamic memory management, reduced development time, reduced dependence on hardware, distributed and agent-oriented features of the hybrid RTAPS facilitate realization of management related objectives like scalability (vertical and horizontal), cost effectiveness, reduced memory requirements, reusability and easier maintenance.

References

Khosla, R. and Dillon, T. (1993), "Combined Symbolic-Artificial Neural Net Alarm Processing System,"- (deals with power system aspects of alarm processing), *Eleventh Power Systems Computation Conference (PSCC)*, Avig non, France, pp. 259-266. Khosla, R. and Dillon, T. (1994), "A Distributed Real-Time Alarm Processing System with Symbolic-Connectionist Computation," in *Proceedings of the IEEE Workshop in Real Time Applications*, Washington D.C., USA .

Khosla, R. and Dillon, T. (1997), "Learning Knowledge and Strategy of a Generic Neuro-Expert System Architecture in Alarm Processing," to appear in *IEEE Transactions in Power Systems*.

Kohonen, T. (1990), "The self-organizing map," *Proceedings of the IEEE*, vol. 78, no. 9, pp. 1464-1480. Khosla, R. and Dillon, T. (1997), "Learning Knowledge and Strategy of a

Nexpert Object Manuals (1989), Neuron Data Inc. CA, USA.

13 FROM DATA REPOSITORIES TO KNOWLEDGE REPOSITORIES

13.1 INTRODUCTION

In the 1970's computerized information (collected set of data items) was used for general management support in the form of management information systems. Since the 1980's computerized information has been regarded as a strategic resource. The 1990's decade can be labeled as the era of globalization of economies. Organizations in the developed world are shifting from purely information based to knowledge based organizations. Today knowledge is being regarded as a strategic resource for gaining competitive advantage. The emphasis today is on intelligent decision making. This shift in emphasis can be seen with development of intelligent systems at various levels of an organization. The emergence of knowledge discovery and data mining area for extraction of intelligent meaningful knowledge from millions of records of data is an evidence of this shift or change in emphasis. This shift or change in emphasis can be called as process of building knowledge repositories from data repositories. The transition from data repositories to knowledge repositories has three primary ingredients, namely, data, information, and knowledge. These three ingredients are used for system modeling in the data engineering, information engineering, and knowledge engineering layers of an organization.

Given this context, in this penultimate chapter of the book the role of IMAHDA is extended to building knowledge systems in general. In the context of building enterprise-wide knowledge systems, IMAHDA is looked in terms of its five information processing phases and the tasks associated with these phases. The research done in the area of information systems is a useful starting point for determining the role of knowledge systems at different organization levels. Keeping this in mind, the first part of this chapter looks at different organizational levels, and the different types of information systems which serve these organizational levels. An application of IMAHDA in developing a knowledge system in the sales and marketing function area at is described in this chapter.

Further, role of objects, agents, and IMAHDA is extended to the data and information engineering layers of system modeling in an organization. A framework for enterprise-wide systems modeling is outlined in the second half of this chapter.

13.2 INFORMATION SYSTEMS AND ORGANIZATIONAL LEVELS

An information system can be defined as a set of interrelated components that collect/retrieve, process, store, and distribute information to support decision making and control in an organization. In addition to supporting decision making, coordination, and control, information systems may also help managers and workers analyze problems, visualize complex subjects, and create new products. From a business perspective, an information system is an organizational and management solution, based on information technology, to a challenge posed by the environment (Laudon and Laudon, 1996).

Most organizations can be divided into five function areas, namely, sales and marketing, manufacturing, finance, accounting, and human resources. Information systems in an organization can be distinguished from each other based on function areas and organizational levels as shown in Figure 13.1. As can be seen in Figure 13.1, the organizational levels are operational, knowledge, management, and strategic.

Four major categories of information systems which serve these organization levels and function areas are:

- Operational level information systems.

- Knowledge level information systems.

- Management level information systems.

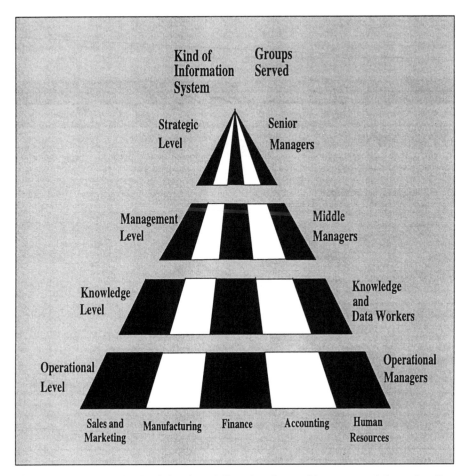

Figure 13.1. Organizational Levels and Function Areas. Management Information Systems 4TH/E. by Laudon/Laudon, © 1996. Reprinted by permission of Prentice-Hall, Inc., Upper Saddle River, NJ.

■ Strategic level information systems.

The descriptions of the information systems at the four levels and their characteristics have been derived from Laudon and Laudon (1996).

13.2.1   Operational Level Information Systems

Operational level information systems support operational managers by keeping track of the elementary activities and transactions of the organization, such as sales, cash deposits, payroll, credit decisions, and flow of materials in a factory.

13.2.2   Knowledge Level Information Systems

Knowledge level systems support knowledge and data workers in an organization. Knowledge level systems support knowledge and data workers help an organization to integrate new knowledge into the business and to help them the organization restrict the flow of paperwork in an organization.

13.2.3   Management Level Information Systems

Management level systems are built to serve the decision making, monitoring, controlling, and administrative activities of middle managers. They may involve hiring decisions, performance monitoring, and cost control in order to ensure that things are going as per the business plans laid down by the senior management. Unlike operational level systems, the management level systems do not use instant information. Instead they use summarized or periodic reports.

13.2.4   Strategic Level Information Systems

Strategic level information systems, on the other hand enable senior management to address strategic issues and long term trends, both in the firm and in the external environment. Their principal concern is matching changes in the external environment with existing capabilities. The information systems at this level facilitate answers to questions like what is the electricity demand trend in the next five years, and how can our organization maintain a competitive advantage given this trend.

13.3   CHARACTERISTICS OF INFORMATION SYSTEMS

The characteristics of information systems are determined by the organizational level at which they are designed. There are six specific types of information systems associated with the four organizational levels.

The operational level is supported by Transaction Processing Systems (TPS). The knowledge level is supported by Knowledge Work Systems (KWS) and Office Automation Systems (OAS). The management level is supported by Management Information Systems (MIS) and Decision Support Systems (DSS). The strategic level is supported by Executive Support Systems (ESS). Figure 13.2 shows the applications of these information systems with respect to the function area and the organizational level.

13.3.1   Transaction Processing Systems

At the operational level tasks, resources, and outcomes are predefined. In other words, the problems are highly structured and decisions are made based a predefined criteria. For example, in a payroll system, method to determine an employee's salary is predefined and simple. Transaction Processing Systems (TPS) which serve the operational level process day to day data like sale receipts, number of cash deposits, and number of computers shipped. They interface with the external environment of the organization and process the raw data coming into the organization. Payroll systems, airline reservation systems, and hotel reservation systems.

13.3.2   Knowledge Work and Office Automation Systems

Knowledge work systems serve knowledge workers at the knowledge level of an organization. Knowledge workers in an organization can be scientists, engineers, designers, lawyers, and doctors. Knowledge work systems like Intelligent Computer Aided Design (ICAD) systems help to create new knowledge like new car design or new product promotion display system. on the other hand, Office Automation Systems (OAS) like electronic mail, video conferencing, news management, and word processing systems are used by data workers (e.g. secretaries) and knowledge workers to control paper flow and increase productivity.

13.3.3   Management Information Systems

As the title suggests, Management Information Systems (MIS) serve the management level of an organization. Unlike the transaction processing systems, MIS are based on internal not external events or data. They operate on weekly, monthly, quarterly and yearly data. They provide summarized reports to the

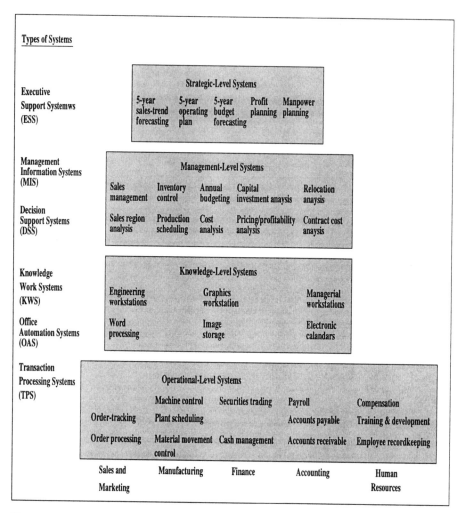

Figure 13.2.   Six Major Types of Information Systems and their Applications. Management Information Systems 4TH/E. by Laudon/Laudon, © 1996. Reprinted by permission of Prentice-Hall, Inc., Upper Saddle River, NJ.

managers and answer questions like: What were the sales in the same quarter last year? What were the overheads in the same quarter last year?, What was the customer base in the same quarter last year? What were the profit margins in the same quarter last year. Thus MIS have a marginal analytical content as compared to TPS. They serve the functions of planning, controlling and decision making at the management level.

13.3.4   Decision Support Systems

Decision Support Systems (DSS) also like the MIS serve the management level but with a few differences. Unlike the MIS, they rely not only on internal data but also on external data. They help managers make decisions which are semistructured, unique, or rapidly changing, and not easily specified in advance. DSSs allow the the managers to directly interact with system and play the *What-If* games. To that extent DSSs are more analytical than the MISs.

13.3.5   Executive Support Systems

Decision support systems address semistructured decision. Executive Support Systems (ESS) address unstructured decisions. They serve the senior management at the strategic level of an organization. ESSs incorporate data from external events like new tax laws, government polices, and new competitors as well as use summarized information from MISs and DSSs. They assist the senior management to answer questions like: What new products or services should be introduced to enhance the existing product line? What are competitors doing? what are the sales projections for the next five years? What new acquisitions will benefit the company?. ESSs filter, compress, and track critical data, emphasizing the reduction of time and effort required to obtain information useful to executives.

The characteristics of the six different types of information systems described in this section are shown in Table 13.1 has been made based on the type of data used by these systems. The information systems at the lower (i.e. TPS) and higher (i.e. ESS) end of the organization interface with the external data, whereas the information systems in between these the two ends of the spectrum rely on the internal data generated from the external data.

13.4   INFORMATION SYSTEMS AND KNOWLEDGE SYSTEMS

The characteristics of information systems shown in Table 13.1 indicate that information systems are largely based on system generated data rather than user data or user intelligence. The informations systems treat information as

Table 13.1. Characteristics of Information Systems. Management Information Systems 4TH/E. by Laudon/Laudon, © 1996. Reprinted by permission of Prentice-Hall, Inc., Upper Saddle River, NJ.

Type of System	Information Inputs	Processing	Information Outputs	Users
ESS	Aggregate data; external, internal	Graphics; simulations; interactive	Projections; responses to queries	Senior managers
DSS	Low-volume data; analytic models	Interactive; simulations, analysis	Special reports; decision analysis; responses to queries	Professionals; staff managers
MIS	Summary transaction data; high-volume data; simple models	Routine reports; simple models; low-level analysis	Summary and exception reports	Middle managers
KWS	Design specifications; knowledge base	Modeling; simulations	Models; graphics	Professionals; technical staff
OAS	Documents; schedules	Document; management; scheduling; comunication	Documents; schedules; mail	Clerical workers
TPS	Transactions; events	Sorting; listing; merging; updating	Detailed reports; lists; summaries	Operations personnel; supervisors

a strategic resource. The information systems do not make a serious attempt to involve the user as a source of data/information/knowledge generation. If the emerging intelligent systems at different levels of an organization are anything to go by, this is clearly not enough. The emerging intelligent systems or knowledge systems treat knowledge as a strategic resource.

Another implicit distinction made between the six types of information systems is based on the structured and unstructured problems they are dealing with. The transaction processing systems because they are based on predefined criteria, procedures, and policies of a company address structured problems. On the other hand, unstructured problems are largely handled at the strategic level by executive support systems. The distinction is not true in its absolute sense when the user is involved as a source of data, information, and knowledge in developing these systems. Intelligent or knowledge systems are known to deal with semi-structured and unstructured problems. These semi-structured and unstructured problems are user related problems at different organization levels. Today one can find intelligent airline reservation systems at the operational level, intelligent e-mail (Maes 1994; Dinh 1995) and news management (Maes 1994) systems at the knowledge level, intelligent production scheduling systems (Hamada et. al 1995) at the management level, and intelligent forecasting and prediction systems (as described in chapter 5) at the strategic level. These systems in which user is a major source of information and knowledge are called knowledge systems. In the era of affordable computer technology, globalization, deregulation, competition, acquisitions and layoffs the need to get more out of your human resource and information resource is ever so great. Knowledge systems have a critical role to play in order make organizations more effective and competitive in this environment.

In the next section, it is shown how IMAHDA can be used in developing a knowledge system, namely, intelligent hiring system in the sales and marketing function area.

13.5 IMAHDA AND ORGANIZATIONAL KNOWLEDGE SYSTEMS

In chapters 5 to 11 the architectural theory of IMAHDA and its application to real time complex alarm processing domain have been described. IMAHDA can also be used to develop organization wide knowledge systems. From a problem solving viewpoint, the five information processing phases and tasks associated with them can be usefully employed in developing knowledge systems. The degree of complexity of a knowledge system will determine whether the control and decision phase are merged with one another or whether they remain distinct from one another. The symbolic and soft computing methods used to accomplish the tasks are based on constraints like high level symbolic reasoning, adaptability, learning, incomplete information, and optimization which are applicable in most knowledge systems.

For purpose of illustration, in this section application of IMAHDA for developing a salesperson hiring knowledge system in the sales & marketing function is briefly described.

13.6 APPLICATION OF IMAHDA IN SALES & MARKETING FUNCTION

A sales manager has many responsibilities including forecasting demand, managing salespersons, and establishing quotas as shown in Figure 13.3.

Forecasting demand involves such activities as estimating future company sales based upon historical trends, knowledge of present market and company conditions, experience, and estimates of future market trends.

Managing salespersons as shown in Figure 13.4 involves such activities as hiring and training the sales team, establishing territories, evaluating performance, and supporting them in their work.

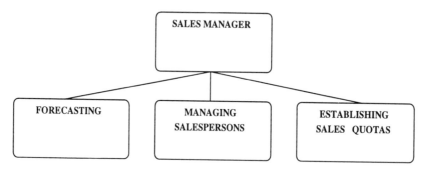

Figure 13.3.   Overview of Sales Management Function

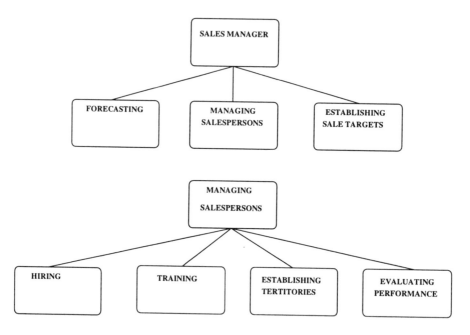

Figure 13.4.   Overview of Managing Salespersons

Establishing sales quotas involves studying a wide range of strategic and operational factors, including past sales in the territory, competitive factors, company's overall strategic and marketing plans, local and national economic conditions, demographic trends, and seasonal sales fluctuations.

The sub-function, hiring under managing salespersons has been chosen for illustrative purposes.

13.6.1   Salesperson Hiring Knowledge System

In the sales management area where performance can be assessed in tangible terms, hiring salespersons is an important decision making activity.

Most organizations rely on short listing and interview as the main strategy for hiring salespersons. Product knowledge, verbal skills, hard work, self-discipline, are generally assumed to be well taken care of in the interviewing process. However,

- the sales manager spends limited time with the salesperson during the interview. In this limited time, it is difficult to realistically assess all the factors which affect the selling behavior of a salesperson.

- organizations do not have any objective skill base or model/s of their salespersons which can help them in evaluation, relocation (to customer service or sales support), and benchmarking new or existing sales staff, and

- salesperson turnover continues to be high in most organizations, thus wasting investments made in them in terms of training, managerial time, and other development activities.

Given that there are enough good reasons for developing a salesperson hiring knowledge system, in the next section brief description of the problem solving agents of IMAHDA as applied to the hiring decision is provided.

13.6.2   IMAHDA and Salesperson Hiring Knowledge System

The five problem solving agents applied in the hiring decision activity are pro-processing agent, decomposition agent, control agent, decision agent, and post-processing agent.The main purpose of this example is to illustrate how the five information processing phases play a role in the hiring decision process.

13.6.2.1   Hiring Proprocessing Agent.   The tasks undertaken by the *Hiring Preprocessing Agent* shown in Figure 13.5 are input conditioning, and noise filtering. The input conditioning task in context of this application checks whether the resumes contain all the information, and extracts the relevant information based on the job description. This may involve scanning the candidate resume. The noise filtering task involves filtering out those resumes which do not meet the selection criteria.

In fact Resumex Inc., USA have built a knowledge based system precisely for this purpose. It preprocesses the resumes and provides a short list of the candidates to be called for a job interview.

13.6.2.2 Hiring Decomposition Agent. The *Hiring Decomposition Agent* decomposes the hiring decision into three abstract agents namely, *General Character Evaluation Agent, Selling Behavior Evaluation Agent*, and *Product Knowledge Evaluation Agent* as shown in Figure 13.5. The *General Character Evaluation Agent* evaluates the general personality, self-discipline, and verbal skills of the candidate. The responsibility for doing this evaluation may be shared between the human agent, namely the sales manager or the human resource manager and the *General Character Evaluation Agent* . The *Selling Behavior Evaluation Agent* and its collaborative control and decision agents have the responsibility of evaluating the selling behavior profile of the candidate and determining whether the candidate's profile meets the requirements of the organization. Finally, the *Product Knowledge Evaluation Agent* and its collaborative agents evaluate the technical aspects of the candidate's product knowledge.

Unlike, the alarm processing application, at this level, the three abstract agents are not independent of each other. In other words, a sales manager makes a hiring decision based on the results provided by the decision agents of these three abstract agents. Thus, at the global level, the hiring decomposition agent exercises global control and has the task of resolving conflicts if any between the abstract agents. For example, the decision agent/s of the *General Character Evaluation Agent* may make a *No* hiring decision, the decision agent/s of the *Selling Behavior Evaluation Agent* may make a *Yes* hiring decision, and decision agent/s of *Product Knowledge Evaluation Agent* may make a *Yes* hiring decision. Then *Hiring Decomposition Agent* based on its previous experience has the responsibility of resolving the conflict and making the final hiring decision.

In the remaining parts of this section, the application of IMAHDA is outlined for the *Selling Behavior Evaluation Agent*.

13.6.2.3 Selling Behavior Evaluation Control Agent. The *Selling Behavior Evaluation Control Agent* determines the decision categories of the selling behavior evaluation. A sales manager often opinionizes on the behavioral categories of their salespersons and match them with different types of customers in order to maximize sales results. These decision categories or decision agents in the decision phase are derived from their experience and the behavioral models developed by psychologists.

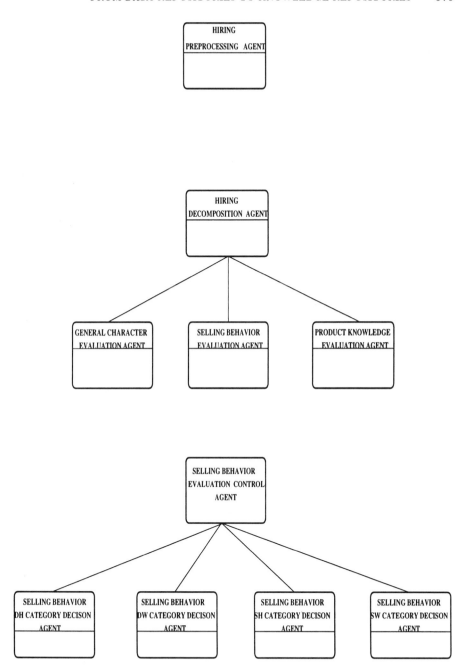

Figure 13.5.   Overview of Information Processing Phases in Hiring Decision Making

The control agent also determines the predominant selling behavioral category of a candidate. This is it does based on category scores computed by the decision category agents.

13.6.2.4   Selling Behavior Evaluation Decision & Postprocessing Agent.
Four decision category agents are shown in Figure 13.5. These are *Dominant Hostile Category Agent, Dominant Warm Category Agent, Submissive Hostile Category Agent* and *Submissive Warm Category Agent* respectively. These decision agents evaluate the candidate on various areas related to selling, customer service, and sales support. They compute their category scores based on the candidate answers, and relay back their scores to the control agent. The candidate's profile is based on scores in various categories.

Some aspects of the *Dominant Hostile Decision Agent* are shown in Figure 13.6. Once the final behavioral category has been determined it is validated by the human agent, i.e. the sales manager as shown in Figure 13.7. For example, if the final category is dominant hostile, then the *Dominant Hostile Decision Agent* communicates with the human agent for validation. Thus the final behavioral category is an objective and independent feedback to the management to facilitate intelligent and smarter decision making. The decision agent provides feedback on a number of other aspects of a candidate as shown in Figure 13.8.

The knowledge explanation agent provides an explanation to the sales manager on the behavioral category of the candidate and the reasons which lead to that category. It produces number of reports for the sales manager in the process.

A Salesperson Hiring knowledge system based some of the concepts described in this section has been developed by Intelligent Software Systems Pty. Ltd., Melbourne, Australia and is being used in the industry. The hiring decision is made based on the recommendations made by the system and comparison of the candidate's profile with a number of benchmark profiles (stored in the system). Figure 13.9 shows a salesperson results screen of candidate by the salesperson hiring knowledge system developed by Intelligent Software Systems Pty Ltd., Melbourne, Australia.

13.7   UNIFIED APPROACH TO ENTERPRISE-WIDE SYSTEM MODELING

The business plans of an enterprise are realized through people, structure and systems. From a systems perspective, it needs to be understood that there are three primary inputs to the systems modeling process in an enterprise. These are data, information and knowledge. As is the practice today these three

NAME:	SELLING BEHAVIOR CAETGORY 1 DECISION AGENT
PARENT AGENT:	BB, ES, FUZZY LOGIC(FL), SANN, IMAHDA DECISION AGENT
COMMUNICATES WITH:	SELLING BEHAVIOR CONTROL AGENT CATEGORY 1 BELIEF BASE AGENT, CATEGOR 1 EXPLN. AGENT CATEGORY 1 DECISION VALIDATION AGENT,
COMMUNICATION CONSTRUCTS:	INFORM : INITIAL CATEGORY DECISION/S REQUEST: CANDIDATE ANSWERS, REQUEST : VALIDATE CATEGORY DECISION
GOALS:	CONTEXT VALIDATION DETERMINE CANDIDATE CATEGORY ESTABLISH COMMUNICATION CHANNELS WITH CONTROL AGENT & OTHER DECISION AGENTS FORMULATE ACTION EXECUTION SEQUENCE RECORD INTERNAL STATE
PERCEPTS:	CANDIDATE ANSWERS ON QUESTIONS RELATED TO SELLING, COMMANDS FROM CONTROL AGENT RESULTS FROM OTHER DECISION AGENTS, VALIDATED RESULTS FROM CATEGORY 1 VALIDATION AGENT,
ACTIONS:	INITIAL CATEGORISATION CONTEXT VALIDATION RULES FINAL CATEGORISATION CONTEXT VALIDATION RULES COMMUNICATION RULES INITIAL CATEGORISATION RULES FINAL CATEGORISATION RULES TRAIN, TEST & VALIDATE CATEGORY 1 NEURAL NETWORK (OPTIONAL)
ENVIRONMENT:	NON-DETERMINSTIC

Figure 13.6.   Dominant Hostile Category Decision Agent

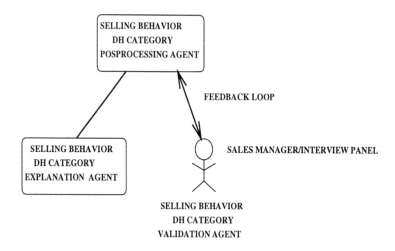

Figure 13.7.   Knowledge Explanation & Decision Validation Agents

primary inputs are modeled separately as if they are independent from each other. This can be seen from the fact that systems (e.g. database systems, information systems, and knowledge or intelligent systems) built around these three primary inputs use a different set of methodologies which have very little in common. However, the reality is quite different. Data is an atomic element in the system building process. It represents raw facts which on their own have no meaning (e.g. Age: 35 years). Information is collection of facts in one context which provides meaning to data (Pressman, 1992). For example, in the context of a bank teller system, customer related information may be a simple collection of facts like Name: Carol Witten, PIN No: 9897 and Cash Balance: $120000. In the context of a loan approval system, customer related information can be a collection of facts like Name: Carol Witten, Age: 35 years, Profession: lawyer, Income: $100000 per annum and Credit Rating: high. On the other hand knowledge is derived from associating information in multiple contexts. For example, in a loan approval system information related to lawyers and gamblers can be associated by a heuristic that "lawyers repay their mortgage faster than gamblers". People are an important element in making this association. It is quite clear from this example that data, information and knowledge are strongly bound to each other and in fact emerge from one another. There is another reality which organizations are facing today. This reality is that we have moved from era of data and information being a strategic to a knowledge being a strategic resource. The lack of accountability of these two realities has resulted in number of problems. At a higher level, the traditional enterprise models

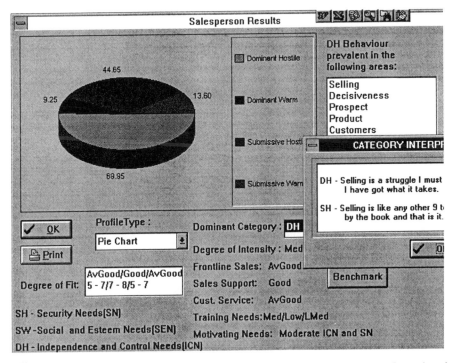

Figure 13.8. A Result Screen of the Salesperson Hiring Knowledge System. Reproduced with Permission from Intelligent Software Systems Pty Ltd., Melbourne, Australia.

concerned with modeling of data and/or information only have been found to be insufficient to effectively realize the business plans of organizations in the 90's. At a lower level because different set of methodologies have been used to model the three system categories, the problems which have emerged. These include little or no knowledge sharing/reuse of problem solving approach in system development resulting in more system development time and cost, lack of interoperability between these three categories of systems, and impediments in forward or backward integration of systems.

In the preceding sections the role of knowledge systems and IMAHDA has been discussed at the strategic, management, knowledge and operational levels of a business enterprise. This section discusses the how the various methodologies used and developed in this book can be effectively used in a consistent manner to develop as well as integrate enterprise-wide systems in the three categories.

13.7.1 Objects, Agents and IMAHDA

The Figure 13.10 shows a multi-layered system modeling structure based on first principles that is, the three primary system development ingredients data, information and knowledge. This multilayered structure is applicable at each organizational level, and in each functional area or business unit. Figure 13.10 also shows the methodologies used to develop the three categories of systems, namely, database systems, information systems, and knowledge or intelligent systems. It can also be seen in Figure 13.10 that a consistent set of methodologies have been proposed for data engineering, information engineering and knowledge engineering. The consistent set of methodologies have been used to address the problems mentioned in the preceding paragraph.

The use of the set of methodologies shown in Figure 13.10 is driven more by their natural modeling inclinations or strengths rather than their extensions and exceptions. The research and practice in artificial intelligence and software engineering has shown that the natural modeling strength of object-oriented models is structural abstraction of data and determining structural relationships. That is, object-oriented models are driven by data and structure and not by tasks, and behavior (especially task-oriented behavior). On the other hand, agents by definition (An entity authorized to act on another's behalf) are task and behavior driven. The PAGE description of agents used in this book indicates that agents involve task and task-oriented behavior abstraction. By incorporating agents as modeling tools in this manner other more sophisticated characteristics of agents like adaptation, learning and others as outlined in chapter 2 can also be modeled incrementally in all three categories of systems. Based on these two contexts, data engineering, information engi-

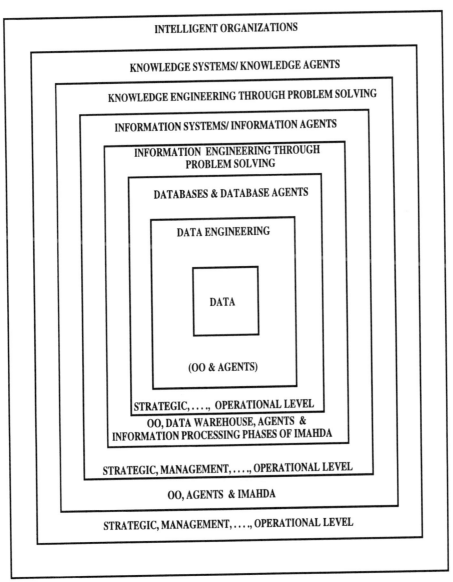

Figure 13.9.   Enterprise-wide System Modeling Framework

neering and knowledge engineering layers of enterprise-wide system modeling are outlined.

In the data engineering stage for building databases, object-oriented methodology can be used for developing structured object-oriented data models of data items and their associated data types (Hughes 1990; Brown 1991; Dillon and Tan 1993). These structured object-oriented data models capture generic representation and operations associated with say, a particular data type. Agents can be used for task-oriented behavior modeling in order to capture user specific operations or actions. This modeling strategy leads to construction of databases and database agents as shown in Figure 13.10. Information engineering through problem solving is based problem solving approach towards building information systems. This approach has been adopted to facilitate reuse of concepts related to the problem solving approach. The five information processing phases of IMAHDA and the tasks associated with them can be used as a problem solving approach for building information systems. Although, IMAHDA's information processing phases and tasks associated with them are useful in the information system layer, intelligent methods used to accomplish the tasks may not be required. Thus the preprocessing, decomposition, control, decision, and postprocessing agents of IMAHDA can be used to model structured and semi-structured problems at the operational, knowledge, management and strategic levels of an enterprise (unstructured problems are best left to the knowledge engineering layer). Agent oriented analysis and design approach as described in chapters 10 and 11 respectively can be used for analysis and design of information systems or information agents. Problem domain structure analysis can be done using object-oriented methodology in conjunction with data warehouse methodology. The data warehouse methodology is useful for abstracting data based on the time dimension which can assist in applications requiring historical and summarized reports or comparisons. Figure 5.2 in chapter 5 shows data model of a forecasting profile system which integrates data on the time dimension (based on data warehouse concepts) as well as the structural dimension (based on object-oriented concepts) which captures the concepts of a forecasting profile system at the operational level and the strategic level. Knowledge engineering through problem solving (e.g. IMAHDA) has been described in this chapter and in chapters 6 to 12 of this book and is not elaborated here.

13.8   SUMMARY

In the 90's knowledge has become an important strategic resource in an organization. This chapter establishes the role of knowledge systems at the operational, knowledge, management, strategic levels of an organization. Knowledge systems are a layer above information systems. Information systems are based on external or raw data coming into an organization and internal data which is built from the external data. Knowledge systems are based on user data or user related data and knowledge used for decision making at four levels of the organization. They deal with semistructured and unstructured problems at operational level, knowledge level, management level, and strategic level of an organization. In the above context IMAHDA is seen as a tool for developing enterprise wide knowledge systems. For illustrative purposes a salesperson hiring knowledge system is modeled using IMAHDA. Further the role of the different methodologies used (e.g. OO and Agents) and developed (IMAHDA) in this is explored for developing a unified approach towards, data engineering, information engineering through problem solving, and knowledge engineering through problem solving on a enterprise-wide basis. This is done to facilitate backward and forward integration of the three categories of systems (i.e. database systems, information systems, and knowledge or intelligent systems) as well facilitate the reuse of the system modeling and problem solving approach adopted by these three categories of systems.

References

Brown, A. W. (1991), *Object-Oriented Databases : Applications in Software Engineering*, McGraw-Hill, New York, USA.

Dillon, T. and Tan, P. L., (1993), *Object-Oriented Conceptual Modeling*, Prentice Hall, Sydney, Australia.

Dinh, S. (1995), "Intelligent E-Mail Agents," *Honors Thesis*, School of Computer Science and Computer Engineering, La Trobe University, Melbourne, Victoria - 3083, Australia.

Hamada, K., Baba, T., Sato, K., and Yufu, M. (1995), "Hybridizing a Genetic Algorithm with Rule-Based Reasoning for Production Planning," *IEEE Expert*, pp. 60-67.

Hughes, J.G. (1990), *Object-Oriented Databases*, Prentice Hall, Englewood Cliffs, NJ, USA.

Laudon, K. and Laudon, J. (1996), *Management Information Systems*, Prentice Hall International Series, Englewood Cliffs, NJ, USA.

Maes, P.(1994), "Agents that Reduce Work and Information Overload," in *Communications of the ACM*, pp. 31-40.

.

14 IMAHDA REVISITED

14.1 INTRODUCTION

In the last thirteen chapters a number of different areas have been covered. It may be useful to recapitulate the role of IMAHDA and its methodologies viz-a-viz these areas. This is to enable the reader emerge from the methodological details, architectural details, application details of IMAHDA and get a broader view of the issues which have been addressed in this book. So in this final chapter IMAHDA and the concepts espoused by it are revisited in the areas of problem solving, hybrid systems, control systems, multi-agent systems, software engineering and reuse, and enterprise-wide system modeling. This chapter in a sense provides backward integration into some of the contributions of this book.

14.2 IMAHDA AND PROBLEM SOLVING

Although IMAHDA deals with intelligent hybrid systems, the architecture is driven by problem solving in a more general sense. The five information processing phases, namely, preprocessing, decomposition, control, and decision can be applied in areas outside hybrid systems for software development. The tasks associated with these phases can also be used, although in some domains it the task set may be larger or smaller. The information processing phases and the tasks associated with them can be used by a problem solver as a guide for dealing with complexity of real world problems.

14.3   IMAHDA AND HYBRID SYSTEMS

This book provides an engineering methodology for building intelligent hybrid systems. It develops an architectural theory of intelligent hybrid systems involving four intelligent methodologies, namely, symbolic AI systems, fuzzy systems, artificial neural networks and genetic algorithms based on a number of perspectives including cognitive science, neurobiolgical control, computational and artificial intelligence, forms of knowledge, learning, physical systems, user and others.. It develops this architectural theory at the task structural level and computational level. The task structural level architecture is defined in terms of the information processing phases, tasks to be accomplished in each phase, task constraints which apply in each phase, knowledge engineering strategy, intelligent methods used to accomplish the tasks, and hybrid arrangement (e.g. combination, transformation or fusion) of the intelligent methods in each phase.

The computational level involves knowledge modeling of the task structure level architecture, developing a computational framework for building real world applications, and a model for dynamic analysis of the computational framework. The knowledge modeling among other aspects includes learning knowledge and strategy of the hybrid architecture. It realizes the computational framework in the form of an Intelligent Multi-Agent Hybrid Distributed Architecture (IMAHDA) and highlights some of its emergent characteristics.

IMAHDA is applied in a complex real time alarm processing application. The agent oriented analysis and design of alarm processing application highlights the association between objects and agents for analysis, design and implementation of software systems.

14.3.1   Software Modeling of Hybrid Systems

In chapter 3, 4 and 5 fusion and transformation, transformation, and combination hybrid systems have been described. These classes of hybrid systems can be modeled based on the vocabulary of generic intelligent agents, software agents, and objects described in this book. For example, neural network and genetic algorithm based fusion and/or transformation systems described in chapters 3 and 5 can be modeled as agents as shown in Figures 14.1, 14.2, and 14.3 respectively. The agent class hierarchy or agent structure shown in Fig-

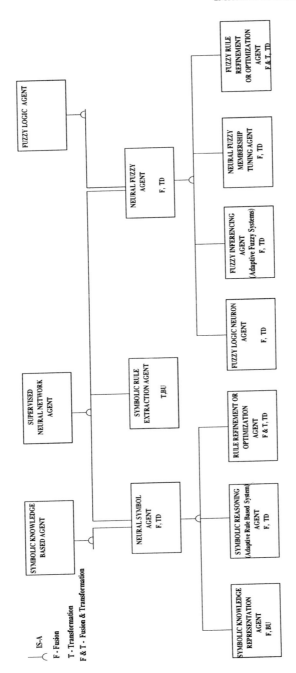

Figure 14.1.   Some Supervised Neural Network Agent Types

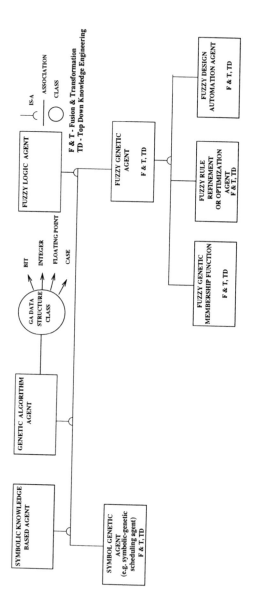

Figure 14.2.   Some Genetic Algorithm Agent Types

ures 14.1, 14.2, and 14.3 also indicates the hybrid configuration and knowledge engineering strategy of the agents.

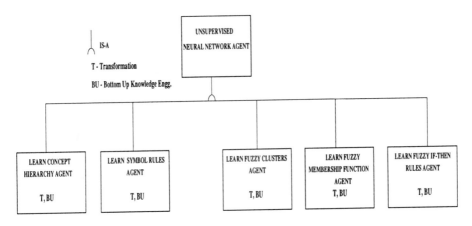

Figure 14.3.   Some Unsupervised Neural Network Agent Types

14.4   IMAHDA AND CONTROL SYSTEMS

In Figure 3.16 of chapter 3 conceptual scheme of a control system is outlined. This conceptual scheme contains generation knowledge, dynamics knowledge, and selection knowledge for modeling a control system. The generation knowledge which generates possible control actions is represented by the control phase of IMAHDA. The dynamics knowledge for predicting the outcome of generated control action is represented by the decision phase of IMAHDA. The selection knowledge which evaluates the predicted outcomes is represented by the post-processing phase. The postprocessing phase also represents the feedback loop from the environment for purpose of adaptation.

14.5   IMAHDA AND MULTI-AGENT SYSTEMS

The classical architectures (e.g. Robo-SOAR (Laird et al. 1991) and THEO (Mitchell 1990) of an agent system or a multi-agent system have been based on explicit deliberation. The principal drawback of these classical architectures as elucidated by Russell and Norvig (1995) is that *explicit reasoning about the effects of low-level actions is too expensive to generate real-time behavior.* In order to respond to this problem, researchers like Rosenschein (1985), and Kaelbling and Rosenschin (1990) working in situated automata have adopted a compilation approach with compiled behavior.

IMAHDA which is built around aspects related to information processing at macrostructure and microstructure level does not totally rely on explicit deliberation. The five information processing phases of IMAHDA represent at a global level a deliberate reasoning structure, whereas reasoning within each phase depending on the granularity of the processes is a mixture (of analytical and automated or compiled methods. Thus it incorporates the strengths of both the classical and situated automata architectures to generate real-time behavior.

14.6   IMAHDA AND SOFTWARE ENGINEERING

Building new software systems today usually entails constructing significant portions of the domain from scratch. The cost of this effort has become prohibitive especially as we attempt to build larger and larger systems. To overcome this problem and advance the state of the art, we must find ways of preserving the existing knowledge and of sharing, reusing, and building on it.

Object-Oriented (OO) software engineering methodology today is one of the premium software engineering methodologies for building software systems. Some of the reasons which have made the OO methodology popular include its intuitive structural modeling approach, strong information hiding properties for software implementation, and reuse. The OO methodology facilitates reuse at the application or problem domain class level and not beyond. That is, it does not facilitate reuse at the task structure level or the problem solving approach level.

IMAHDA facilitates reuse not only at the application level (through use of its OO features) but also at the problem-solving approach level through definition its information processing phases and other aspects related to these phases. Besides the agent oriented (the term "oriented" has been used as the agents nclude object-oriented properties like inheritance) analysis and design described in chapters 9 and 10 illustrate the integration of objects with agents for system modeling and how these two methodologies enrich the software analysis and design process.

14.7   IMAHDA AND ENTERPRISE-WIDE SYSTEMS MODELING

Objects, Agents and IMAHDA can be used as a means for alleviating the problems associated with enterprise modeling, backward and forward integration of three categories of systems, namely, database systems, information systems and knowledge systems or intelligent systems. The chapter 13 of this book describes an enterprise-wide system building approach data engineering, information engineering through problem solving, and knowledge engineering through problem solving using objects, agents and IMAHDA. It does this by distinguishing between structural abstraction, task abstraction, task-oriented behavior abstraction.

References

Laird, J. E., Yager, E.S., Hucka, M., and Tuck, M. (1991), "Robo-Soar: An Integration of External Interaction, Planning, and Learning using Soar," *Robotics and Autonomous Systems*, vol. 8, pp. 113-129.

Mitchell, T. M. (1990), "Becoming Increasingly Reactive (Mobile Robots)," *Proceedings of the Eighth National Conference on Artificial Intelligence (AAAI90)* , vol. 2, pp. 1051-1058.

Rosenchein,S. J. (1985), "Formal Theories of Knowledge in AI and Robotics," *New Generation Computing*, vol. 3 no. 4, pp. 345-357.

Kaelbling, L. P. and Rosenchein, S. J. (1990), "Action and Planning in Embedded Agents," *Robotics and Autonomous Systems*, vol. 6 no. 1-2, pp. 35-48.

Russell, S., and Norvig, P. (1995), *Artificial Intelligence - A Modern Approach*, Prentice Hall, New Jersey, USA, pp. 788-790.

Appendix A
Input Features of the Animal Domain

Problem Space

In this page and the next one the features used for classification of various animals in the animal kingdom and their classes have been tabulated. The input features tabulated in this page forms only a partial description of the animal kingdom. The animal domain has been used for illustrating the learning knowledge and strategy of IMAHDA.

hair	milk	feathers	egglaying	pouches
land	tree	field-meadows	lake-river	large
medium	fly	fly-short-dist	fly-long-dist	five-digit
claws	long-head	odd-toed	even-toed	herbivore
sturdy-feet	web-feet	tail	trunk	long-mane
bird-or-prey	hoofs	climb-trees	brownish-yellow color	long-tail
short-legs	can-leap	long-tail-with -tip-held-up	rosette-shaped -spots	big-paws
small-eyes	yellow-eyes	black-stripes	short- mane	beard-like-growth - on-cheeks
rounded-ears	long-yellowish-thick -hair	white-bellyside-extends-to-flanks	white-bellyside-does not-extend-to-flanks	white-tail
yellowish-color	reddish-ochre-color	light-reddish-ochre-color	narrow-stripes	very-narrow-stripes
closely-spaced-stripes	yellowish-red-color	white-belly-area-restricted	white-color	widely-spaced-stripes
very-light-reddish-ochre-color	tapered-ears	small	dark-line-from-eye -to-ear	small-paws
narrow-head	slender-body	blue-eyes	whiskers	carnivore
long-black-hair	large-head	medium-head	small-head	heavy-cheek-hair
very-heavy-cheek-hair	light-cheek-hair	belly-mane	dig-dens	long-legs

Figure A.1.   Input Features of the Animal Domain

397

Appendix B
Classes in the Animal Domain

Playtus	Anteater	Kangaroo	Koala	Wallaby	Wombat	Bandicoot	Opussum
Whale	Dolphin	Porpoise	Toothed-whale	Seal	Sperm-whale	Gibbon	Orangutun
Chimpanzee	Gorilla	Spider-monkey	Night-monkey	Howler-monkey	Marmoset	White-bear	Brown-bear
Black-bear	Silver-bear	Puma	Lion	Tiger	Leopard	Wild-cat	Wolf
Fox	Coyote	Collie	Jackal	Dingo	Labrador	Bulldog	African-elephant
Indian-elephant	Grevy-zebra	Mountain-zebra	Gaugge-zebra	Brown-horse	White-horse	Black-horse	Cattle
Siberian tiger	Indochina tiger	Bengal tiger	South china tiger	Javan tiger	Caspian tiger	White tiger	Sumartan tiger
Ostrich	Kiwi	Emu	Rhea	Penguin	Chicken	Turkey	Duck
Geese	Guinea-fowl	Pigeon	Pheasant	Stork	Flamingo	Swan	Albatross
Gull	Auks	Wader	Pelican	Hawk	Falcon	Night-hawk	Owl
Eagle	Vulture	Condor	Sparrow	Starling	Finch	Thrush	Robin
Songbird	Nightingale	Bluebird	Warbler	Blackbird	Skylark	Wagtail	Wren
Crow	Blue-jay	Jackdaw	Raven	Magpie	Lyre-bird	Bird-paradise	Bower-bird

Figure B.1.   Classes in the Animal Domain

Appendix C
TTS Power Network

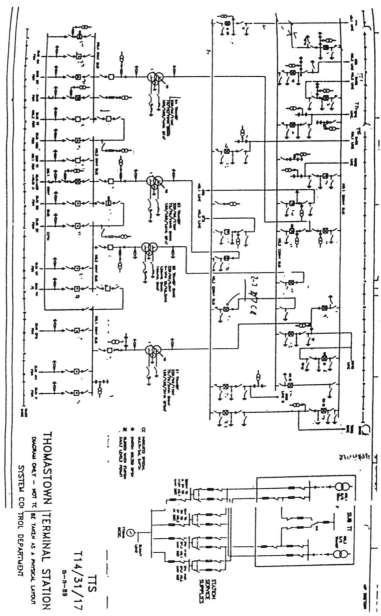

Figure C.1.   TTS Power Network

Appendix D
TTS Substation Power Network

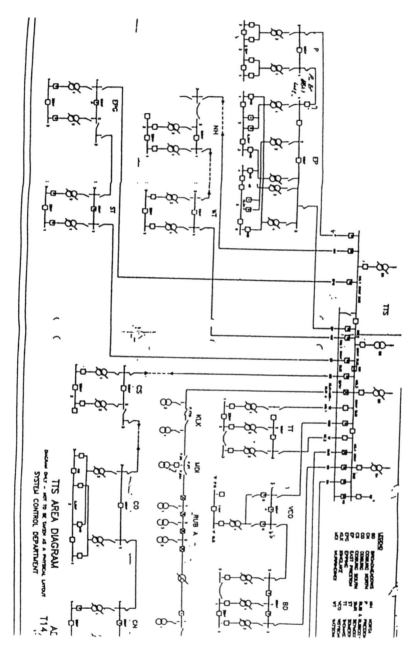

Figure D.1.   TTS Substation Power Network

Appendix E
Real Time Alarm Data

Real Time Alarm Data

8DEC90 1438:16 TTS ON BUILDING FIRE
8DEC90 1438:18 TTS OFF BUILDING FIRE
8DEC90 1438:18 TTS ON FIRE EQPT FAULT
8DEC90 1438:21 TTS OFF FIRE EQPT FAULT
9DEC90 0815:19 TTS OPEN 2-3 220KV B/T CB
9DEC90 0815:26 TTS OPEN 2-3 220KV B/T CB ROI
9DEC90 0815:26 TTS OPEN 2-3 220KV B/T CB 3 BUS ROI
9DEC90 0815:49 TTS CLOSE KTS 2 220KV CB ROI
9DEC90 0816:03 TTS CLOSE KTS 2 220KV CB
9DEC90 0816:11 TTS ON ALARM EQPT DC FAIL
9DEC90 0816:12 TTS OFF ALARM EQPT DC FAIL
9DEC90 0816:15 TTS OPEN KTS 1/1 BUS 220KV CB
9DEC90 0816:22 TTS ON ALARM EQPT DC FAIL
9DEC90 0816:24 TTS OPEN KTS 1/1 BUS 220KV CB ROI
9DEC90 0816:24 TTS OFF ALARM EQPT DC FAIL
9DEC90 0817:56 TTS OPER 2 220/66KV TR GAS
9DEC90 0818:56 TTS ON 220/66KV TR TEMP
9DEC90 1115:40 TTS CLOSE KTS 1/1 BUS 220KV CB ROI
9DEC90 1116:02 TTS CLOSE KTS 1/1 BUS 220KV CB ROI
9DEC90 1116:17 TTS OPEN KTS 2 220KV CB
9DEC90 1116:26 TTS OPEN KTS 2 220KV CB ROI
9DEC90 1731:15 TTS ON 220/66KV TR OLTC EQPT
9DEC90 1731:24 TTS OFF 220/66KV TR OLTC EQPT
10DEC90 1321:20 TTS ON 220/66KV TR TEMP
10DEC90 1321:26 TTS ON ALARM EQPT DC FAIL
10DEC90 1321:37 TTS OFF ALARM EQPT DC FAIL
10DEC90 1321:56 TTS OPER 2 220/66KV TR GAS
10DEC90 1321:56 TTS ON 220/66KV TR TEMP
10DEC90 1324:08 TTS RESET 2 220/66KV TR GAS
10DEC90 1324:08 TTS OFF 220/66KV TR TEMP
10DEC90 1326:20 TTS OPER 2 220/66KV TR GAS

```
10DEC90 1326:20 TTS ON 220/66KV TR TEMP
10DEC90 1327:11 TTS RESET 2 220/66KV TR GAS
10DEC90 1327:11 TTS OFF 220/66KV TR TEMP
10DEC90 1328:32 TTS OPER 2 220/66KV TR GAS
10DEC90 1328:32 TTS ON 220/66KV TR TEMP
10DEC90 1330:08 TTS ON 415V SELECTED SUPPLY FAIL
10DEC90 1330:08 TTS ON COMM EQPT
10DEC90 1330:08 TTS ON ALARM EQPT DC FAIL
10DEC90 1331:00 TTS RESET 2 220/66KV TR GAS
10DEC90 1331:00 TTS OFF 220/66KV TR TEMP
10DEC90 1332:03 TTS OPER 4 220/66KV TR GAS
10DEC90 1332:03 TTS ON 220/66KV TR TEMP
10DEC90 1332:42 TTS RESET 4 220/66KV TR GAS
10DEC90 1332:42 TTS OFF 220/66KV TR TEMP
10DEC90 1334:53 TTS OFF ALARM EQPT DC FAIL
10DEC90 1335:00 TTS OFF COMM EQPT
10DEC90 1337:38 TTS OPER 2 220/66KV TR GAS
10DEC90 1337:38 TTS ON 220/66KV TR TEMP
10DEC90 1338:59 TTS RESET 2 220/66KV TR GAS
10DEC90 1338:59 TTS OFF 220/66KV TR TEMP
10DEC90 1340:57 TTS OPER 2 220/66KV TR GAS
10DEC90 1340:57 TTS ON 220/66KV TR TEMP
10DEC90 1341:52 TTS RESET 2 220/66KV TR GAS
10DEC90 1341:52 TTS OFF 220/66KV TR TEMP
10DEC90 1349:52 TTS OPER 2 220/66KV TR GAS
10DEC90 1349:52 TTS ON 220/66KV TR TEMP
10DEC90 1352:06 TTS RESET 2 220/66KV TR GAS
10DEC90 1352:06 TTS OFF 220/66KV TR TEMP
10DEC90 1356:59 TTS OPER 2 220/66KV TR GAS
10DEC90 1356:59 TTS ON 220/66KV TR TEMP
10DEC90 1357:52 TTS RESET 2 220/66KV TR GAS
10DEC90 1357:52 TTS OFF 220/66KV TR TEMP
10DEC90 1403:17 TTS OPER 2 220/66KV TR GAS
10DEC90 1403:17 TTS ON 220/66KV TR TEMP
10DEC90 1404:42 TTS RESET 2 220/66KV TR GAS
10DEC90 1404:42 TTS OFF 220/66KV TR TEMP
10DEC90 1416:01 TTS OFF 415V SELECTED SUPPLY FAIL
10DEC90 1716:56 TTS OPER CB TRIP
10DEC90 1716:57 TTS OPER CB TRIP
10DEC90 1716:58 TTS OPEN KLK-MDI-RUB A 66KV CB
10DEC90 1716:58 TTS OPER CB TRIP
```

```
10DEC90 1716:59 TTS OPEN VCO 66KV CB
10DEC90 1717:00 TTS OPER CB TRIP
10DEC90 1717:01 TTS OPEN 2-3 B/T 66KV CB
10DEC90 1717:01 TTS OPER CB TRIP
10DEC90 1717:01 TTS OPER CB TRIP
10DEC90 1717:01 TTS OPER CB TRIP
10DEC90 1717:02 TTS OPEN B3 TR 66KV CB
10DEC90 1717:02 TTS OPEN CS 66KV CB
10DEC90 1717:03 TTS OPEN CN 66KV CB
11DEC90 1410:26 TTS OPER CAPACITOR FAULT
11DEC90 1417:22 TTS RESET CAPACITOR FAULT
11DEC90 1420:12 TTS OPER NH - STATION GENERAL
11DEC90 1421:32 TTS RESET NH - STATION GENERAL
11DEC90 1422:06 TTS OPER NH - TR FAULT
11DEC90 1422:16 TTS RESET NH - TR FAULT
11DEC90 1422:30 TTS OPER P - STATION GENERAL
11DEC90 1422:41 TTS RESET P - STATION GENERAL
11DEC90 1422:53 TTS OPER P - TR FAULT
11DEC90 1423:02 TTS RESET P - TR FAULT
11DEC90 1433:23 TTS OPER ST - STATION GENERAL
11DEC90 1430:21 TTS RESET ST - STATION GENERAL
11DEC90 1431:52 TTS OPER CN - STATION GENERAL
11DEC90 1432:03 TTS RESET CN - STATION GENERAL
11DEC90 1432:18 TTS OPER CN - TR FAULT
11DEC90 1432:36 TTS RESET CN - TR FAULT
11DEC90 1433:08 TTS OPER TT - STATION GENERAL
11DEC90 1433:47 TTS RESET TT - STATION GENERAL
11DEC90 1433:58 TTS OPER TT - INTERTRIP NO VOLT
11DEC90 1434:16 TTS RESET TT - INTERTRIP NO VOLT
11DEC90 1434:26 TTS OPER CS - STATION GENERAL
11DEC90 1434:25 TTS RESET CS - STATION GENERAL
11DEC90 1435:06 TTS OPER CS - TR FAULT
11DEC90 1435:15 TTS RESET CS - TR FAULT
11DEC90 1435:47 TTS OPER EP - STATION GENERAL
11DEC90 1436:20 TTS RESET EP - STATION GENERAL
11DEC90 1436:37 TTS OPER EP - TR FAULT
13DEC90 1339:36 TTS OPEN CN 66KV CB
13DEC90 1339:53 TTS CLOSE CN 66KV CB
13DEC90 1344:41 TTS OPEN BD 66KV CB
13DEC90 1344:44 TTS CLOSE BD 66KV CB
13DEC90 1344:47 TTS OPEN SMR 66KV CB
```

13DEC90 1344:49 TTS CLOSE SMR 66KV CB
13DEC90 1344:52 TTS OPEN SMR 66KV CB
13DEC90 1344:54 TTS CLOSE SMR 66KV CB
13DEC90 1345:55 TTS OPEN SMR 66KV CB
13DEC90 1402:30 TTS CLOSE SMR 66KV CB
13DEC90 1408:09 TTS OPEN B3 TR 66KV CB
13DEC90 1408:10 TTS CLOSE B3 TR 66KV CB
13DEC90 1408:53 TTS CLOSE KLK-MDI-RUB A 66KV CB
13DEC90 1409:45 TTS OPEN KLK-MDI-RUB A 66KV CB
19DEC90 1449:12 TTS ON COMM EQPT
19DEC90 1532:34 TTS OFF COMM EQPT
19DEC90 1532:46 TTS ON COMM EQPT
19DEC90 1532:51 TTS OFF COMM EQPT
19DEC90 1533:07 TTS ON COMM EQPT
19DEC90 1533:16 TTS OFF COMM EQPT
19DEC90 1533:16 TTS OFF COMM EQPT
19DEC90 1533:28 TTS ON COMM EQPT
19DEC90 1541:18 TTS OFF COMM EQPT
19DEC90 1533:28 TTS ON COMM EQPT
19DEC90 1541:18 TTS OFF COMM EQPT
3JAN91 1316:58 TTS ON OSCILLO OPERATED
3JAN91 1317:00 TTS OFF OSCILLO OPERATED
3JAN91 1317:08 TTS ON COMM EQPT
3JAN91 1317:09 TTS OFF COMM EQPT
5JAN91 1317:15 TTS ON 415V SELECTED SUPPLY FAIL
5JAN91 1317:22 TTS OFF 415V SELECTED SUPPLY FAIL
5JAN91 1317:24 TTS ON DIESEL RUNNING
5JAN91 1317:32 TTS ON DIESEL RUNNING
5JAN91 1317:45 TTS ON 415V SELECTED SUPPLY FAIL
5JAN91 1317:45 TTS ON COMM EQPT
5JAN91 1317:45 TTS ON ALARM EQPT DC FAIL
5JAN91 1319:20 TTS OFF DIESEL RUNNING
5JAN91 1322:34 TTS OFF COMM EQPT
5JAN91 1323:01 TTS ON STATION GENERAL
5JAN91 1323:42 TTS RESET CB TRIP
5JAN91 1323:43 TTS OPER CB TRIP
5JAN91 1323:44 TTS OPEN BD 66KV CB
5JAN91 1323:05 TTS OPER CB TRIP
5JAN91 1323:15 TTS OPEN KTS 1/1 BUS 220KV CB
5JAN91 1323:20 TTS OPER CB TRIP
5JAN91 1323:25 TTS OPEN SMTS 2 220KV CB

5JAN91 1323:26 TTS OPER CB TRIP
5JAN91 1323:27 TTS OPER CB TRIP
5JAN91 1323:28 TTS OPEN ROTS 220KV CB
5JAN91 1323:29 TTS OPEN B4 TR 1 BUS 220KV CB
5JAN91 1323:30 TTS OPER CB TRIP
5JAN91 1323:31 TTS OPER CB TRIP
5JAN91 1323:39 TTS OPEN TSTS 220KV CB
5JAN91 1323:33 TTS OPER CB TRIP
5JAN91 1323:35 TTS OPEN RWTS 1/1 BUS 220KV CB
5JAN91 1323:39 TTS OPEN B1 TR 220KV CB
5JAN91 1323:42 TTS OPER CB TRIP
5JAN91 1323:44 TTS OPEN KTS 2 220KV CB
22FEB91 0836:44 TTS ON COMM EQPT URGENT
22FEB91 0836:48 TTS OFF COMM EQPT URGENT
22FEB91 0836:48 TTS ON VF GUARD LOSS
22FEB91 0837:49 TTS OFF VF GUARD LOSS
22FEB91 0837:39 TTS OPER CB TRIP
22FEB91 0837:40 TTS OPEN 2-3 B/T 66KV CB
22FEB91 0837:40 TTS OPER CB TRIP
22FEB91 0837:41 TTS ON ST 66KV CB
22FEB91 0837:41 TTS OPER CB TRIP
22FEB91 0837:42 TTS ON WT 66KV CB
22FEB91 0837:43 TTS OPER CB TRIP
22FEB91 0837:44 TTS ON EP 66KV CB
22FEB91 0837:45 TTS OPER CB TRIP
22FEB91 0837:45 TTS ON B2 TR 66KV CB
22FEB91 0837:46 TTS OPER CB TRIP
22FEB91 0837:47 TTS ON 1-2 B/T 66KV CB

Index